Economics and the Philosophy of Science

ECONOMICS
AND THE
PHILOSOPHY
OF SCIENCE

Deborah A. Redman

New York   Oxford
OXFORD UNIVERSITY PRESS
1991

Oxford University Press

Oxford   New York   Toronto
Delhi   Bombay   Calcutta   Madras   Karachi
Petaling Jaya   Singapore   Hong Kong   Tokyo
Nairobi   Dar es Salaam   Cape Town
Melbourne   Auckland

and associated companies in
Berlin   Ibadan

Copyright © 1991 by Oxford University Press, Inc.

Published by Oxford University Press, Inc.,
200 Madison Avenue, New York, New York 10016

Oxford is a registered trademark of Oxford University Press

Library of Congress Cataloging-in-Publication Data
Redman, Deborah A.
Economics and the philosophy of science /
Deborah A. Redman. p.   cm.   Includes Bibliographical references.
ISBN 0-19-506412-7
1. Economics—Philosophy.   2. Science—Philosophy.
I. Title.
HB71.R35   1991   330′.01′5—dc20
90-32099

2 4 6 8 9 7 5 3 1

Printed in the United States of America
on acid-free paper

We all have our philosophies, whether or not we are aware of this fact, and our philosophies are not worth very much. But the impact of our philosophies upon our actions and our lives is often devastating. This makes it necessary to try to improve our philosophies by criticism. This is the only apology for the continued existence of philosophy which I am able to offer.

KARL R. POPPER. *Objective Knowledge* (Oxford: Oxford University Press, 1972), p. 33.

[T]he ideas of economists and political philosophers, *both when they are right and when they are wrong* [my emphasis], are more powerful than is commonly understood. Indeed the world is ruled by little else.

JOHN MAYNARD KEYNES. *The General Theory of Employment, Interest and Money.* Vol. 7 of *The Collected Writings of John Maynard Keynes* (London: Macmillan/ New York: St. Martin's Press for The Royal Economic Society, 1973), p. 383.

A person is not likely to be a good political economist who is nothing else.

J. S. MILL, "Auguste Comte and Positivism." In *Essays on Ethics, Religion and Society.* Vol. 10 of *Collected Works of John Stuart Mill* (Toronto: Toronto University Press, 1969), p. 306.

[I]n the study of society exclusive concentration on a speciality has a peculiarly baneful effect: it will not merely prevent us from being attractive company or good citizens but may impair our competence in our proper field—or at least for some of the most important tasks we have to perform. The physicist who is only a physicist can still be a first-class physicist and a most valuable member of society. But nobody can be a great economist who is only an economist—and I am even tempted to add that the economist who is only an economist is likely to become a nuisance if not a positive danger.

FRIEDRICH A. VON HAYEK, "The Dilemma of Specialization." In L. D. White, ed. *The State of the Social Sciences* (Chicago: University of Chicago Press, 1956), pp. 462–63.

It is by failure that we learn, and perhaps the greatest service is to point out where the failures lie.

KENNETH E. BOULDING, "What Went Wrong with Economics?" *American Economist* 30, no. 1 (Spring 1986): 12.

Preface

In this work I have tried to provide an accurate critical survey of the philosophy of science, beginning with the Vienna Circle, and to trace modern economists' relationship with the philosophy of science. Because I wanted to represent these ideas as accurately as possible, I have made extensive use of citations from original sources. The reproduction of authors' own words—although at times cumbersome—seemed to me the most honest and accurate way to set forth the history of their ideas while at the same time lending support to my own interpretation. Economists often eschew good citation practice; my use of quotations and notes prevents the reader from reaching the conclusion that I simply developed my arguments in a vacuum and also lessens possible criticism of the ideas presented in this manner. Moreover, few social scientists take the time to read works in the philosophy of science, although many have heard about them. If they merely read the present work, at least they will have been exposed to a small portion of some important works in the philosophy of science and economics. Letting the authors of these works speak for themselves may spark interest in reading more of the original sources.

Through the use of the first-person pronoun "I," an attempt has been made to distinguish my insights from those of others, which will—it is hoped—contribute to the book's clarity. A good part of the survey of philosophic and economic thought reflects a rather standard interpretation that can be found in the literature. Where I, however, hope to add something new or provide a view that perhaps deviates from the view represented in the literature, I have employed "I."

In the section on economics I attempt to represent as many methodological positions as possible and to draw on the ideas of economists from all areas of the discipline, both mainstream and nontraditional. In this sense I wanted to introduce readers to the modern methodology of economics and to familiarize them with modern arguments in economics that are supported by positions grounded in the philosophy of science. Part of my aim has been prohibitive. For instance, I hope to show that falsification (as conclusive disproof) does not exist in science and that defending a theory because it has not yet been "falsified" may sound sophisticated and scientific but is in reality indefensible. Hence, continued reference to falsification itself becomes an "immunizing strategy." The second major prohibitive contribution deals with the interpretation of Kuhn's philosophical position that science is consensus, or the adher-

ence to one "paradigm" or "disciplinary matrix." The danger this view presents to science and intellectual activity in general is discussed within the
framework of its significance for economics. The chief goal of the work is to
give economists and other social scientists the necessary background to discuss methodological matters with authority.

I have no doubt "picked" on Professors Blaug and Hutchison in the
section on economics. I hope they and others will not have forgotten Sidney
Schoeffler's lament (1955, p. vii) that the methodologist is caught in a dilemma in which he

> often antagonizes the very people whom he respects and admires the most;
> and that he often wins the approval of at least some people for whose ap-
> proval he does not care. *And this somewhat melancholy situation is even
> aggravated by the fact that the methodologist will usually "pick" on the most
> creative and most productive scholars in his field—since their work is the most
> important, it is also the most "worthy" of methodological analysis.* There is a
> distinct element of intellectual patricide in methodological work, and this
> aspect of it does not help to win it social approval. (my emphasis)

Both Blaug and Hutchison are unique in that they were the first to apply the
basic concepts of twentieth-century philosophy of science to economics.
Those who explore new pathways face a much greater challenge than those
who follow. I have learned from them both. I am especially indebted to T.
W. Hutchison, who most graciously consented to read and comment on the
manuscript.

Where there was no official translation available or in existence, I have
provided my own; the original text appears in the notes. I have attempted to
assemble as comprehensive a bibliography as possible for purely pedantic
reasons. Where German or French works and their English translations differ
substantially, I have cited versions in both languages. Good bibliographies are
an invaluable research tool: I hope this one will serve the autodidact as well as
the methodologist. Nonetheless, my attempt to provide a useful, comprehensive bibliography failed here because of the sheer number of works involved
and the need for a meaningful organization of these works. Thus, out of this
work grew a companion volume, *Economic Methodology: A Bibliography
with References to Works in the Philosophy of Science (1860–1988)* (1989),
which is a fairly comprehensive bibliography of works on economic methodology drawn from the writings of economists and philosophers of science, as
well as relevant works in the philosophy of science.

D. A. R.

Acknowledgments

Acknowledgment is made to the following individuals and organizations for their kind permission to quote, sometimes at length, from the works listed below:

"The American Tradition in Economics," by H. G. Johnson. Repr. by permission of the *Quarterly Journal of Business and Economics* (formerly the *Nebraska Journal of Economics and Business*) 16, no. 3 (Summer 1977): 17–20, 22–26. Copyright © 1977 by the *Quarterly Journal of Business and Economics*.

"Book Review: Toward an Historiography of Science," by W. A. Smeaton and Stafford Finlay. *Annals of Science* 18 (1962): 125, 128–29. Repr. by permission of Taylor & Francis Ltd. and W. A. Smeaton. Copyright © 1962 by Taylor & Francis Ltd.

"Comment 2: Mark Blaug," by Mark Blaug. Repr. by permission of Basil Blackwell Publishers Ltd. from *Economics in Disarray.* Ed. Peter Wiles and Guy Routh (Oxford: Basil Blackwell, 1984), pp. 32–33. Copyright © 1984 by Peter Wiles and Guy Routh.

"Descriptive, Predictive and Normative Theory," by Oskar Morgenstern. Repr. by permission of *Kyklos,* Helbing & Lichtenhahn Verlag AG, Basel, from *Kyklos* 25, no. 4 (1972), 706–7. Copyright © 1972 by Helbing & Lichtenhahn Verlag AG.

"Economics: History, Doctrine, Science, Art," by Jürg Niehans. Repr. by permission of *Kyklos,* Helbing & Lichtenhahn Verlag AG, Basel, from *Kyklos* 34, no. 2 (1981), 171–72. Copyright © 1981 by Helbing & Lichtenhahn Verlag AG.

"The Feigl diagram," by Herbert Feigl. Repr. by permission of the University of Minnesota Press from "The 'Orthodox' View of Theories: Remarks in Defense as Well a Critique," in *Analyses of Theories and Methods of Physics and Psychology,* ed. Michael Radner and Stephen Winokur, vol. 4 of the Minnesota Studies in the Philosophy of Science, p. 6. Copyright © 1970 by the University of Minnesota.

"Die 'Feyerbendglocke" des Szientismus," by Hans Lenk. Repr. by permission of the Verband der wissenschaftlichen Gesellschaften Österreichs from *Conceptus* 16 (1982): 4, 6–8, 10. Copyright © 1982 by the Verband der wissenschaftlichen Gesellschaften Österreichs.

"Isaac Newton," by Dudley Shapere. Repr. by permission of Macmillan Publishing Company from the *Encyclopedia of Philosophy,* ed. Paul A. Edwards, vol. 5, p. 490. Copyright © 1967 by Macmillan, Inc.

"The Logic of the Social Sciences," by Karl R. Popper. Repr. by permission of Karl R. Popper from *The Positivist Dispute in German Sociology* (New York: Harper and Row, 1976), pp. 89–96, 102–4. Copyright © 1976 by Sir Karl R. Popper and © 1969 by Hermann Luchterhand Verlag.

"Normal Science and Its Dangers," by Karl R. Popper. Repr. by permission of Cambridge University Press from *Criticism and the Growth of Knowledge,* ed. Imre Lakatos and Alan Musgrave. (London: Cambridge University Press, 1970), pp. 52–53. Copyright © 1970 by Cambridge University Press.

"N. R. Hanson." Repr. by permission of *Time* Magazine from *Time* 89, April 28, 1967, p. 103. Copyright © 1967 by *Time* Inc.

"On Imre Lakatos," by William W. Bartley III. Repr. by permission of Kluwer Academic Publishers from *Essays in Memory of Imre Lakatos,* ed. R. S. Cohen, P. K. Feyerabend, and M. W. Wartofsky, vol. 39 of Boston Studies in the Philosophy of Science (Dordrecht: D. Reidel, 1976), pp. 37–38. Copyright © 1976 by D. Reidel Publishing Company.

"On the History and Philosophy of Science and Economics," by T. W. Hutchison. Repr. by permission of Basil Blackwell and The University of Chicago Press from chapter 3 of *Knowledge and Ignorance in Economics* (Oxford/Chicago: Basil Blackwell/University of Chicago Press, 1977), pp. 40, 50–51, 58, 61, 159. Copyright © 1977 by Basil Blackwell and The University of Chicago Press.

The Philosophy of Karl Popper, ed. Paul Schilpp. Repr. by permission of Open Court Publishing Company, La Salle, Illinois, and The Library of Living Philosophers. Copyright © 1974 The Library of Living Philosophers, Inc.

"Pursuing Scientific Truth," by Dexter B. Northrop. Repr. by permission of the author from the editorial page of *Chemical & Engineering News* (published by the American Chemical Society), 58, no. 2, May 19, 1980: 4.

"Rational Reconstructions," by Noretta Koertge. Repr. by permission of of Kluwer Academic Publishers from *Essays in Memory of Imre Lakatos,* ed. R. S. Cohen, P. K. Feyerabend, and M. W. Wartofsky, vol. 39 of Boston Studies in the Philosophy of Science (Dordrecht: D. Reidel, 1976), p. 362. Copyright © 1976 by D. Reidel Publishing Company.

"The Scope and Method of Economic Science," by Phyllis Deane. Repr. by permission of Basil Blackwell Ltd. on behalf of The Royal Economic Society from the *Economic Journal* 93 (1983): 11–12. Copyright © 1983 by Basil Blackwell Ltd.

"The Status of Master's Programs in Economics," by Robert J. Thornton and Jon T. Innes. Repr. by permission of the American Economic Association and the authors from the *Journal of Economic Perspectives* 2, no. 1 (Winter 1988): 178. Copyright © 1988 by the American Economic Association.

"Theories of Rationality," by William W. Bartley III. Repr. by permission of Open Court Publishing Company, La Salle, Illinois, from *Evolutionary*

Epistemology, Rationality, and the Sociology of Knowledge, ed. Gerard Radnitzky and W. W. Bartley III (La Salle: Open Court, 1987), p. 211. Copyright © 1987 by Gerard Radnitzky and W. W. Bartley III.

"Writing, Typing and Economics," by John Kenneth Galbraith. Repr. by permission of The Atlantic from *The Atlantic Monthly* 241 (March 1978): 105. Copyright © 1978 by The Atlantic.

Contents

PART II   ECONOMICS AND THE PHILOSOPHY OF SCIENCE

I

PHILOSOPHY OF SCIENCE:
Rationality, Growth, Ignorance, Objectivity, and Criticism

1

The Problems

The three principal problems of philosophy, according to the philosopher of science William Bartley, are (1) the problem of knowledge, (2) the problem of rationality, and (3) the reconciliation of the first problem with the second (Bartley, 1964, p. 3). Certainly, rationality and the growth of knowledge have become central issues in the philosophy of science since Thomas Kuhn and others—especially those paying attention to the history of science—have shown that the methods producing major scientific advances historically do not jibe with the accounts given by philosophers of science.

The roots of the philosophy of science lie in two opposed streams of philosophical thought: in seventeenth-century rationalism (the theory that reason rather than experience is the source of knowledge [Descartes, Spinoza, Leibniz]) and empiricism (the theory that experience is the source of knowledge [Locke, Berkeley, Hume]). During the Scientific Revolution these two strands of thought became fused, yielding a view of scientific method that is still popular today: the scientist proposes a theory on the basis of inductive logic and confirms or refutes the theory by experimental test of the predictions deductively derived from the theory. It is assumed that appraisal of theories or hypotheses follows from an application of scientific rules; that is, the justification of a statement is either inferred from principles (rationalism) or inferred from evidence (empiricism). New theories are adopted because of their greater explanatory power, and thus science progresses ever closer to the truth. This view of scientific method is outdated and mistaken. One reason that it is no longer valid is that the concepts of "science" and "method" have changed, no longer being solely associated with physics:

> . . . in Newton's time there was only one developed science, physics; the late eighteenth century saw the emergence of chemistry, the nineteenth of biology and psychology, the twentieth of the social sciences, and these have brought with them new problems of method not amenable to the comparatively simple analysis which is appropriate to physics. The range of meaning of "science" has been widened and is a matter of dispute not to be settled by methodological prescriptions. The question whether or not history is a sci-

3

ence, for example, cannot be arbitrarily decided by pointing out that it is
neither experimental nor predictive. (Caws, 1967, p. 342)

This method grew up with a science that was incredibly successful. The
great esteem accorded modern science led people to believe that there must
be something special about the scientific method that guaranteed these superior achievements. This, in turn, spurred a greater interest in methodology.
The scientific community no doubt viewed its labors as the fruits of a superior
method: scientific enterprise represents rationality par excellence, and is perhaps the noblest pursuit of mankind. Consider the characterization of science
made by Karl Popper (1972a, p. 216) in 1963:

> The history of science, like the history of all human ideas, is a history of
> irresponsible dreams, of obstinacy, and of error. But science is one of the very
> few human activities—perhaps the only one—in which errors are systemati-
> cally criticized and fairly often, in time, corrected.

This corrective attribute of science is the mechanism guaranteeing progress.
But expecially since the 1960s the rational image of science has become
tarnished. The seventeenth-century foundations of science—that theories are
derived through observation and experiment—is not reliable. Science as a
cumulative process and hence as a progressive enterprise has been challenged,
generating an immense literature on the growth of knowledge. A clean-cut
observation–theory distinction (that is, that observation reports and theory
are distinct processes) was proven invalid, as was a strict distinction between
the context of justification of scientific theories and the context of their discovery. It was found that scientific concepts were not as precise as reputed. The
pursuit of a science unified by a universal method has been all but abandoned.
Similarly, construing science to mean a single category whose legitimacy and
progress are measured by the extent to which the methods of physics are
employed has been rejected as naive by most philosophers.
Not every philosopher of science has concentrated on all of these problems. For instance, Popper has dedicated much of his research to the demarcation problem, Carnap to foundations, Kuhn to the growth of knowledge.
Likewise, each problem has evolved. The demarcation problem, for example,
has evolved from the positivists, who set out to separate the meaningful from
the meaningless, to Popper, who seeks to demarcate science from nonscience,
to Bartley, who believes that separating the critical from the uncritical or
pseudocritical is the most fruitful approach to minimizing and driving out the
ignorant, superstitious, ideological, and pseudoscientific elements in science.
Why is a deviation of substantial length into the modern developments in
the philosophy of science justified when the subject matter here is concerned
chiefly with the social sciences? First, the naive, seventeenth-century view of
science remains influential in the social sciences. One might venture to remark: "So what? The natural sciences have achieved awesome successes despite adherence to a naive philosophy of science." The social sciences, however, have a different relationship with the philosophy of science: the social

sciences—especially economics—claim scientific status based, to a great extent, on their use of "the scientific method," that is, the mistaken view of science rooted in seventeenth-century physics. This claim is not tenable.

A second reason for an in-depth examination of this body of philosophic thought is that social science literature is replete with allusions to some aspect of modern philosophy of science: everyone has seen arguments supported by Kuhnian paradigms, Popperian falsification, Lakatosian research programs, and related jargon. Philosophical ideas are obviously having a major impact on science. As a result, there is a great need for a concise, neutral, accurate introduction to the vast philosophy of science literature so that economists and other social scientists can become well informed with the least possible expenditure of energy and time. Indeed, there is a major dilemma involved with this transfer of knowledge from the philosophy of science: an outsider cannot evaluate the arguments because he or she does not have the necessary background knowledge. The historian of science I. Bernard Cohen (1977, pp. 312–13) describes the dilemma in the following way:

> It often seems as if a historian would have to become aware of all the major developments in philosophy of science before he could make a valid judgement as to the possible ways in which any or all parts of the philosophy of science might be useful for his problems. This task seems well beyond the philosophical capacities of almost all historians. Whereas the philosopher of science can very easily improve the historical level of his discussions, a similar change (and even "improvement") of the level of philosophy of science used by historians seems to me far more difficult to achieve.

It is unfortunate that very few surveys of the philosophy of science or textbooks exist.[1] The literature in the philosophy of science is an ongoing dialogue among its members; a reading of a few major works by the uninitiated can be misleading because the reader comes in in the middle of the conversation, so to speak. In addition, many of the short introductions to the topic offered by economists—usually by historians of economic thought—or by other social scientists are often highly oversimplified and in some ways quite unsophisticated. The survey here is an attempt to bridge the gap in knowledge between historians, social scientists, and philosophers by providing a critical summary of recent developments in the nature of science in easily readable form. It does not and cannot substitute, however, for a reading of the actual works.

Notes

1. Chalmers (1982) is a popular elementary introduction to the philosophy of science; the intellectual debt to Popper is quite apparent. H. I. Brown (1977) is very readable and recommended as a balance to Chalmers. In part I Brown reviews the doctrines of logical empiricism; in part II he discusses the "new philosophy of science." Newton-Smith (1981), Suppe (1977b), and two volumes of Stegmüller's three-volume *Hauptströmungen der Gegenwartsphilosophie* (1978 and 1987; currently not available in English) provide sophisticated and more demanding discussions of the

philosophy of science. Hacking (1981b) and Gutting (1980) collect many key readings dealing with Kuhn, revolutions, and the sociological view of science. A few surveys of the literature exist but are of limited usefulness, for example, Braun (1977); Danto (1967); Harré (1967); Kisiel with Johnson (1974); Laudan (1979, 1968); and McMullin (1970).

Four economists who have provided summaries of the developments in the philosophy of science with applications to economics are Blaug (1980b), Caldwell (1982), Klant (1984), and Pheby (1988). Philosophy of science has become so popular that *The Economist* recently ran an article reviewing some of the problems of the discipline (25 April 1987); the year before (26 April 1986) a general survey of Anglo–American philosophy appeared.

2

The Decline of Logical Positivism

Modern philosophy of science grew out of a movement called logical positivism, which emerged in the early twentieth century. The movement was highly complex; it should be cautioned that any brief attempt to characterize the movement, such as what follows, cannot do it justice.[1] Logical positivism, a term coined by A. E. Blumberg and Herbert Feigl in 1931, is the name given to the philosophical ideas put forth by a mixed group of academicians who referred to themselves as members of the Vienna Circle, the Wiener Kreis. Logical positivism is not to be confused with the positivism of August Comte (1798–1857) or with logical empiricism, the more mature form of logical positivism that evolved after the Vienna Circle dissolved with the Nazi Anschluß in 1938. Logical empiricism as a school of philosophical thought continues today. The clear distinction made here is not always observed in discussions on logical positivism and empiricism.

Logical positivism as a movement was essentially Germanic in its inception, growing out of evening meetings of the Vienna Circle. Not well known is the fact that its members were not just philosophers and scientists; the Vienna Circle membership was quite mixed. Led by Moritz Schlick (physicist [1882–1936]), the Vienna Circle was in part composed of members Rudolf Carnap (philosopher [1891–1970]), Friedrich Waismann (at that time a student, later philosopher [1891–1959]), Herbert Feigl (then a student, later philosopher [1902–]), Felix Kaufmann (lawyer [1895–1949]), Victor Kraft (historian [1880–1975]), Kurt Gödel (mathematician [1906–1978]), Philipp Frank (physicist [1884–1966]), and Otto Neurath (sociologist and economist [1882–1945]). Karl Popper (1902–), Ludwig Wittgenstein (1889–1951), and Karl Menger (1902–) spoke at some of these sessions but were not regular members. Also noteworthy is Hans Reichenbach's (1891–1953) Berlin School (where Carl Hempel [1905–] was a member), which also served as a locus for the development and dissemination of the ideas of logical positivism.

Because the members of the movement reacted strongly against the romantic, irrational, and ideological climate of nineteenth- and early-twentieth-century Germany, logical positivism is most popularly associated with its arduous pursuit of the elimination of the ideological and metaphysical elements of science and culture (the latter being included because their program also embraced the social sciences). Frederick Suppe warns in his work on positivism and its legacy, *The Structure of Scientific Theories* (1977b, pp. 6–7), that overemphasis of this single component of the positivist program may lead to a vulgarized understanding of the movement:

> It is often said that positivism emerged as a response to the metaphysical excesses of Hegel and his neo-Hegelian successors (for example, McTaggert, Bradley and others) who sought to explain reality in terms of abstract metaphysical entities (for example, Entelechy and the Absolute) which did not admit of empirical specification. While it certainly was an important goal (pursued with messianic fervor) of logical positivism to eliminate such metaphysical entities from philosophy and science, it seems to me that its role in the inception of the movement tends to be exaggerated, and that giving it too central an importance tends to obscure the true origins of the positivist doctrines.

(Likewise, the attempt to attribute all of the problems of modern economics to the excesses of logical positivism, as is sometimes argued in some histories of economic thought, is rather rash.)

The first and perhaps most significant component of the positivist program was a view that knowledge is grounded in experience and that meaningful statements are verifiable by observation and experiment. The verifiability criterion was used to exclude metaphysical statements—hence the reference to "Entelechies" and the "Absolute." (The entelechy was the invention of a German biologist and philosopher, Hans Driesch, who spent much of his life explaining biological processes; those that he could not explain, however, he attributed to an entelechy.) When no observation reports or laws supported the existence of entities (such as the entelechy or the Absolute), the entity could not be verified and was therefore rejected by positivists as metaphysical. Although the verifiability criterion excluded metaphysical propositions, the positivists soon discovered that it was so strict that it also ruled out most scientific laws, which naturally cannot be conclusively verified, provoking Karl Popper's famous comment (1972b, p. 36) that "positivists, in their anxiety to annihilate metaphysics, annihilate natural science along with it." Verifiability was modified to confirmability, a weaker criterion indicating that it is possible to derive true propositions from a statement. Confirmability turned out to be too weak: it allowed in some nonsense statements. The logical positivists never resolved this problem.

The logical positivist program was dedicated to clarity, rigor, and attention to detail. Many of the logical positivists believed that all significant problems could and should be reducible to problems formalizable in logic. In addition, they aimed to form an *Einheitswissenschaft,* an all-encompassing science joined by one method: the logical method of analysis. They assumed that science is rational and progressive. The spreading of science meant, in their view, extending rationality to the culture. Interestingly, many logical positivists believed that they were not doing philosophy; they advocated a strict division between science and philosophy, if not the destruction of philosophy.[2] This is more understandable if one recalls that German idealism was an antiscience movement. Hence, bringing science to the world meant not only liberation from the speculative and metaphysical but also exclusion of philosophical (that is, idealist), historical, psychological, and sociological factors, which cannot be confirmed or tested.

In 1929 the Vienna Circle published a position paper called "Scientific World-View: The Vienna Circle," *Wissenschaftliche Weltanschauung: der Wiener Kreis*. It began holding international congresses and put out the journal *Erkenntnis* from 1930 to 1939. In 1939 *Erkenntnis* was replaced by the *Journal of Unified Science,* and monograph publication began in the Circle's International Encyclopedia of Unified Science series. The Institute for the Unity of Science was established in 1936 in the Hague, and later moved with Philipp Frank to Boston.

The Vienna Circle dispersed during World War II for strictly political reasons. Schlick was murdered in 1936 by a deranged student, and as Ayer notes (1959, pp. 6–7), the Austrian press came close to implying that Schlick's death should be condoned because logical positivists deserved to be shot. Under the Naxi government, the sale of logical positivist literature was prohibited. The logical positivist "diaspora" was the result.

Waismann and Neurath emigrated to England. Carnap, Feigl, Gödel, Hempel, Kaufmann, K. Menger, Reichenbach, and Richard von Mises emigrated to the United States.[3] Logical positivism became an international movement and took on a new nature, as well as a new appellation: logical empiricism. The logical empiricists found they had much in common with the American pragmatists and operationalists, among whom C. S. Peirce (1839–1914) and P. W. Bridgman (1882–1961) were especially influential. Logical empiricism concerned itself less with the metaphysical and became markedly less extreme and dogmatic in nature. Logical empiricism was also quite compatible with the Anglo-American tradition of analytic philosophy, and in many ways has become absorbed into that tradition.

Logical empiricsts' analysis of theories has become known as the "received view," or sometimes also the orthodox or standard view.[4] The logical positivists' view of science was inductivist: science (verification) was grounded in experience (that is, in observation) and hence believed to be objective. A problem arose when the logical positivists realized that they needed to show that one person's experience was identical to another's. If that could not be proved, the observation statements that the logical positivists asserted give science its meaning would really be subjective. The answer was that propositions, not experience, verify other propositions—the crudest formulation of the received view. In its more mature form, theories are viewed as "axiomatic systems, wherein theoretical assertions are given a particular observational interpretation via correspondence rules" (Suppe, 1979, p. 317). The received view alternatively treats theories as "a deductively connected collection of laws" (Suppe, 1979, p. 331). More clearly, by the received view in its "original version" (Suppe, 1977b, p. 12), a theory is to be axiomatized in axiomatic logic. The axioms of the theory are scientific laws. Three types of terms are involved: logical and mathematical terms, theoretical terms, and observation terms. The axiomatizations include explicit definitions for theoretical terms called correspondence rules (sometimes called operational rules or a dictionary), which link them to observation terms. Carnap (1966, p. 266) gives an example of a correspondence rule: the temperature of a gas (an observable) is proportional to the mean kinetic energy (nonobservable, theoretical term) of its molecules. From this, empirical laws can be derived.

In the 1960s the received view came under increasing attack, as John Passmore (Australian philosopher [1914–]) notes ironically, "not from metaphysicians but from philosophers who would in a general sense be happy enough to describe themselves as 'logical empiricists' " (1967a, p. 56). In the same breath, Passmore pronounced logical positivism dead as a movement (1967a, p. 56). Indeed, by 1969 even the received view was succumbing to attacks. Modifications and attempted resuscitations of the received view, continue, however.[5]

The attacks on logical empiricism are of two sorts. There are criticisms of specific features of the received view, as well as offers of wholly different alternative philosophies. Among those who offer new philosophies of science are Paul Feyerabend, Norwood Hanson, Thomas Kuhn, Michael Polanyi, Stephen Toulmin, and others. Emphasizing the role of history of science, their views make methodology a part of the historical process. The consequence for the philosophy of science is that they show that major advances, such as those made by Galileo, Newton, Darwin, and Einstein, did not occur in accordance with the orthodox account. That is, the philosophy of science is at odds with the history of science. In addition, the history of science indicates that advances in science, that is, the growth of knowledge, are not a function of scientists' adherence to methodological rules. The upshot is that science, then, may not be viewed as a strictly rational process. I want to begin with this more extreme view.

Notes

1. For a treatment of logical positivism see Achinstein and Barker (1969); Ashby (1964); Ayer (1959*—this is a classic); Feigl (1981b, c, d)†; Ganslandt (1984); Hanson (1969a); Joergensen (1970)†; Juhos (1971)†; Kraft (1974)†; Lorenz (1984); Passmore (1967a); Schleichert (1975)*; Stegmüller (1987, vol. 1, chap. 9); Suppe (1977b). The economist Bruce Caldwell (1982, pp. 11–35) provides a short history of logical positivism and empiricism. Alexander (1964) discusses the philosophy of science between 1850 and 1910, Laudan (1968) theories of scientific method from Plato to Mach.

2. Carnap and Mach especially denied that they were philosophers and renounced philosophy as meaningless. Schlick, on the other hand, referred to himself as

*Collection of original works.
†Former member of the Vienna Circle.

a philosopher and lectured that philosophy was taking a new turn. See Schlick's "Wende der Philosophie" in Schleichert (1975, pp. 12–19), where he explains (p. 13): "Ich bin nämlich übberzeugt, daß wir in einer durchaus endgültigen Wendung der Philosophie mitten darin stehen und daß wir sachlich berechtigt sind, den unfrucht-baren Streit der Systeme als beendigt anzusehen." Schlick's ideas were, however, by far the least extreme of the ideas of all the members of the Vienna Circle.

3. Carnap went to Chicago; Frank to Boston; Gödel to Princeton; Feigl to Minnesota; Tarski (a Polish logician often associated with the Vienna Circle [1901–]) to Berkeley; Hempel to Yale and then to Princeton; Reichenbach to UCLA; Wais-mann to Cambridge and Oxford.

4. See Suppe (1977b, pp. 6–61).

5. The tie between the work of the logical positivists, logical empiricists, and analytical philosophers is their emphasis on language, and especially on the probem of meaning in the use of language. (For example, the positivists held that all meaningful statements are of two sorts: analytical [= tautologous] and empirical.) See H. I. Brown's part I (1977) for an introduction to logical empiricism, which has been ne-glected here due to space considerations.

3

The Pendulum Swings in the Other Direction: Sociological Explanations of Science

I have lumped together Polanyi, Fleck, and Kuhn for two reasons. They have in common a view that science cannot be rationally justified and hence belongs to the sociology of science. Theirs is an elitist view: appraisal of theories is the job of scientists only. The scientific community is a closed community. They all agree that science is a special enterprise. On what does it base its authority, then? For Polanyi, the authority of science lies in the judgment of scientists—in the "tacit dimension" of science, shared and understood only by scientists. For Fleck, it rests in the *Denkkollektiv;* for Kuhn, in the leap of faith to a paradigm.

The second reason for the organization is that the view is in many ways diametrically opposed to that of the positivists. Like positivism, it has its excesses. But this overreaction is useful in illuminating the extreme poles of the spectrum of views and hence the positivists' problems, as well as many of the problems in modern philosophy of science.

Michael Polanyi: Personal and Tacit Knowledge

Michael Polanyi (1891–1976) was born in Budapest, Hungary. In 1913 he finished his medical degree at the University of Budapest, and then joined the army as a medical officer. Shortly thereafter he contracted diphtheria, putting an end to his military career. While recuperating, Polanyi wrote his Ph.D. dissertation in chemistry by corresponding with Albert Einstein (1879–1955). He taught briefly at the University of Budapest, and then moved to the Kaiser Wilhelm Institute in Berlin in 1920. There he worked with Max Planck (1858–1947), Albert Einstein, and Irwin Schroedinger (1887–1961), among others. In 1929 he was appointed a life member of the institute, but he resigned in

1933 in protest of the Hitler regime. He then accepted the chair in physical chemistry at Victoria University in Manchester, England, where his research interest shifted from chemistry to the philosophy of science. In 1948 he exchanged his chair in chemistry for one in the social sciences, from which he retired in 1958. In retirement he was elected a fellow of and resided at Merton College, Oxford.[1]

Michael Polanyi develops a thesis that "into all acts of judgement there enters, and must enter, a personal decision which cannot be accounted for" (1962, p. 1). Hence, all knowledge is personal knowledge—the title of perhaps his best-known work (1958). In *Personal Knowledge* (1958, p. vii), Polanyi begins by rejecting the Laplacian ideal of scientific detachment:

> In the exact sciences, this false ideal is perhaps harmless, for it is in fact disregarded there by scientists. But we shall see that it exercises a destructive influence in biology, psychology and sociology, and falsifies our whole outlook far beyond the domain of science. I want to establish an alternative ideal of knowledge, quite generally.

The emphasis on "personal" knowledge is meant to correct the positivist view that scientific knowledge is impersonal and thus objective. Polanyi (1958, pp. vii, viii) does not wish to convey the idea that personal knowledge implies subjective knowledge:

> Comprehension is neither an arbitrary act nor a passive experience, but a responsible act claiming universal validity. Such knowing is indeed *objective* in the sense of establishing contact with a hidden reality; a contact that is defined as the condition for anticipating an indeterminate range of yet unknown (and perhaps inconceivable) true implications. It seems reasonable to describe this fusion of the personal and the objective as Personal Knowledge.

The dichotomy between fact and value set forth in positivist doctrine is specious, for distinguishing between facts of scientific value and those not "depends ultimately on a sense of intellectual beauty," or on an "emotional response which can never be dispassionately defined" (1958, p. 135). The stubborn perpetuation of the myth of objectivity threatens science, argues Polanyi (1958, p. 142), because it is a

> misguided intellectual passion—a passion for achieving absolutely impersonal knowledge which, being unable to recognize any persons, presents us with a picture of the universe in which we ourselves are absent. In such a universe there is no one capable of creating and upholding scientific values; hence there is no science.

The thrust of Polanyi's argument is that science does not rest on any purely objective method but is "a system of beliefs to which we are committed" (Polanyi, 1958, p. 171). Polanyi (1952, pp. 218–19) has the following to say:

> I hold that the propositions embodied in natural science are not derived by any definite rule from the data of experience, and that they can neither be

verified nor falsified by experience according to any definite rule. Discovery, verification and falsification proceed according to certain maxims which cannot be precisely formulated and still less proved or disproved, and the application of which relies in every case on a personal judgement exercised (or accredited) by ourselves. These maxims and the art of interpreting them may be said to constitute the premisses of science, but I prefer to call them our scientific beliefs. These premisses or beliefs are embodied in a tradition, the tradition of science.

Because science is a system of beliefs, it cannot be accounted for by experience or by reason alone and "it cannot be represented in noncommittal terms" (Polanyi, 1958, p. 171). "[O]nce you face up to the ubiquitous controlling position of unformalisable mental skills, you do meet difficulties for the justification of knowledge that cannot be disposed of within the framework of rationalism" (Polanyi, 1962, p. 2). From what does Polanyi believe science derives its authority, then? Essentially, it is drawn from convention, that is, scientific opinion or the "tacit understanding" between scientists "on the trustworthiness of the method to which they are setting out to analyse" (Polanyi, 1952, p. 231).

The link between Polanyi's work and that of the positivists and logical empiricists is threefold. Polanyi's work focused chiefly on theory discovery and not on theory choice. The positivists and logical empiricists believed that discovery belonged to the psychological realm, unlike the logic of justification. In this sense Polanyi's emphasis on the sociological and psychological aspects of discovery were not new.

His portrait of science as a nonrational, "personal" (as opposed to a purely objective) activity, however, sets him in direct opposition to the positivists and logical empiricists. His work can quite clearly be seen to draw to its logical conclusions the positivists' dilemma of experience: How can we claim that science is really not grounded in the personal experience and perception of each individual scientist? Those philosophers of science who maintain that science is a rational activity cannot accept his argument.

Unlike most philosophers of science, Polanyi does discuss in detail the social sciences and biology. In fact, his purpose, in part, is to show that the logical positivists' view that the scientific method is objective is not only false but dangerous when applied to the social sciences (although innocuous with respect to the natural sciences). Implicit in this distinction is a rejection of the unity-of-science thesis. For an Anglo-Saxon philosopher of science living in the first half of the twentieth century to hold this view is most unusual because most of them are/were physicists and assume that the prototype method is the method of physics.[2]

Ludwik Fleck: *Denkkollektiv* and *Denkstil*

Ludwik Fleck (1896–1961) is an almost unknown philosopher of science whose major work, *Entstehung und Entwicklung einer wissenschaftlichen*

Tatsache, published in 1935, was largely ignored because of the unreceptive political climate in Europe. Fleck is enjoying a posthumous revival, or first appreciation, after having been recognized by Thomas Kuhn in his *Structure of Scientific Revolutions.* Indeed, a reading of Fleck's *Entstehung* reveals just how indebted Kuhn is to him.

Fleck, like Polanyi, was educated in medicine. He grew up in Lwów, Poland, whose science and culture were much influenced by the Vienna Circle. He studied philosophy on the side, and then not particularly systematically. Fleck was acquainted with the philosophy of the Vienna Circle;[3] the theme of his work on the philosophy of science published in 1935 stands in stark contrast to that of the Vienna Circle.

The thirties proved to be an especially inopportune time for Fleck to bring out the book. Because Fleck was Jewish, he and his family were arrested by the Nazis in 1942. Shortly thereafter he was deported to Auschwitz, where he worked at first as a nurse, then was moved to the serology unit, where he was to work on vaccine production. In 1943 the Nazis founded a laboratory for manufacturing typhus vaccine at the concentration camp in Buchenwald; Fleck was deported there to work on a more productive way to manufacture the vaccine. He, his wife, and his son all managed to survive the war. After the war Fleck worked as an academician and doctor. In 1957, in failing health, he moved to Israel to be near his son. He died there in 1961, months before Kuhn's work appeared and his own work enjoyed a revival.

In Fleck's view, science is not a rational enterprise. He characterizes science as an historical, psychological, sociological, closed system of beliefs (*Meinungssystem*). The holding of these beliefs is quite dogmatic (*Beharrungstendenz von Meinungssystemen*). The initiation into a discipline—the student's training—is, in his view, indoctrination: "Each didactic introduction is literally a guiding into the subject, a soft coercion" (Fleck, 1980, p. 137).[4]

Fleck's philosophy of science revolves around the concepts of *Denkstil* and *Denkkollektiv.* It is not rigorously set forth; he admits that there are gaps but argues that the logical positivists' belief in the absolute is their Achilles' heel. A *Denkkollektiv* refers to the scientific community and corresponds roughly to Kuhn's "disciplinary matrix." A stable *Denkkollektiv* is clearly the forerunner of Kuhn's normal science. The *Denkstil,* similar to Kuhn's "exemplars," is "a guided perception, with the corresponding cognitive and actual digestion of that perceived" (Fleck, 1980, p. 130).[5] The *Denkstil* includes the problems and judgments common to a discipline, and methods used in problem solving. The relationship between *Denkstil* and *Denkkollektiv* is described by Fleck (1980, p. 135):

> We call the collective carrier of the *Denkstil* the *Denkkollektiv.* We are using the concept of the *Denkkollektiv* as a way of examining the social conditioning of thinking and it is not meant to be the values of a fixed group or social class. The concept is so to speak more functional than substantive in nature, similar for example to the concept of a force field in physics. A *Denkkollektiv* is always in existence if two or more people exchange ideas: these are fleeting, arbitrary *Denkkollektive,* which in a moment exist and pass. Yet even in

them a particular attitude is manifested which takes possession of none of the participants, but which often returns when the particular persons meet again.[6]

In Fleck's system, science does not accept new ideas through revolutions, as Kuhn suggests, but rather through mutations (*Mutationen*) of the *Denkstil*. Mutations either adapt (*anpassen*) or do not adapt. Fleck essentially adapts Charles Darwin (1809–1882) to the sociology of knowledge. Here Kuhn does not follow him; but Stephen Toulmin's thesis in his *Human Understanding* (1972) (see part I, chapter 5 in the present volume) expands on this idea.[7]

It should be kept in mind that Fleck generalizes what he knew of the medical scientific world to all of the sciences. It is unfortunate that no group—neither the Vienna Circle nor the Twardowski (Kamimierz Twardowski [1866–1938]) school of philosophy in Poland—took an interest in his work during his lifetime, for it is highly probable that he would have developed his work further had a sounding board existed.[8]

Thomas Kuhn's Scientific Revolutions

The key features of Kuhn's work are its emphasis on revolutions and the incorporation of the role of the sociology of knowledge into the philosophy of science. Kuhn (1922–) received his Ph.D. in physics from Harvard, then made a career change. The reason that Kuhn (1970c, p. v) gives for this "shift in career plans" is that

> (a) fortunate involvement with an experimental college course treating physical science for the non-scientist provided my first exposure to the history of science. To my complete surprise, that exposure to out-of-date scientific theory and practice radically undermined some of my basic conceptions about the nature of science and the reasons for its special success.

So Kuhn went on to develop an account of science that he believed was more in keeping with the history of science than the positivist or falsificationist views. The result was his *Structure of Scientific Revolutions,* probably the best-known work in the philosophy of science—ironically, written by a historian of science.

Crucial to understanding Kuhn's analysis is his concept of a "paradigm." According to the broad sense in which he employs the concept, a paradigm is composed of general theoretical assumptions, laws, techniques, and metaphysical principles that guide scientists in their work and members of a particular scientific community. Because Kuhn used *paradigm* in no fewer than twenty-one different senses (Masterman, 1970, p. 61), it has become the source of considerable misunderstanding; thus, Kuhn has essentially abandoned his usage of *paradigms* and replaced it with *disciplinary matrices* and *exemplars,* which capture the two main meanings of the word.[9]

A disciplinary matrix corresponds to the broad definition of paradigm mentioned above. " 'Disciplinary' because it is the common possession of the practitioners of a professional discipline; 'matrix' because it is composed of ordered elements of various sorts, each requiring further specification" (Kuhn, 1977l, p. 463). Three constituents of a disciplinary matrix are symbolic generalizations, models, and exemplars. Symbolic generalizations (Kuhn, 1977l, p. 463) are "expressions . . . which can readily be cast in some logical form like $(x)(y)(z)\phi(x,y,z)$. They are the formal, or the readily formalizable, components of the disciplinary matrix." Models are "analogies," which Kuhn discusses only in passing. Exemplars are "concrete problem solutions, accepted by the group as, in a quite usual sense, paradigmatic" (Kuhn, 1977l, p. 463). Exemplars are the second major sense in which Kuhn uses *paradigms.* A scientist comes to know and understand the disciplinary matrix from the study of "exemplars," which are examples showing the apprentice how the laws of the theories apply to phenomena. Exemplars appear in textbooks and present an informal description of an experimental setup and how this can be generalized to nature. From the study of exemplars, the student develops a "learned perception of similarity." To clarify this latter concept, Kuhn (1977l) gives an example of a father who takes his son, Johnny, to a zoological garden where Johnny learns, through the constant correction of his father, how to use the words *ducks, geese,* and *swans* properly.

Kuhn's example fails in its simplicity, for exemplars, or the comprehension of textbook examples, are not analogous to definitions, and Johnny is learning to apply definitions. Kuhn (1977l, p. 470) perhaps better illustrates what he means by learned perception of similarity in the following passage:

> Students of physics regularly report that they have read through a chapter of their text, understood is perfectly, but nonetheless had difficulty solving the problems at the chapter's end. Almost invariably their difficulty is in setting up the appropriate equations, in relating the words and examples given in the text to the particular problems they are asked to solve. Ordinarily, also, those difficulties dissolve in the same way. The student discovers a way to see his problem as like a problem he has already encountered. Once that likeness or analogy has been seen, only manipulative difficulties remain.

Although Kuhn has abandoned his usage of *paradigms,* the reason for introducing the term should not be lost. He introduced the term to show that the scientific community has certain things in common: exemplars, values, teaching methods, metaphysical principles, and so on. Kuhn puts the matter as follows in his "Second Thoughts on Paradigms" (1977l, p. 482):

> Return, finally, to the term 'paradigm.' It entered *The Structure of Scientific Revolutions* because I, the book's historian-author, could not, when examining the membership of a scientific community, retrieve enough shared rules to account for the group's unproblematic conduct of research. Shared examples of successful practice could, I next concluded, provide what the group lacked in rules. Those examples were its paradigms, and as such essential to its continued research. Unfortunately, having gotten that far, I allowed the term's applications to expand, embracing all shared group commitments, all components of what I now wish to call disciplinary matrix. Inevitably, the result was confusion, and it obscured the original reasons for introducing a

special term. But those reasons still stand. Shared examples can serve cognitive functions commonly attributed to shared rules. When they do, knowledge develops differently from the way it does when governed by rules. This paper has, above all, been an effort to isolate, clarify, and drive home those essential points. If they can be seen, we shall be able to dispense with the term 'paradigm,' though not with the concept that led to its introduction.

And this represents his attack on positivism. By introducing paradigms Kuhn makes the unit of appraisal broader than a single theory. Moreover, he believes he provides a better account of science than that of the received view, which employs correspondence rules.

> Johnny, in short, has to apply symbolic labels to nature without anything like definitions or correspondence rules. In their absence he employs a learned but nonetheless primitive perception of similarity and difference. While acquiring the perception, he has learned something about nature. This knowledge can thereafter be embedded, not in generalizations or rules, but in the similarity relationship itself. (Kuhn, 1977l, pp. 475–76)

In many respects, Kuhn's learned similarity relationship resembles Polanyi's "tacit knowledge" because the emphasis is on knowledge gained "by doing science rather than by acquiring rules for doing it" (Kuhn, 1970c, p. 191).

The main thrust of Kuhn's work is that major changes in science occur through revolutions, "those non-cumulative developmental episodes in which an older paradigm (= disciplinary matrix) is replaced in whole or in part by an incompatible new one" (Kuhn, 1970c, p. 92). Logically, if there are revolutions, there must be nonrevolutionary periods. Kuhn calls these periods "normal science." During times of normal science the members of the scientific community refine the disciplinary matrix. If the science is mature, it is governed by one disciplinary matrix during times of normal science. (Kuhn asserts that physics is mature; the social sciences are not, in his opinion.[10])

Revolutions begin with a crisis. A crisis arises when an anomaly (that is, an unsolved scientific puzzle whose solution is crucial to the maintenance of a healthy disciplinary matrix) cannot be compromised with the prevailing disciplinary matrix. Proponents of the disciplinary matrix slowly lose confidence in it, and eventually a new matrix will emerge whose features seem attractive enough to cause most of the proponents of the old disciplinary matrix to switch to the new one. According to Kuhn (1970c, p. 77), "[T]he decision to reject one theory for another is always simultaneously the decision to accept another," and it is much like a "religious conversion" (1970c, chap. 12) or a "gestalt switch" (1970c, chap. 10). Like suddenly viewing the other side of the reversible figures so often appearing in the philosophy literature, the gestalt switch approximates switching views from bird to rabbit, or viewing a box first as facing east and then as facing west.[11] Important for understanding Kuhnian revolutions are the sociological and psychological components of science involved with the timing of the shift (Kuhn, 1970c, pp. 67–68):

> Because it demands large-scale paradigm destruction and major shifts in the problems and techniques of normal science, the emergence of new theories is

generally preceded by a period of pronounced professional insecurity. As one might expect, that insecurity is generated by the persistent failure of the puzzles of normal science to comme out as they should.

So it is clear that the choice between theories and disciplinary matrices cannot be based on logic alone.

It is worth considering this aspect of Kuhn's theory more closely, for it has led others to charge him with irrationalism and relativism. If the concept of the gestalt switch is drawn to its logical conclusions, it leads to what Kuhn and Feyerabend both refer to as the incommensurability problem. No one can justify whether seeing the duck or rabbit is better, and both duck and rabbit cannot be seen at the same time. Because of this, disciplinary matrices or theories are incommensurable; that is, there is no rational basis by which theories can be compared. Kuhn (1970b, p. 266) puts this problem into historical context when he writes:

> The point-by-point comparison of two successive theories demands a language into which at least the empirical consequences of both can be translated without loss or change. That such a language lies ready to hand has been widely assumed since at least the seventeenth century when philosophers took the neutrality of pure sensation-reports for granted and sought a 'universal character' which would display all languages for expressing them as one. Ideally the primitive vocabulary of such a language would consist of pure sense-datum terms plus syntactic connectives. Philosophers have now abandoned hope of achieving any such ideal, but many of them continue to assume that theories can be compared by recourse to a basic vocabulary consisting entirely of words which are attached to nature in ways that are unproblematic and, to the extent necessary, independent of theory. . . . Feyerabend and I have argued at length that no such vocabulary is availabile. In the transition from one theory to the next words change their meanings or conditions of applicability in subtle ways.

Moreover, the new disciplinary matrix that emerges need not necessarily better explain the world than the old one. That is to say, in his *Structure of Scientific Revolutions* Kuhn challenges the notion that science is not only rational but also a continuous, cumulative activity.

There is a final point to be made about Kuhn's philosophy of science before turning to the modifications Kuhn has made of his work, as well as criticisms. For Kuhn, like Polanyi and Fleck, scientific knowledge is the collective opinion of the scientific community: "[T]he responsibility for applying shared scientific values, must be left to the specialists' group" (Kuhn, 1970b, p. 263). Kuhn (1970a, p. 21) describes what this means for scientific explanation:

> Already it should be clear that the explanation must, in the final analysis, be psychological or sociological. It must, that is, be a description of a value system, an ideology, together with an analysis of the institutions through which that system is transmitted and enforced. Knowing what scientists value, we may hope to understand what problems they will undertake and what choices they will make in particular circumstances of conflict. I doubt that there is another sort of answer to be found.

Since the second edition of *The Structure of Scientific Revolutions* has
appeared, Kuhn has so drastically compromised his initial thesis that there is
little left that could be called revolutionary. It is certainly a major drawback in
his work that it has been so modified by qualifications of terminology—
especially of his key concept of a paradigm—in later editions, postscripts, and
replies to critics that it is now almost impossible to understand where he
stands. Adding to this difficulty is the problem with external interpretations of
his work: Kuhn (1977l, p. 459) has expressed disappointment over the indis-
criminate way in which his work has been interpreted by others:

> One aspect of the response [to my book] does, however, from time to time
> dismay me. Monitoring conversations, particularly among the book's enthusi-
> asts, I have sometimes found it hard to believe that all parties to the discussion
> had been engaged with the same volume. Part of the reason for its success is, I
> regretfully conclude, that it can be too nearly all things to all people.

Moreover, Kuhn's *new* concepts—disciplinary matrix, exemplar, and the
learned similarity relationship—also do not entirely escape this problem.
"Like the paradigm they have an intuitive appeal which prompts uncritical
acceptance and invites the sort of self-defeating plastic application which
Kuhn deplores" (Suppe, 1977a, p. 484).

Certainly because critics have charged Kuhn with irrationality and relativism, he has dedicated much of his recent work to proving that incommensurability is compatible with rationality, that is, that theories can be compared
even though a neutral language cannot be found. Kuhn (1977i, pp. 321–22)
recently argues that there are five standard criteria for evaluating the adequacy of a theory:

> First, a theory should be accurate within its domain, that is, consequences
> deducible from a theory should be in demonstrated agreement with the
> results of existing experiments and observations. Second, a theory should be
> consistent, not only internally or with itself, but also with other currently
> accepted theories applicable to related aspects of nature. Third, it should
> have broad scope: in particular, a theory's consequences should extend far
> beyond the particular observations, laws, or subtheories it was initially de-
> signed to explain. Fourth, and closely related, it should be simple, bringing
> order to phenomena that in its absence would be individually isolated and,
> as a set, confused. Fifth—a somewhat less standard item, but one of special
> importance to actual scientific decisions—a theory should be fruitful of new
> research findings: it should, that is, disclose new phenomena or previously
> unnoted relationships among those already known. These five character-
> istics—accuracy, consistency, scope, simplicity, and fruitfulness—are all stan-
> dard criteria for evaluating the adequacy of a theory.

In a similar way Kuhn denies that he is a relativist. He (1970b, p. 264)
asserts instead that he is an evolutionist: "For me, therefore, scientific development is, like biological evolution, unidirectional and irreversible. One scientific theory is not as good as another for doing what scientists normally do. In
that sense I am not a relativist."

Israel Scheffler has emphasized the weakness in Kuhn's employment of the words *revolution* and *vision* to mean a gestalt switch. The adoption of a disciplinary matrix is not an instantaneous or individual affair such as Kuhn supposes when he discusses the reversal of visual doubles. "[V]ision may perhaps approximately serve as a metaphor for *comprehension* of a paradigm or theory, though not for its testing and acceptance or rejection" (Scheffler, 1972, pp. 373–74). To reduce a revolution to a gestalt switch is, moreover, to omit many of the sociological and psychological aspects of science that Kuhn wishes to emphasize.

Kuhn's notion of normal science has been scathingly attacked by Popper and Feyerabend, among others. The reason for this is discussed quite fully and lucidly by Popper (1970, pp. 52–53):

> 'Normal' science, in Kuhn's sense, exists. It is the activity of the non-revolutionary, or more precisely, the not-too-critical professional: of the science student who accepts the ruling dogma of the day; who does not wish to challenge it; and who accepts a new revolutionary theory only if almost everybody else is ready to accept it—if it becomes fashionable by a kind of band-wagon effect. To resist a new fashion needs perhaps as much courage as was needed to bring it about. . . . In my view the 'normal' scientist, as Kuhn describes him, is a person one ought to be sorry for. . . . The 'normal' scientist, in my view, has been taught badly. I believe, and so do many others, that all teaching on the University level (and if possible below) should be training and encouragement in critical thinking. The 'normal' scientist, as described by Kuhn, has been badly taught. He has been taught in a dogmatic spirit: he is the victim of indoctrination. He has learned a technique which can be applied without asking for the reason why (especially in quantum mechanics). As a consequence, he has become what may be called an *applied* scientist, in contradistinction to what I should call a *pure scientist*. He is, as Kuhn put it, content to solve 'puzzles'. The choice of this term seems to indicate that Kuhn wishes to stress that it is not a really fundamental problem which the 'normal' scientist is prepared to tackle: it is, rather, a routine problem, a problem of applying what one has learned: Kuhn describes it as a problem in which a dominant theory (which he calls a 'paradigm') is applied. The success of the 'normal' scientist consists, entirely, in showing that the ruling theory can be properly and satisfactorily applied in order to reach a solution of the puzzle in question.

Despite the enormous problems with *The Structure of Scientific Revolutions,* Kuhn's work has been quite influential in his own field and in the philosophy of science in the sense that it has been widely read by professionals and has generated considerable controversy, as well as at least two colloquia.[12] Historians of science, including Kuhn, however, have not adopted the framework of revolutions to interpret history.[13] As the above-mentioned criticisms clearly suggest, the result of the modifications of the theory is that the original work loses much of its revolutionary character.[14] Sociologists, biologists, economists, and other social scientists have nonetheless blithely applied Kuhn's framework to the "softer" fields. Ironic is the fact that Kuhn regards all of these fields as prescientific disciplines not subject to the "normal science–crisis–revolution–normal science with a new disciplinary matrix" pattern.

Suppe (1977b, pp. 647–48) has summed up Kuhn's contribution in the following way:

> *First*, as Kuhn has modified and attentuated his views, it has seemed to many that he was retreating toward a neopositivistic view. . . . *Second,* there is the growing perception, defended briefly above, that Kuhn's account unnecessarily but essentially shortchanges the role of rationality in the growth of science. *Third,* there is a growing skepticism over Kuhn's claims that the history of science exemplifies his views on how science oscillates between normal and revolutionary science, despite brief attempts of his to counter such doubts. . . . *Fourth,* there is a growing realization that Kuhn's position commits him to a metaphysical and epistemological view of science which is fundamentally defective since it makes discovering how the world really is irrelevant to scientific knowledge, reducing scientific knowledge to the collective beliefs of members of scientific disciplines.

Newton-Smith (1981, p. 116) perhaps offers a more balanced view:[15]

> Kuhn's over-reaction has served to counter the once popular position on the opposite extreme of the spectrum. For many philosophers of science thought that while the process of theory discovery was not rule-governed, the process of theory choice in the face of a given body of evidence was capable of being represented in a system of binding rules.

Notes

1. See Baker (1979) and Manno (1977) for further biographical information on Polanyi.

2. As mentioned in chapter 1, this view that the method of physics is the measure of all science is no longer accepted by most philosophers today. In his elementary textbook on the philosophy of science (1982, p. 166), A. F. Chalmers reaches the following conclusion:

> . . . I suggest that the question that constitutes the title of this book [What Is This Thing Called Science?] is a misleading and presumptuous one. It presumes that there is a single category "science", and implies that various areas of knowledge, physics, biology, history, sociology and so on, either come under that category or do not. I do not know how such a general characterization of science can be established or defended. Philosophers do not have resources that enable them to legislate on the criteria that must be satisfied if an area of knowledge is to be acceptable or "scientific". Each area of knowledge can be analyzed for what it is.

See also note 10.

3. It is uncertain how Fleck learned of the writings of the Vienna Circle. Schäfer and Schnelle, editors of the 1980 reprint of Fleck's *Entstehung,* assume that Fleck learned of the positivist philosophy through the meetings of the Lwów-Warszawa philosophers that he attended:

> Von Twardowskis Schülern ausgehend, entstand in Polen zu jener Zeit die "Lwów-Warszawa"-Schule (vgl. Zamecki 1977), eine vom Wiener Kreis stark

beeinflußte neopositivistische Strömung. Auch dieser Kreis war an interdisziplinären Kontakten sehr interessiert und bot entsprechende Diskussionsrunden. Auch hier war Fleck regelmäßiger Teilnehmer. Vermutlich ist Flecks Vertrautheit mit der Philosophie des Wiener Kreises, gegen die er sein Buch richtet, über die Verbindung mit der Twardowski-Schule zu erklären (Fleck, 1980, p. xviii).

4. The passage in Fleck (1980, p. 137) reads: "Jede didaktische Einführung ist also wörtlich eine Hineinführung, ein sanfter Zwang."

5. The original German text (Fleck, 1980, p. 130) is "Wir können also Denkstil als gerichtetes Wahrnehmen, mit entsprechendem gedanklichen und sachlichen Verarbeiten des Wahrgenommenen, definieren."

6. The German passage (Fleck, 1980, p. 135) follows:

Den gemeinschaftlichen Träger des Denkstiles nennen wir: das Denkkollektiv. Dem Begriff des Denkkollektives, wie wir ihn als Untersuchungsmittel sozialer Bedingtheit des Denkens verwenden, kommt nicht der Wert einer fixen Gruppe oder Gesellschaftsklasse zu. Es ist sozusagen mehr funktioneller als substanzieller Begriff, dem Kraftfeldbegriffe der Physik z.B. vergleichbar. Ein Denkkollektiv ist immer dann vorhanden, wenn zwei oder mehrere Menschen Gedanken austauschen: dies sind momentane, zufällige Denkkollektive, die jeden Augenblick entstehen und vergehen. Doch auch in ihnen stellt sich eine besondere Stimmung ein, der keiner der Teilnehmer sonst habhaft wird, die aber oft wiederkehrt, wenn die bestimmten Personen zusammenkommen.

7. I do not mean to imply that Toulmin is intellectually indebted to Fleck. Nothing appears in Toulmin's *Human Understanding* to indicate that he is familiar with Fleck's *Entstehung*. (Fleck's name does not appear in the index.) Toulmin, however, does draw from Darwin. It is certainly true that Kuhn refers to his philosophy as "evolutionary," and in this way has much in common with Fleck and Toulmin, and even with the contemporary Popper. (See note 8.)

8. There are some similarities between the lives and works of Karl Popper and Ludwik Fleck. Popper's *Logik der Forschung* was first published in 1934, and like Fleck's *Entstehung,* was a refutation of certain aspects of the philosophy of the Vienna Circle. It, too, was ignored, in part due to an unreceptive political climate. In addition, Popper's more modern work (for example, 1987b) deals increasingly with evolutionary aspects of the philosophy of science and the significance of Darwin. Compare note 2, chapter 4.

9. The first point about Kuhn's work is that no one can understand it by simply reading his *Structure of Scientific Revolutions,* which first appeared in 1962 and was followed in 1970 by a second, enlarged edition. In the "Postscript 1969" (pp. 174–210) of the second edition Kuhn discusses the problems with his usage of the word *paradigm,* and attempts to meet criticisms of his critics in general. Further modifications of his position appear in Kuhn 1970a, pp. 1–23 (repr. in his 1977b, pp. 266–92); Kuhn 1970b, pp. 231–78; Kuhn 1977l, pp. 458–82 (repr. in his 1977b, pp. 293–319); Kuhn 1977i, pp. 320–39.

Second, usually when Kuhn uses the term *disciplinary matrix* (or *paradigm*), he means a theory or complex of theories: "Elsewhere I use the term 'paradigm' rather than 'theory' to denote what is rejected and replaced during scientific revolutions" (Kuhn, 1970a, p. 2 n. 1).

Finally, Kuhn's change of terminology—to *disciplinary matrix* and *exemplars*—in fact corresponds quite closely to Fleck's discussion of the *Denkkollektiv* (the scientific community limited to one particular discipline) and the *Denkstil* (the scientific material or substance carried by the *Denkkollektiv*). Kuhn acknowledges Fleck on pp. vi–vii of the preface to *The Structure of Scientific Revolutions;* the intellectual debt to Fleck has undoubtedly grown with Kuhn's modification of "paradigms."

10. Kuhn considers physics to be the prototype science (as do most philosophers of science). To a great extent he identifies the maturity of a science with how closely the methods correspond to those of physics. That is reflected in the following passage (Kuhn, 1970b, pp. 245–46):

> Fortunately, though no prescription will force it, the transition to maturity does come to many fields, and it is well worth waiting and struggling to attain. Each of the currently established sciences has emerged from a previously more speculative branch of natural philosophy, medicine, or the crafts at some relatively well-defined period in the past. Other fields will surely experience the same transition in the future. Only after it occurs does progress become an obvious characteristic of a field. And only then do those prescriptions of mine which my critics decry come into play. . . . Such a field first gains maturity when provided with theory and technique which satisfy the four following conditions. First is Sir Karl's demarcation criterion without which no field is potentially a science: for some range of natural phenomena concrete predictions must emerge from the practice of the field. Second, for some interesting subclass of phenomena, whatever passes for predictive success must be consistently achieved. . . . Third, predictive techniques must have roots in a theory which, however metaphysical, simultaneously justifies them, explains their limited success, and suggests means for their improvement in both precision and scope. Finally, the improvement of predictive technique must be a challenging task, demanding on occasions the very highest measure of talent and devotion.
>
> These conditions are, of course, tantamount to the description of a good scientific theory.

There is much that speaks against this view. See, for example, note 2.

11. The reversible figures, or visual doubles, which are scattered across the pages of modern philosophy of science works, do appear in some economic textbooks, for example, in Paul Samuelson's elementary economic texts. Two of the most common visual doubles are reproduced below. This is what Kuhn is referring to when he writes about gestalt switches from duck to rabbit or from boxes viewed from above to boxes viewed below. The point of this exercise is to demonstrate that the positivists' efforts to produce a neutral observation language must fail. N. R. Hanson devoted much of his life work to this theme.

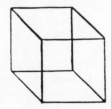

FIGURE 3–1. Visual Doubles or Reversible Figures

12. Musgrave and Lakatos, eds., *Criticism and the Growth of Knowledge* (1970) grew out of an international colloquium held in London on Kuhn's work in 1965. The follow-up to that volume is *Progress and Rationality in Science,* volume 58 of the Boston Studies in the Philosophy of Science series, edited by Gerard Radnitzky and Gunnar Andersson (1978).

13. Although historians of science have not used Kuhn's framework of revolutions to explain other aspects of the history of science, other historians have. According to the historian David Hollinger (1973, pp. 371–72), Kuhn's framework has been employed "explicitly by historians of art, religion, political organization, social thought, and American foreign policy, in addition to their more predictable use by historians of the social sciences and the natural sciences." He adds (p. 372):

> The result is that many "applications" of Kuhn take the somewhat incongruous form of analogies between science and non-science; a Kuhnian "scientific revolution" is explicitly compared to the American decision to withdraw from Vietnam under the pressure of anti-war demonstrations at the Pentagon and at Chicago.

Faber (1977) and Veit-Brause (1985) criticize historians' use of Kuhn's framework.

14. One truly revolutionary aspect of Kuhn's work is that *The Structure of Scientific Revolutions* was published as one of the final monographs in the International Encyclopedia of Unified Science series, the mouthpiece of logical positivism.

I found Kuhn's work difficult to represent fairly because Kuhn has altered his position so extensively. Kuhn admits errors and is willing to modify his position; his intellectual honesty is really very refreshing. Newton-Smith (1981, p. 124) explains that there are actually two Kuhns:

> One, the temperate non-rationalist, takes it that while there is agreement on the factors which guide theory choice, these factors cannot be justified. The other, the embryonic rationalist, takes it that the five ways (the five criteria of theory choice) can be justified as the criteria to be used in achieving progress in science; that is, in increasing puzzle-solving capacity.

Criticisms of Kuhn's work have under no circumstances been exhausted. The criticisms that appear in the text seem to me to be the most damaging criticisms.

For a fuller treatment of Kuhn and his work, see Ben-David (1964); Böhler (1972); H. I. Brown (1983); Diederich (1986, 1974); Glymour (1980, chap. 4)*; Gutting (1980)†; Grünfeld (1979); Hacking (1981b); King (1971); Kockelmans (1972); Krüger (1974); Lakatos and Musgrave (1970); Newton-Smith (1981, chaps. v–vi); Scheffler (1972); Shapere (1971, 1964); Shapin (1982); Stegmüller (1987, vol. 3, chap. 3); Stove (1982); Suppe (1977a, 1978b, pp. 135–51); Toulmin (1972b, pp. 98–130). See also the works of Karl Popper, Imre Lakatos, and Paul Feyerabend.

I. B. Cohen (1976, 1985) treats the origins of the concept scientific revolution. On the history of science, see Kuhn (1977g and 1977j) and Ravetz (1985).

Something that I have omitted but that Kuhn is known for, is his advocacy of a balanced use of the external and internal history of science. Internal history of

*Glymour's "New Fuzziness and Old Problems" (1980, chap. 4) treats variations on Kuhn's philosophy, e.g., the work of Laudan, Sneed, and Stegmüller, which I have not been able to discuss here due to space considerations..

†Pages 321–39 of Gutting's work contain a useful bibliography on Kuhn and applications of his philosophy to other fields.

science—often associated with the work of Alexandre Koyré—focuses on the intellectual aspects of the development of science. The historiography is to refer to the textbooks and journals of the particular period under study. This is the dominant approach in the history of science. The external history of science considers institutional factors. For instance, social values become important because they determine which questions scientists focus on. Kuhn advocates the external historiographical approach be used, not exclusively but in complement to the internal approach, and maintains that "[p]utting the two together is perhaps the greatest challenge now faced by the profession, and there are increasing signs of a response" (Kuhn, 1977g, p. 110).

15. This comment that Newton-Smith's view may be more balanced than Suppe's has been made in order to bring to the reader's attention the fact that philosophers of science have a particular bias. Both Suppe and Newton-Smith continue in the tradition of logical empiricism. Both are rationalists. Suppe is, however, more inclined to dismiss Kuhn's work as psychology rather than philosophy of science. Suppe concludes in his "Exemplars, Theories and Disciplinary Matrixes" (1977a, p. 499):

> But as one has also come to expect, Kuhn has clouded these insights by his insistence on reifying his insights into shared entities—paradigms in *The Structure of Scientific Revolutions* and now disciplinary matrixes and shared resemblance relations—which are supposed to account for and explain the features of the scientific enterprise he discusses. His insights are too important, and potentially too valuable, to be clouded or obscured by such exercises in reification—and I wish he would stop populating science with new entities so that we more easily can have access to his insights.

Suppe's critique is correct but ignores the fact that Kuhn's work is valuable because of its mistakes.

4

The Popperian School

There is a certain risk involved in grouping Popper, Lakatos, Feyerabend, and Bartley together under the heading "the Popperian school." Their philosophies are in some ways glaringly divergent. Despite their differences, their works do have common features: perhaps most important, they take Popper's philosophy as their starting point. Lakatos, Feyerabend, and Bartley were all Popper's students and have achieved a reputation of their own. The Popperian school is hardly limited to the small group that will be discussed here. Among Popper's former students and sympathizers are Joseph Agassi (student), Ian C. Jarvie (student), J.W.N. Watkins (student; he received Popper's chair at the London School of Economics [LSE] after Popper retired in 1959), A. I. Sabra (student), Preston King and J. O. Wisdom (philosophers at the LSE), Donald T. Campbell (psychologist), F. A. von Hayek (economist at the LSE), Sir John Eccles (neurophysicist), Sir Peter Medawar (biologist), Jacques Monad (biochemist), Noretta Koertge (student), A. F. Chalmers (student), Hans Albert (economist), and Helmut Spinner (philosopher). The Popperian school has a closer tie to the positivists and logical empiricists than the group of sociological philosophers of science already examined. This is especially true in Popper's case.

Sir Karl Popper

Sir Karl Popper was born in Vienna in 1902, and received his Ph.D. from the University of Vienna in 1928.[1] Popper attended some of the meetings of the Vienna Circle; he, however, had differences with the circle's members and thus prefers not to be called a positivist. He was in fact nicknamed "the official opposition" by Otto Neurath.

Popper is perhaps best known for his *Logik der Forschung* (English version, *The Logic of Scientific Discovery* published in 1959), an antipositivist work that first came out in 1934 as number 9 of the Vienna Circle's Schriften zur wissenschaftlichen Weltauffassung. Although read by some of the mem-

27

bers of the Vienna Circle, Popper's *Logik der Forschung* was not very warmly received by the German-speaking world in general.[2]

Correctly predicting that war would break out in Europe, Popper left Vienna in 1937 for New Zealand, where he taught at Canterbury University College in Christchurch until 1945.[3] His war work (against totalitarianism and the irrationality of war) is *The Open Society and Its Enemies* (1943). That and the accompanying piece, *The Poverty of Historicism* (1957, 1960), established his reputation as a social and political philosopher.

After the war Popper settled in England and took up his work at the London School of Economics, where he taught until his retirement in 1969. In many ways Popper has gone against the mainstream of philosophical thought. Consequently, the academic world has not always been appreciative of his contributions. His work, however, has been recognized by the British society in particular (he was knighted in 1965) and by philosophers around the world in general.

Popper's relationship with the Vienna Circle is worth investigating because it has been the source of various squabbles. For instance, his opponents, especially members of the Frankfurt school, label him a positivist. As Victor Kraft (1974, p. 185), the circle's historian, notes, "Popper never belonged to the Vienna Circle, never took part in its meetings, and yet cannot be thought of as outside it."[4] As has been mentioned, in the sixties John Passmore pronounced positivism in its classical form dead. Popper (1974a, p. 69) takes responsibility for its death. Popper, however, has been criticized by several members of the Vienna Circle for overemphasizing his differences with them (Popper, 1974b, p. 971). Popper insists he was not a member of the circle, yet admits that many members "accepted most of [his] ideas," namely Carnap, Feigl, Frank, Gomperz, Hahn, Hempel, Kraft, Menger, R. von Mises, and Waismann (Popper, 1974b, p. 970). Popper (1974b, p. 970) names Neurath, Reichenbach, and Schlick as those who felt his ideas were "dangerous and to be combated." Certainly, Popper's arguments transformed the ideas of some of the members of the Vienna Circle and vice versa, but the dispersion of circle members because of the war, as well as changing times, also contributed to the intellectual evolution of the positivists.[5]

Popper and the positivists have much in common. They are empiricists, although Popper emphasizes the rational component of knowledge.[6] Popper (1974a, p. 70) reveals perhaps the major unifying factor:

> But what attracted me perhaps most to the Vienna Circle was the "scientific attitude" or, as I now prefer to call it, the rational attitude. . . . It is in this general attitude, the attitude of the enlightenment, and in this critical view of philosophy—of what philosophy unfortunately is, and of what it ought to be—that I still feel very much at one with the Vienna Circle and with its spiritual father, Bertrand Russell. This explains perhaps why I was sometimes thought by members of the Circle, such as Carnap, to be one of them, and to overstress my differences with them.

As Kraft (1974, p. 200) summarizes, "There was no unbridgeable opposition (between them); rather a common basis."[7]

Hume's Problem, or the Problem of Induction

Perhaps the most radical aspect of Popper's philosophy of science is his complete rejection of induction (Popper, 1974a, p. 116): "As for induction (or inductive logic, or inductive behaviour, or learning by induction or by repetition or by "instruction") I assert that there is no such thing." Popper's is a purely deductive philosophy.[8] This rejection of induction is one of the distinguishing marks of many of the members of the Popperian school.[9] It represents perhaps the greatest division between Popper and the works of the positivists and logical empiricists.

Popper's rejection of induction rests on a belief that the prior probability of any law must equal zero.[10] The intuitive proof of this runs as follows. Popper argues that no matter how often one encounters white swans and only white swans, the universal statement "All swans are white" can never be confirmed, as the logical positivists had believed: the future could yield a black swan. Because the future could yield infinitely many black swans, the prior probability of "All swans are white" (and other universal generalizations) being true must be zero.

Although the above reasoning makes good sense, Popper's position must be viewed as an extreme one. One philosopher (Andersson, 1986, p. 242) comments humorously that "[a]fter Hume and Popper it [induction] is a 'mortal sin of the first order.' " And, indeed, as Popper (1974b, p. 1015) himself points out, his position is far more extreme than Hume's:

> I agree with Hume's opinion that induction is invalid and in no sense justified. Consequently neither Hume nor I can accept the traditional formulations which uncritically ask for the justification of induction; such a request is uncritical because it is blind to the possibility that induction is invalid *in every sense,* and therefore *unjustifiable.*
>
> I disagree with Hume's opinion (the opinion incidentally of almost all philosophers) that induction is a fact and in any case needed. I hold that neither animals nor men use any procedure like induction, or any argumennt based on the repetition of instances. The belief that we use induction is simply a mistake. It is a kind of optical illusion.

Critical Rationalism and the Tie to Falsification

Popper argues that he replaces the positivists' justification of theories with their criticism, the positivists' criterion of meaning with a criterion of demarcation.[11] Victor Kraft (1974, pp. 188–89) relates that

> [t]he problem, which formed the starting point for Popper's epistemological investigations did not come to him in reading philosophical literature, but forced itself upon him in his own thinking. It grew out of his study of theories which were current at that time: Marx's philosophy of history, Freud's psychoanalysis, and Alfred Adler's individual psychology. Because of his doubts about these theories he found himself confronted by these questions: How

can one decide whether a theory is correct? How do scientific statements get their validity? How can we distinguish between scientific and unscientific assertions?

Popper is fighting intellectual relativism; it was in part his war experience, his realization that the Viennese version of Marxism was indefensible and dangerous, that sparked his work.[12] To Popper, critical rationalism is the Socratic ideal of rational discussion,[13] and he (1976c, p. 292) concludes, "If the method of rational critical discussion should establish itself, then this should make the use of violence obsolete: *critical reason is the only alternative to violence so far discovered.*"

Intertwined with his critical method and his answer to induction is his criterion of demarcation, falsification:

> This solution to the problem of induction gives rise to a new theory on the method of science, to an analysis of the *critical method,* the method of trial and error: the method of proposing bold hypotheses, and exposing them to the severest criticism, in order to detect where we have erred.
>
> From the point of view of this methodology, we start our investigation with *problems.* We always find ourselves in a certain problem situation; and we choose a problem which we hope we may be able to solve. The solution, always tentative, consists in a theory, a hypothesis, a conjecture. The various competing theories are compared and critically discussed, in order to detect their shortcomings; and the always changing, always inconclusive results of the critical discussion constitute what may be called "the science of the day."
>
> Thus *there is no induction:* we never argue from facts to theories, unless by way of refutation or "falsification." This view of science may be described as selective, as Darwinian. By contrast, theories of method which assert that we proceed by induction or which stress *verification* (rather than *falsification*) are typically Lamarckian: they stress *instruction* by the environment rather than *selection* by the environment. (Popper, 1974a, p. 68)

Popper's method of trial-and-error elimination, the *modus tollens,* takes a dialectic form that he (1974a, pp. 105–6) describes in the following passage:

> . . . all scientific discussions start with a problem (P_1), to which we offer some sort of tentative solution—a *tentative theory* (TT); this theory is then criticized, in an attempt at *error elimination* (EE); and as in the case of dialectic, this process renews itself: the theory and its critical revision give rise to new *problems* (P_2).
>
> Later I condensed this into the following schema:
>
> $$P_1 \rightarrow TT \rightarrow EE \rightarrow P_2 ,$$
>
> a schema which I often used in lectures.
>
> I liked to sum up this schema by saying that science *begins with problems, and ends with problems.* But I was always a little worried about this summary, for every scientific problem arises, in its turn, in a theoretical context. It is soaked in theory. So I used to say that we may begin the schema at any place: we may begin with TT_1 and end with TT_2; or we may begin with EE_1 and end with EE_2. However, I used to add that it is often from some *practical problem*

that a theoretical development starts; and although any formulation of a practical problem unavoidably brings in theory, the practical problem itself may be just "felt.". . .

Popper's account of science, then, starts with *problems* and not observations, as the positivists had postulated. In simple form, Popper's message: falsify theories, don't confirm them. And falsification is at first glance an intuitively appealing view of science. Whereas a universal statement cannot be confirmed, it can be deduced as false. Consider, for instance, the observation "A black swan was seen at place x in Germany at time t." From this we can easily deduce that the universal statement "All swans are white" is false.

Falsifiability requires that an observation statement can be formulated that, if found true, would contradict the hypothesis, thus falsifying it. For instance, the statement "Either the sun will or won't shine on August 15th" is not falsifiable; no observation statement can be formulated that would falsify the statement. (As a consequence, tautologies and definitions are also not falsifiable.) All scientific theories or "conjectures" should be submitted to crucial tests or experiments that may falsify them. Falsification occurs if the hypothesis fails to stand up to the observational tests and experiments.[14]

Popper's view of the growth of knowledge and of objectivity both stem from his critical rationalism. An objective theory is one that "is arguable, which can be exposed to rational criticism, preferably a theory which can be tested; not one which merely appeals to our subjective intuitions" (Popper, 1974a, p. 110). The growth of science, contends Popper, was largely ignored by the positivists and is central to his work. The critical method becomes the "instrument of growth" (Popper, 1974a, p. 91):

> In *Logik der Forschung* I tried to show that our knowledge grows through trial and error-elimination, and that the main difference between its pre-scientific and its scientific growth is that on the scientific level we consciously search for our errors: *the conscious adoption of the critical method* becomes the main instrument of growth. It seems that already at that time I was well aware that the critical method—or the critical approach—consists, generally, in the search for difficulties or contradictions and their tentative resolution, and that this approach could be carried far beyond science, for which alone *critical tests* are characteristic.

Not only should science proliferate theories as much as possible, subjecting them to crucial tests so that science grows, it should propose "bold conjectures," that is, conjectures that run a great risk of being false. Highly falsifiable theories (those that rule out more) are preferred to less falsifiable ones. (Hence, for example, "All swans are white" is preferred to "All swans are white, grey, or black.") Unlike the inductivists, scientists should not be cautious. Science, according to Popper, progresses by trial *and* error; scientists learn from their mistakes, from the falsifications of theories. As science grows, theories become more falsifiable and accumulate ever higher levels of information content. The criterion specifying that conjectures be highly falsifiable helps ensure that they are sharply formulated. Once a theory has withstood severe tests, it is said to have been "corroborated" or "confirmed" (which should under no circumstance be confused with the positivists' usage of *confirmation* to mean established as true). The history of a mature science, for Popper, is the piecemeal approximation of a group of theories ever closer to the truth.

Why Falsification Fails

There seem to be five very convincing reasons that falsification fails. First, theories, unlike the single statement "All swans are white," are complex webs of assumptions, laws, and various conditions. Since the unit of appraisal is in practice not a simple statement, the scientist cannot know which assumption of the theory is causing the problem. The scientist can only conclude that at least one of the many assumptions is false; hence, the theory cannot be conclusively falsified. This is the Duhem problem, which is treated more fully in the next section, where Lakatos is discussed.

Second, Popper fails to develop a fully noninductive schema; the induction problem resurfaces. Popper does not believe we can ever know the truth; the goal of science is to obtain not the truth but increasing verisimilitude, or increasing "truth" content. Then how does one know when one theory is better than another? Theory comparison depends on the degree of corroboration, or how well a theory has stood up to severe tests (Popper, 1974a, p. 82). So if theory A has passed one hundred severe tests, we infer that it will pass more and is hence reliable: induction. Popper realizes induction reappears, but does not modify his extreme anti-inductivist position. The following discussion, which appears in the Schilpp volume at footnote 165b (Popper, 1974b, pp. 1192–93), is lengthy but worthy of reproduction in full:

> Truthlikeness or verisimilitude is very important. For there is a probabilistic though typically noninductivist argument which is invalid if it is used to establish the probability of a theory's being true, but which becomes valid (though essentially nonnumerical) if we replace truth by verisimilitude. The argument can be used only by realists who not only assume that there is a real world but also that this world is by and large more similar to the way modern theories describe it than to the way superseded theories describe it. On this basis we can argue that it would be a highly improbable coincidence if a theory like Einstein's could correctly predict very precise measurements not predicted by its predecessors unless there is "some truth" in it. This must not be interpreted to mean that it is improbable that the theory is not true (and hence probable that it is true). But it can be interpreted to mean that it is probable that the theory has both a high truth content and a high degree of verisimilitude; which means here only, "a higher degree of verisimilitude *than those of its competitors* which led to predictions that were less successful, and which are thus less well corroborated".
>
> The argument is typically noninductive because in contradistinction to inductive arguments such as Carnap's the probability that the theory in question has a high degree of verisimilitude is (like degree of corroboration)

inverse to the initial probability of the theory, prior to testing. Moreover, it only establishes a probability of verisimilitude relative to its competitors (and especially to its predecessors). In spite of this, *there may be a "whiff" of inductivism here* [my emphasis]. It enters with the vague realist assumption that reality, though unknown, is in some respects similar to what science tells us or, in other words, with the assumption that science can progress towards greater verisimilitude.

Newton-Smith (1981, p. 68) analyzes Popper's reference to a "whiff" of inductivism in the following way; "On one meaning of the word 'whiff' a whiff is 'a kind of flatfish', and certainly this argument is kind of fishy. On another construal 'whiff' is a puff of air. But it is just false to say that there is a whiff of inductivism here—there is a full-blown storm."

Popper's treatment of *ad hoc* theories and growth of science, then, is doomed. For Popper, a theory is said to be *ad hoc* if it cannot be "independently tested." If it is independently testable, the theory's truth content increases because of the *ad hoc* modification; if it is not independently testable, Popper treats the *ad hoc* modification as mere deflection of criticism. But insistence on increasing truth content fails in a purely deductive system. Popper's method also rests on being able to establish that progress can be made—something that he cannot do without induction.

The third way in which falsification is inadequate is that the history of science indicates that the best theories would have been rejected had falsification really been the method followed.[15] History shows that the practice of science has not been one of rejecting theories when observation conflicts with the theory. Popper (1974b, p. 984) explains this problem in his "Replies to My Critics" under §6, entitled "Difficulties of the Demarcation Proposal":

> (8) But I was yet a little more self-critical in *Logik der Forschung:* I first noticed that such a rule of method is, necessarily, somewhat vague—as is the problem of demarcation altogether. Clearly, one can say that if you avoid falsification *at any price,* you give up empirical science in my sense. But I found that, in addition, supersensitivity with respect to refuting criticism was just as dangerous: there is a legitimate place for dogmatism, though a very limited place. He who gives up his theory too easily in the face of apparent refutations will never discover the possibilities inherent in his theory. *There is room in science for debate:* for attack and therefore also for defence. Only if we try to defend them can we learn all the different possibilities inherent in our theories. As always, science is conjecture. You have to conjecture when to stop defending a favourite theory, and when to try a new one.
>
> (9) Thus I did not propose the simple rule: "Look out for refutations, and never dogmatically defend your theory." Still, it was much better advice than dogmatic defence at any price. The truth is that we must be constantly critical; self-critical with respect to our own theories, and self-critical with respect to our own criticism; and, of course, we must never evade an issue.

This, then, is roughly the *methodological form* of (D), of the criterion of demarcation. Propose theories which can be criticized. Think about possible decisive falsifying experiments—crucial experiments. But do not give up your theories too easily, not, at any rate, before you have critically examined your criticism.

A related matter, and the fourth point, is that Popper overestimates the objectivity of scientists. As Lakatos once remarked, "You know a scientist who wants to falsify his theory?" (Newton-Smith, 1981, p. 52). Falsification can in fact be "immunized" (Popper, 1974b, p. 983):

(6) As I explained in the very first chapter of *Logik der Forschung,* we can always adopt evasive tactics in the face of refutations. I called these tactics (for historical reasons) "conventionalist stratagems (or twists)", but my friend Professor Hans Albert has found a much better term for them. He calls them *"immunizing tactics or stratagems"*: we can always immunize a theory against refutation. There are many such evasive immunizing tactics; and if nothing better occurs to us, we can always deny the objectivity—or even the existence—of the refuting observation. (Remember the people who *refused* to look through Galileo's telescope.) Those intellectuals who are more interested in being right than in learning something interesting but unexpected are by no means rare exceptions.

(7) None of the difficulties so far discussed is terribly serious: it may seem that a little intellectual honesty would go a long way to overcome them. By and large this is true. But how can we describe this intellectual honesty in logical terms? I described it in *Logik der Forschung* as *a rule of method,* or a *methodological rule:* "Do not try to evade falsification, but stick your neck out!

The final difficulty is that observation statements (what Popper refers to as "basic statements") are fallible. Since it is to observation statements that we turn to falsify a theory conclusively, Popper's empirical basis for science must be absolute. But this is not the case. Popper (1972b, p. 111) discusses this problem with the use of a lovely metaphor in a now famous passage of his *Logic of Scientific Discovery:*

The empirical basis of objective science has thus nothing 'absolute' about it. Science does not rest upon solid bedrock. The bold structure of its theories rises, as it were, above a swamp. It is like a building erected on piles. The piles are driven down from above into the swamp, but not down to any natural or 'given' base; and if we stop driving the piles deeper, it is not because we have reached firm ground. We simply stop when we are satisfied that the piles are firm enough to carry the structure, at least for the time being.

Popper argues in *The Logic of Scientific Discovery* (1972b, §30) that acceptance and rejection of basic statements ultimately rest on a decision reached through a process much like a trial by jury. (In chapter 5 in the section on N. R. Hanson the fallibility of observation statements is discussed in greater detail.) The fact that science rests on a decision brings Newton-Smith (1981, p. 64) to the conclusion that we have returned to Kuhn's irrationalism: hence, he dubs Popper "the irrational rationalist." He also reminds us that many philosophers and academicians do not reject Popper's philosophy because "they do not succeed in thinking themselves into the system" (Newton-Smith, 1981, p. 62).[16] I think the real reason for this is that they sympathize with his metaphysical position: it is difficult not to appreciate this stance, which Popper (1966b, p. 225) characterizes as "fundamentally an attitude of admitting that *'I may be wrong and you may be right, and by an effort, we may get nearer to the truth.'* "

But Popper's critical rationalism—the Socratic element of his philosophy of science—need not be renounced because falsification fails. They are mutually exclusive. Certainly, Popper's critical rationalism is attractive, and then not only with respect to the natural sciences. Bartley further develops the school in this direction, while Lakatos develops the idea that empirical problems—when evidence conflicts with theory—cannot lead to the overthrow or refutation of a theory.

Imre Lakatos: Metaphysics Transformed into the Methodology of Scientific Research Programs

Imre Lakatos was born in Budapest on 5 November 1922. Schooled as a philosopher of mathematics, he became famous for his philosophy of science. He is often deemed second to Popper as the leading representative of critical rationalism.

Lakatos studied mathematics, physics, and philosophy in Debrecen (in eastern Hungary) and in Budapest from 1944 to 1948; in 1948 he received his Ph.D. He was politically active and a member of the anti-Nazi resistance movement during the German occupation. In 1947 he became a high official in the Hungarian Ministry of Education; his revisionary ideas, however, led to his imprisonment from 1950 to 1953. During the Hungarian uprising of 1956 Lakatos was threatened again with the prospect of imprisonment and consequently fled to Vienna. From there he made his way to Britain, where he commenced his study of the philosophy of mathematics at Cambridge. In 1960 he received his Ph.D. from Cambridge and took up a post at the London School of Economics. There he became famous for his methodology of scientific research programs (generally abbreviated to MSRP). In addition, Lakatos was responsible for organizing several famous philosophy of science colloquia; the one in 1965 brought Rudolf Carnap and Karl Popper together after decades of estrangement. Lakatos died suddenly and prematurely (at age fifty-one) in 1974, at the zenith of his career.[17]

Lakatos (1974b, p. 318) believed he was correcting deficiencies in Popper's work:

> I have tried to amend the falsificationist definition of science so that it no longer rules out essential gambits of actual science. I tried to bring about such an amendment, *primarily by shifting the problem of appraising theories to the problem of appraising historical series of theories, or, rather, of 'research programmes', and by changing the falsificationist rules of theory rejection.*

Although the reader is barraged by new terminology—research programs, hard core, protective belt, positive and negative heuristic, progressive and degenerating programs—a basic understanding of the working of Lakatos's MSRP is easily acquired. Lakatos (1974c, p. 8; 1978b, p. 4) utilizes a broad framework called a research program for the following reasons:

> [T]he typical descriptive unit of great scientific achievements is not an iso-
> lated hypothesis but rather a research programme. Science is not simply trial-
> and-error, a series of conjectures and refutations. 'All swans are white' may
> be falsified by the discovery of one black swan. But such trivial trial and error
> does not rank as science. Newtonian science, for instance, is not simply a set
> of four conjectures—the three laws of mechanics and the law of gravitation.
> These four laws constitute only the 'hard core' of the Newtonian programme.
> But this hard core is tenaciously protected from refutation by a vast 'protec-
> tive belt' of auxiliary hypotheses. And, even more importantly, the research
> programme also has a 'heuristic', that is, a powerful problem-solving machin-
> ery, which, with the help of sophisticated mathematical techniques, digests
> anomalies and even turns them into positive evidence.

The hard core mentioned above by Lakatos is the framework of general hypotheses that make up the theory. The hard core will not be falsified by the adherents of the programme; it is, for all purposes, taken as given. The negative heuristic is the condition that the hard core remain unmodified. If a scientist opts to modify the hard core, the scientist automatically chooses to work on a new research program. A protective belt contains the set of auxil-iary hypotheses, conditions, and observation statements lying outside the hard core but that, unlike the hard core, may be falsified. The positive heuris-tic, which is not altogether unambiguously defined by Lakatos (1970, p. 135; 1978b, p. 50), "consists of a partially articulated set of suggestions or hints on how to change, develop the 'refutable variants' of the research-programme, how to modify, sophisticate, the 'refutable' protective belt."

With this new apparatus Lakatos restricts Popper's concept of falsifica-tion. This point is summed up in Lakatos's key article "Falsification and the Methodology of Scientific Research Programmes" (1970, p. 133; 1978b, p. 48):

> All scientific research programmes may be characterized by their 'hard core'.
> The negative heuristic of the programme forbids us to direct the *modus*
> *tollens* at this 'hard core'. Instead, we must use our ingenuity to articulate or
> even invent 'auxiliary hypotheses', which form a *protective belt* around this
> core, and we must redirect the *modus tollens* to *these*.

But when falsification is restricted to the protective belt, the whole concept changes. Essentially, when Lakatos says a theory is falsified, he does not mean rejected but that most scientists quit working with it. Although Lakatos notes this, and in fact labels Popperian falsificationism "naive falsificationism" com-pared with his "sophisticated falsificationism," this change in usage cannot be stressed enough. Lakatos (1970, p. 122; 1978b, p. 37) remarks that

[t]he problem-shift from naive to sophisticated falsification involves a seman-
tic difficulty. For the naive falsificationist a 'refutation' is an experimental
result which, by force of his decisions, is made to conflict with the theory
under test. But according to sophisticated falsificationism one must not take
such decisions before the alleged 'refuting instance' has become the confirming instance of a new, better theory. Therefore whenever we see terms like
'refutation', 'falsification', 'counterexample', we have to check in each case
whether these terms are being applied in virtue of decisions by the naive or by
the sophisticated falsificationist.

Then according to the sophisticated falsificationist, "falsify" does not mean to
disprove conclusively as ordinary usage would suggest.[18] Why this is has to do
with Lakatos's theory of the comparison of research programs.

How, then, does a sophisticated falsificationist compare or appraise research programs? Progressive programs are preferable to degenerating ones,
asserts Lakatos. By progressive is meant that the program is coherent (that is,
provides for future research) and leads to the discovery of novel phenomena.
In contrast, explains Lakatos (1974c, p. 8), "[i]n degenerating programmes
however, theories are fabricated only in order to accommodate *known* facts."

The problem is that history shows that programs have not always pro-
gressed in linear fashion, that is, they have exhibited periods that could be
designated as degenerating (for instance, no novel predictions, increased
truth content, coherency, and so on). Thus, Lakatos is forced to admit that no
program may ever completely degenerate because the right modification of
the protective belt could always potentially revive the it. Lakatos (1970, p.
155; 1978b, p. 69) tries to escape this problem by grounding his MSRP in
"heuristic power":

The idea of competing scientific research programmes leads us to the problem: *how are research programmes eliminated?* It has transpired from our
previous considerations that a degenerating problemshift is no more a suffi-
cient reason to eliminate a research programme than some old-fashioned
'refutation' or a Kuhnian 'crisis'. *Can there be any objective* (as opposed to
socio-psychological) *reason to reject a programme, that is, to eliminate its hard
core and its programme for constructing protective belts?* Our answer, in
outline, is that such an objective reason is provided by a rival research programme which explains the previous success of its rival and supersedes it by a
further display of *heuristic power.*

The footnote belonging to the end of the passage above (1970, p. 155) is well
worth reproducing: "I use *'heuristic power'* here as a technical term to characterize the power of a research programme to anticipate theoretically novel
facts in its growth. I could of course use *'explanatory power'*. . . ." This does
not help Lakatos, for how can one know without hindsight whether program
X has more explanatory power than program Y? His criterion is undermined
because even powerfully heuristic programs undergo more stagnant and un-
productive periods.

Lakatos alludes to Kuhn in the passage above, and they do share a most interesting relationship. It is important to keep in mind that Lakatos sets out to refute not only Popper's notion of falsification but Kuhn's theory of revolutions. Lakatos calls his MSRP "normative," Kuhn's methodology "socio-psychological."[19] Lakatos wants his model to explain scientific change as an activity that is governed by and grows according to rules and standards of reason. He asserts that his programs can be tested against the history of science (which he, interestingly enough, equates solely with the history of physics). For Lakatos, Kuhn's depiction of scientific change is relativist anathema: "a kind of religious change," or "a mystical conversion that is not and cannot be governed by rules of reason and that falls totally in the realm of the (*social*) *psychology of discovery*" (Lakatos, 1970, p. 93; 1978b, p. 9).

Despite Lakatos's antagonism toward Kuhn's theory of scientific change, the MSRP bears a striking resemblance to Kuhn's methodology of paradigms. Lakatos admits (1970, p. 155; 1978b, p. 69) that "indeed, what he [that is, Kuhn] calls 'normal science' is nothing but a research programme that has achieved monopoly." And from Kuhn's side (1970b, p. 256) we learn that "[t]hough his terminology is different, his analytic apparatus is as close to mine as need be: hard core, work in the protective belt, and degenerative phase are close parallels for my paradigms, normal science, and crisis."

In summary, Lakatos wants his MSRP to achieve the following: (1) to evaluate SRPs, that is, to act as an account of theory appraisal, (2) to explain scientific change as largely rational, and (3) to act as a demarcation criterion between science and nonscience. Before evaluating whether he has achieved these goals, we will turn briefly to a related issue: the Duhem thesis.

The Revival of the Duhem Thesis

Lakatos, unlike Popper, builds upon the work of Pierre Duhem (1861–1916). The significance of the Duhem thesis is one of the central themes in the modern philosophy of science, and one that remains controversial.[20]

Pierre Duhem's *The Aim and Structure of Physical Theory* first appeared in 1906. There Duhem argues that the falsification (that is, rejection) of theories is ambiguous. "Crucial experiments" do not let us distinguish between theories, Duhem suggests, because "a 'crucial experiment' is impossible in physics" (Duhem, 1962, p. 188). Because "an experiment in physics can never condemn an isolated hypothesis but only a whole theoretical group" (Duhem, 1962, p. 183), we cannot locate the faulty assumption involved. In short, Duhem denies that theory complexes can be falsified conclusively by a crucial experiment.[21] Falsification could exist only if we knew that the other assumptions and theories upon which the model is based are true, but this is the failure of inductive logic. W. V. Quine has developed a view similar to Duhem's.

Popper (1972a, p. 239) does not believe the Duhem "holistic view of tests" presents "a serious difficulty for the fallibilist and falsificationist," although he acknowledges that the problem exists.

Now it has to be admitted that we can often test only a large chunk of a theoretical system, and sometimes perhaps only the whole system, and that, in these cases, it is sheer guesswork which of its ingredients should be held responsible for any falsification; a point which I have tried to emphasize—also with reference to Duhem—for a long time past. Though this argument may turn a verificationist into a sceptic, it does not affect those who hold that all our theories are guesses anyway.

Lakatos's view of science is not so sanguine. He (1974c, p. 317) asserts that "both Popper and Grünbaum stubbornly overestimate the immediate striking force of purely negative criticism, whether empirical or logical."

Yet, Lakatos insists he is a falsificationist—a view seemingly inconsistent with the Duhem thesis, which he incorporates into his philosophy. He (1970, pp. 184–85; 1978b, p. 97) accomplishes this by altering the meaning of *falsification,* and by arguing that there are two versions of the Duhem thesis:[22]

> In its *strong interpretation* the Duhem–Quine thesis excludes any *rational* selection rule among the alternatives; this version is inconsistent with all forms of methodological falsificationism. The two interpretations have not been clearly separated, although the difference is methodologically vital. Duhem seems to have held only the weak interpretation: for him the selection is a matter of 'sagacity': we must always make the right choices in order to get nearer to 'natural selection'.

How does Lakatos deal with the Duhem problem, that falsification of the problematic component cannot occur because that part of the theoretical labyrinth cannot be identified? Lakatos insists that the hard core of the program be inviolable, thus restricting falsification (= rejection) of an *ad hoc* hypothesis to the protective belt. In addition, he specifies that *ad hoc* hypotheses be independently testable. These two conditions should make it easier to identify problematic hypotheses: if an *ad hoc* hypothesis withstands testing it is accepted; if not, it is rejected. Of course, Lakatos restricts rejections to *ad hoc* hypotheses: scientific theories are never falsifiable in the sense of being conclusively rejected.

Not all philosophers of science agree that the Duhem problem is capable of being solved. Lakatos does not lay claim to having solved the problem, despite the construction of a rather artificially "hard" core.[23]

Behind the Methodology of Scientific Research Programs

In this section I would like to show that Lakatos's MSRP fails. In particular, I will treat the following issues: (1) that rationality as Lakatos defines it fails, (2) that the demarcation criterion fails, (3) that his relationship to the Popper school of thought is dubious, and (4) that Lakatos's "rational reconstructions" of the history of science—his theory of history—grossly pervert history.

One of the major criticisms of the MSRP is that Lakatos has failed to show that science is rational. It is certainly true that Lakatos runs into the same problem with inductivism as Popper: he needs inductive argument to establish that one program has more truth content (verisimilitude) than another and to establish positive evidence (corroboration). Hence, Feigl (1971, p. 147) labels Lakatos "a second-level inductivist." Yet Lakatos's work contains a new dimension: he cannot rebuke a scientist for sticking with a degenerating program. Lakatos (1974b, p. 319) describes this dilemma as follows:

> It is very difficult to decide, especially if one does not demand progress at each single step, when a research programme has degenerated hopelessly; or when one or two rival programmes has achieved a decisive advantage over the other. In this sense there can be no 'instant rationality'. *Neither the logicians's [sic] proof of inconsistency nor the experimental scientists's [sic] verdict of anomaly can defeat a research program at one blow.* The falsificationist can be 'wise' only after the event if he wants to apply falsificationism to research programmes rather than to isolated theories.

This, of course, leaves the scientist with nothing by which to judge the merits of a research program; it is rational to persist with a program even if the evidence does not appear to support it. Only with hindsight can we judge whether a program was progressive. Here and now, then—that is, without the advantage of hindsight—any program becomes as good as the other. For this reason Lakatos has been accused of "epistemological anarchism," a stance usually associated with Paul Feyerabend.[24] Indeed, Feyerabend embraces Lakatos as a fellow anarchist, and he (1970a, p. 215) labels the Lakatosian methodology a "verbal ornament" and "a memorial to happier times when it was still thought possible to run a complex and often catastrophic business like science by following a few simple and 'rational' rules."

It should be clear that if one cannot distinguish between progressive and degenerating research programs, then Lakatos does not have a criterion of demarcation between science and nonscience. Newton-Smith (1981, p. 90) rightly poses the next logical question: Why is it so important for Lakatos (and Popper as well) to establish a demarcation criterion? The answer lies in the fact that "[f]or Lakatos and Popper, the polemical tone of their discussion reveals that the point is simply to condemn certain forms of activity. As with Popper, the pseudo-scientists who are to be condemned are Freud and Marx. . . ." Indeed, scattered through Lakatos's writings are comments such as "What *novel* fact has Marxism *predicted* since, say, 1917?" (1970, p. 176; 1978b, p. 88).

As a final comment on the demarcation criterion, it seems safe to say that most philosophers would consider Popper and Lakatos's attempts to erect a demarcation criterion not only a failure but also an impossible goal.[25] To classify it solely as a failure seems to indicate that demarcation is possible—a dubious propostion.

A third and related conclusion drawn here is that Lakatos's philosophy shares a dubious tie to that of Popper. Koertge (1978, pp. 269–70) discusses the relationship between Popper's and Lakatos's philosophies so clearly that her analysis is reproduced in full:

. . . Lakatos' position is in fact an *inversion* of Popper's basic views. The hyphen in 'Popper-Lakatos' should be read like the sign of opposition in 'acid-base', not like the glide in 'Marxist-Leninist'.

Briefly, the situation as I see it is this: Lakatos has moved science from the falsifiable side of Popper's line of demarcation. His method of appraising science is nothing but an adaptation of Popper's method of appraising metaphysics. Lakatos finds the Popperian method of bold conjectures–severe testing applicable only in pre-scientific trial-and-error learning.

So, very roughly their theories are related as follows:

Lakatos' theory of the appraisal of = Popper's theory of the appraisal of
science   metaphysics.

Lakatos' theory of pre-scientific = Popper's theory of science.
activity

Elsewhere Koertge (1979, p. 69) elaborates on this theme.[26]

Nevertheless according to LAKATOS all scientific theories are to be treated as if they were unfalsifiable frameworks of explanation or metaphysical principles. Ironically, LAKATOS of all people transformed Popper's analysis of the social sciences into a normative methodological paradigm for the physical sciences, although he had little respect for the social sciences!

Because of the Duhem problem—when experiments and the theoretical system contradict one another, which part of the system should then be rejected?—LAKATOS decided that scientific theories can never be falsifiable. And so the demarcation between science and metaphysics disappeared.

This position is affirmed elsewhere. For instance, Bartley (1976b, pp. 37–38) describes how Lakatos came to name his methodology the methodology of scientific research programs:[27]

His fascination with research programmes in the history of science reflected his interest in the strategies whereby a good idea might come to power. In this connexion he explained to me in 1961—against my considerable scepticism and resistance—that Popper was quite wrong to say that words do not matter. Quite the contrary, Lakatos insisted, ideas are of secondary importance compared to the *names* one gives to them: if you give your ideas good names, they will be accepted—and you will be named the father. . . . It is then appropriate that Lakatos should have acquired his chief fame during the final decade of his life for his 'scientific research programmes.' This was an idea that he took over completely developed from the accounts by Popper, Agassi, and Watkins of 'metaphysical research programmes.' Lakatos had the good sense to see that the word 'metaphysics' presented an insuperable public-relations obstacle to the professional philosophers of scientific bent who lacked his own sense of humour. So he calmly changed the word 'metaphysical' to the word 'scientific' and won the acclaim that he had intended for the notion.

Even more disturbing than Lakatos's deceptive use of *scientific* described in the passage above is his theory of history, or rational reconstructions, the interpretation of the past predicated on assumptions that the scientists acted rationally. Lakatos begins his famous "History of Science and Its Rational

Reconstructions" (1971a, p. 91; 1978b, p. 102) with the following paraphrase of Kant's dictum: "Philosophy of science without history of science is empty; history of science without philosophy of science is blind." This innocent and reasonable-sounding maxim is developed and reinterpreted by Lakatos in such a way as to cause historians of science to react violently.[28] The problem is summed up tactfully by Koertge (1976, p. 362) in the following passage:

> It must certainly be admitted that Lakatos' own historical efforts showed an ever decreasing emphasis on factual accuracy. . . . His early 'Proofs and Refutations' (1963–64) was carefully designed so as to avoid confusion between the reconstruction and what really happened. . . . However, in his 'Falsification and the Methodology of Research Programmes' (1970) he introduced the very misleading convention of linking the names of historical figures, e.g., Prout, with views which the historical characters themselves explicitly denied. Thus on p. 183, the text says that Prout recognized certain anomalies, while in a footnote we read, "Alas, all this is rational reconstruction rather than actual history. Prout denied the existence of any anomalies." . . . The general point is to note that this style of historical writing still permits the attentive reader to distinguish what is *intended* to be fact from what is deliberate fiction. But later in the essay, Lakatos even gives up the practice of providing factual footnotes for the reader.

This same point is made more vigorously by the historian of science Gerald Holton (1974, p. 75). He is citing Lakatos (1970, p. 146; 1978b, pp. 60–61) in the following passage.

> Lakatos then gives examples of what happens to an historical case study when done in this style, including his own reconstruction of "Bohr's plan . . . to work out first the theory of the hydrogen atom (1912–1913)."
>
> > His model was to be based on a fixed proton-nucleus with an electron in a circular orbit . . . ; after this he thought of taking the possible spin of the electron into account. . . . All this was planned right at the start.
>
> As it happens, Bohr's early work has been very carefully studied by historians of science, and this version produced by "rational reconstruction" is an ahistorical parody that makes one's hair stand on end.

A third and final example of Lakatos's abuse of historical fact is given in a footnote (1970, p. 140 n. 4; 1978b, p. 55 n. 3), which states in full: "This section may again strike the historian as more a caricature than a sketch; but I hope it serves its purpose. (Cf. *above,* p. 138) Some statements are to be taken not with a grain, but with tons, of salt."

Why would Lakatos do this? Kuhn (1971, p. 143) believes Lakatos was not really interested in the history of science. Yet, Kuhn must then overlook the fact that Lakatos does claim that Popper's philosophy fails because it clashes with the history of science, and moreover, that Lakatos asserts that his methodology is an improvement over Popper's on historical grounds. Koertge (1976, p. 363) reaches the conclusion that Lakatos is trying to apply ideal laws of physics to history—what she calls "Galilean reconstructions." Koertge is perhaps too generous (she, after all, wrote her dissertation at the LSE). I am not sure whether an answer to this question can be found; perhaps Lakatos's zealousness in fighting Kuhn's alleged irrationalism plays a role.[29]

There is a related issue that needs to be treated: the problem of the internal versus the external history of science. Because Kuhn (1971, p. 140; see also his discussion in 1977g) is the expert on this distinction, his definition follows:

> In standard usage among historians, internal history is the sort that focuses primarily or exclusively on the professional activities of the members of a particular scientific community: What theories do they hold? What experiments do they perform? How do the two interact to produce novelty? External history, on the other hand, considers the relations between such scientific communities and the larger culture. The role of changing religious or economic traditions in scientific development thus belongs to external history, as does its converse. Among other standard topics for the externalist are scientific institutions and education, as well as the relations between science and technology.

One task of historians of science is to characterize the transitions in the history of science. Whether internal changes are characterized as rational, nonrational, or irrational varies by scholar. Weighting of the importance of external history also varies. This choice forces historians to embrace a specific philosophical position. (Of course, historians are forced into a philosophical position simply by virtue of the fact that they must possess some conception of what science and scientific concepts are. By forced I mean that a philosophical stance cannot be avoided. And often this stance is taken for granted, that is, is unconscious.) Indeed, here the historical and philosophical spheres mesh because the philosopher must also determine which events are transitions, forcing him or her to make a historical interpretation. Newton-Smith (1981, p. 93) puts the matter as follows:

> If the historian of science subscribes as some do to the general thesis that rational transitions are to be given exclusively internal explanations and that non-rational transitions are to be given external explanations, he will have to employ a theory of rationality, the discussion of which is traditionally the province of the philosopher. Even if one does not subscribe to this controversial thesis (to be discussed in Chapter X) that different kinds of explanation, which would give philosophy a direct and exceedingly important bearing on the history, the connection between these disciplines is intimate enough for us to accept Lakatos' dictum that 'philosophy of science without history is empty; history of science without philosophy of science is blind.'

The best methodology in Lakatos's view is the one that minimizes the role of external factors in its rational reconstructions. He argues that his MSRP does this. But as Newton-Smith (1981, p. 94) notes, to define what is good as minimization of external factors, and to assert that the MSRP does this is to beg the question. How important are external factors to the history of science? If external factors are important, should methodology ignore them?

This overlapping of the philosophical and historical strongly suggests that a good philosopher needs, at the same time, to be a good historian and vice versa: Lakatos's rephrasing of Kant's dictum. Yet, when Lakatos consciously ignores or distorts known historical facts, he perverts history and thereby violates his own value system. As Kuhn (1971, p. 143) concludes, "When one's historical narrative demands footnotes which point out its fabrications, then the time has come to reconsider one's philosophical position." Lakatos (1970, p. 122; 1978b, p. 37) declares that "[s]ophisticated methodological falsificationism offers new standards for intellectual honesty."[30] Given this view, it is a pleasure to conclude with Andersson's (1986, p. 241), and indirectly Agassi's (1986a), final assessment of Lakatos's methodology of scientific research programs.

> Agassi's main thesis [in his review of Radnitzky and Andersson's *Progress and Rationality in Science,* 1978] is that Lakatos' MSRP is a failure. According to Agassi this is the main lesson to be learnt from the (Lakatosian memorial) volume. This is a controversial view and a view that is not admitted in the volume, at least not explicitly, but it is a view which I think is right. Agassi concludes that "the Lakatos era is over", at least in the philosophy of science, and that "the volume is bound to remain noticed, as it signifies the end of an era."

Paul Feyerabend, the "Dadasoph"

Paul Feyerabend (1924–) is a Viennese-born philosopher of science who denies any real expertise in the philosophy of science because he has never studied the subject.[31] During the Nazi occupation of Austria Feyerabend was inducted into the army and was later wounded. After the war he studied theater at the Weimar Institute. In 1947 he studied history, physics, math, and astronomy at the University of Vienna. In 1951 he received his Ph.D. and left for England to study with Wittgenstein. Wittgenstein's death, however, altered his plans and put him into contact with Karl Popper at the London School of Economics. At first embracing Popper's ideas, Feyerabend later became disenchanted and has since spent a good part of his lifetime refuting them. Feyerabend has made an international career of teaching; in the fifties he taught at the University of Bristol and at the Institute of Science and Fine Arts in Vienna, and at one time in the sixties he had simultaneous appointments at Berkeley, Yale, the University of London, and the Free University of Berlin. He has been professor of philosophy at the University of California at Berkeley since 1958.[32]

Newton-Smith (1981, p. 125) begins his chapter on Feyerabend with the following remark: "No more lively or entertaining critique of the scientific method has been provided than that offered by Feyerabend in his *Against Method,* which might well have been called *Against Received Opinion.*" Certainly, Feyerabend has the most controversial, colorful, and flamboyant writing style among philosophers and academicians in general—seemingly a consequence of his love for the theater. Feyerabend (1975, p. 21) refers to himself as an "epistemological anarchist," yet concedes he prefers to call himself a "flippant dadaist," for a dadaist "would never hurt a fly," while anarchism connotes violence.[33] Accordingly, Hans Lenk (1982) has found a befitting appellation for Feyerabend: "Dadasoph."

Feyerabend is against methodology and contemporary philosophy; hence, his slogans "anything goes" and "citizens' initiatives instead of philosophy." He (1978b, p. 117) admits he is suspicious of all intellectuals, delights in showing them that they take themselves far too seriously. His flippant style (he has no intention of being scholarly, he asserts[34]) and his zealous attacks on scientists, intellectuals, and philosophers have won him fame, a fame, however, that has backfired. Some academicians dismiss his work as frivolous; many are outraged by his antics. For instance, Martin Gardner's reaction to him (1982/1983, p. 34), which follows, is typical: "To most philosophers Feyerabend is a brilliant but tiresome, self-centered, repetitive buffoon whose reputation rests mainly on the noise and confusion he generates, and the savagery with which he pummels everybody who disagrees with him." Feyerabend's "noise" has clearly made him noticed, but only now are philosophers and historians of science coming to recognize that Feyerabend's writings do make an important contribution to the contemporary philosophy of science.[35]

"Anything Goes!" and Proliferation

In *Against Method* (1978a), *Science in a Free Society* (1978b), *Erkenntnis für freie Menschen* (1979), and *Wider den Methodenzwang* (1983) Feyerabend makes a case not just for pluralism (proliferation of methods and methodologies) but for the proliferation of traditions.[36] His thesis rests on the arguments that (1) scientific methodologies fail to provide rules adequate to guide science (so progress has not necessarily been the product of rational inquiry), (2) theories are incommensurable, (3) science is not necessarily superior to other types of knowledge, and (4) removal of methodological constraints enhances individual freedoms and creativity.

Feyerabend is not really for "anything goes" but against rules. The slogan is employed humorously against "the rationalists" to show that

> science is always full of lacuna and contradictions, that ignorance, pig-headedness, reliance on prejudice, lying, far from impeding the march of knowledge are essential presuppositions of it and that the traditional virtues of precision, consistency, 'honesty', respect for facts, maximum knowledge under given circumstances, if practised with determination, may bring it to a standstill. It has also emerged that logical principles not only play a much smaller role in the (argumentative and nonargumentative) moves that advance science, but that the attempt to enforce them universally would seriously impede science. (Feyerabend, 1978a, p. 260)

Feyerabend believes adoption of scientific rules causes science to become less adaptable and more rigid and dogmatic. Much of his argument would be taken for granted by historians. For instance, he asserts (1978a, p. 295) that rules result in "too simple a view of the talents of man and the circumstances which encourage, or cause, their development." It also "neglects the complex physical and historical conditions which influence scientific change" (1978a, p. 295). Rules, in fact, would inhibit progress (Feyerabend, 1978a, pp. 23–24):

> [G]iven any rule, however 'fundamental' or 'necessary' for science, there are always circumstances when it is advisable not only to ignore the rule, but to adopt its opposite. For example, there are circumstances when it is advisable to introduce, elaborate, and defend *ad hoc* hypotheses, or hypotheses which contradict well-established and generally accepted experimental results, or hypotheses whose content is smaller than the content of the existing and empirically adequate alternative, or self-inconsistent hypotheses, and so on.

In short, Feyerabend (1978b, p. 13) shows by historical example, "(a) that the rules (standards) *were actually violated* and that the more perceptive scientists were aware of the violations; and (b) that they *had to be violated.* Insistence on the rules would not have improved matters, it would have arrested progress."

So "anything goes" is not to be understood as an appeal to scientists to have eight or nine drinks before they enter the laboratory, or for scientists to forgo education, and so on. Feyerabend (1978b, p. 32) also emphasizes that he does not "intend to replace one set of general rules by another such set." Just as Feyerabend has softened up his terminology "epistemological anarchism," he (1978b, p. 188) also empties "anything goes" of its extreme connotation:

> But 'anything goes' does not express any conviction of mine, it is a jocular summary of the predicament of the rationalist: if you want universal standards, I say, if you cannot live without principles that hold independently of situation, shape of world, exigencies of research, temperamental peculiarities, then I can give you such a principle. It will be empty, useless, and pretty ridiculous—but it will be a 'principle'. It will be the 'principle' 'anything goes'.

Beneath the jokes, wit, and acerbic attacks on the rationalists, Feyerabend is communicating a simple and hardly controversial message: he (1978a, p. 32) wants "to convince the reader that *all methodologies* [he probably means methods], *even the most obvious ones, have their limits.*"

His concept of proliferation is given a political dimension. In *Science in a Free Society* (1978b, p. 9) he

> develops the idea of a free society and defines the role of science (intellectuals) in it. *A free society is a society in which all traditions have equal rights and equal access to the centres of power* (this differs from the customary definition where *individuals* have equal rights of access to positions *defined by a special tradition*—the tradition of Western Science and Rationalism).

Feyerabend hopes to open up science in a way corresponding to the pluralism envisioned by John Stuart Mill in *On Liberty.* Urging removal of rules has often been interpreted as an open invitation to quacks and metaphysicians to invade science. Yet, Feyerabend argues that a proliferation (pluralism) of traditions, theories, methods, and soon would prevent science from becoming a stale defense of an ideology and education from becoming indoctrination. In a pluralistic environment a crank would be exposed immediately for not testing a theory in a way favoring an opponent's view and by refusing to acknowledge the limitations or shortcomings of the theory. The common sense behind this is that commitment to one tradition often involves a certain blindness; proliferation would expose this as one tradition among many—in this way bringing to light its weaknesses and beauty.

Incommensurability of theories strengthens Feyerabend's argument for proliferation. His conception of incommensurability stems from the theory-dependence of observation: it is impossible to formulate basic concepts of one theory completely in terms of another because interpretation of concepts and observation statements that employ them are theory dependent. Feyerabend does not mean that theories cannot be compared. They can be compared, but then the criteria of comparison—for example, linearity, reasonableness of assumptions, and the like—are subjective.

One of the more provocative logical conclusions drawn from Feyerabend's view of incommensurability is that scientific knowledge is not necessarily superior to other types of knowledge. Feyerabend points out that especially because of incommensurability, it is not possible to prove one source of knowledge is better than another. He offers historical cases to support this view. For example, acupuncture and Chinese medicine in general are now accepted as valid knowledge forms, although at first rejected as a primitive, non-Western, unscientific tradition. The tie to pluralism is then facile; one should keep an open mind and investigate *all* forms of knowledge (traditions). Indeed, a common theme in all of Feyerabend's works is that Westerners (ethnocentrically) take science for granted without adequately investigating other forms of knowledge.

Feyerabend arrives at the conclusion that freedom such as Mill defended in his essay *On Liberty* can be achieved by removal of methodological constraints and by giving individuals the choice between scientific and other sources of knowledge. Only in a free society—that is, one in which all traditions are given an equal chance to be entertained by individuals—can an individual render an intelligent decision:

> A mature citizen is not a man who has been *instructed* in a special ideology, such as Puritanism, or critical rationalism, and who now carries this ideology with him like a mental tumour, a mature citizen is a person who has learned how to make up his mind and who has then *decided* in favour of what he thinks suits him best. He is a person who has a certain mental toughness (he does not fall for the first ideological street singer he happens to meet) and who is therefore able *consciously to choose* the business that seems to be most attractive to him rather than being swallowed by it. To prepare himself for his choice he will study the major ideologies as *historical phenomena,* he will study science as a historical phenomenon and not as the one and only sensible way of approaching a problem. He will study it together with other fairy-tales such as the myths of 'primitive' societies so that he has the information needed for arriving at a free decision. An essential part of a general education

of this kind is acquaintance with the most outstanding propagandists in all fields, so that the pupil can build up a resistance against all propaganda, including the propaganda called 'argument'. (Feyerabend, 1978a, p. 308)

What Feyerabend Is Actually Against Is Modern Education

Feyerabend is unquestionably grossly disillusioned with modern education. Why? It does not produce thinkers. Instead, it trains students to be conformists, sycophants, and hypocrites—to regurgitate the ideas of "the masters." Feyerabend harks back to the times before the Vienna Circle in the passage produced below, which appears as the conclusion to a chapter entitled "From Incompetent Professionalism to Professionalized Incompetence—the Rise of a New Breed of Intellectuals" (1978b, pp. 204–5):[37]

> The writers of the Vienna Circle and the early critical rationalists who distorted science and ruined philosophy in the manner just described belonged to a generation still vaguely familiar with physics. Besides, they started a new trend, they did not merely take it over from more inventive predecessors. They *invented* the errors they spread, they had to *fight* to get them accepted and so they had to possess a modicum of *intelligence*. They also suspected that science was more complex than the models they proposed and so they worked hard to make them plausible. They were pioneers, even if only pioneers of simplemindedness. The situation is very different with the new breed of philosophers of science that now populate our universities. They received their philosophy ready made, they did not invent it. Nor do they have much time or inclination to examine its foundations. Instead of bold thinkers who are prepared to defend implausible ideas against a majority of opponents we have now anxious conformists who try to conceal their fear (of failure, of unemployment) behind a stern defence of the status quo. This defence has entered its epicyclic stages: attention is directed to details and considerable work is done to cover up minor faults and deficiencies. But the basic illiteracy remains and it is reinforced for hardly any one of the new breed possesses the detailed knowledge of scientific procedure that occasionally made their ancestors a little hesitant in their pronouncements. For them 'science'is what Popper or Carnap or, more recently, Kuhn say it is—and that is that. It is to be admitted that some sciences, going through a period of stagnation now present their results in axiomatic form, or try to reduce them to correlation hypotheses. This does not remove the stagnation, but makes the sciences more similar to what philosophers of science think science is. Having no motivation to break through the circle and much reason (both emotional and financial) to stay in it philosophers of science can therefore be illiterates with good conscience. Small wonder that intelligent criticism is hard to find. . . .

On the surface this is a complaint against the Popper school, the empiricists, and the contemporary philosophy of science. Underneath runs a deep dissatisfaction with the mediocrity of modern education and with "academic nepotism," the filiopietistic system of favoritism that operates on the always-agree-with-your-supervisors principle. Feyerabend (1978a, p. 46) is frustrated because he is for a truly *critical* rationalism, where unanimity of opinion cannot survive:

> Unanimity of opinion may be fitting for a church, for the frightened or greedy victims of some (ancient, or modern) myth, or for the weak and willing followers of some tyrant. Variety of opinion is necessary for objective knowledge. And a method that encourages variety is also the only method that is compatible with a humanitarian outlook.

Feyerabend is against scholasticism, that is, a restricted criticism of minor details. He is against making a subject unnecessarily difficult and against using forbidding technical terminology where a simple formulation is possible, and so on. He is for using one's head; he (1978b, pp. 139–40) even argues that a lack of respect for the mind ("soul") is analogous to abuse of the body. He goes so far as to compare the neglect of the mind with murder.[38] In brief, Feyerabend (1981a, p. 12) is making the point (which he believes has somehow been lost or forgotten) that all disciplines are academic, that is, use and develop intellectual skills: "[s]trictly speaking, *all sciences are Geisteswissenschaften.*"

Why Feyerabend Is Really a "Rationalist in Disguise," or Why Feyerabend Doesn't Deserve to Be a "Darling of the Left"

Feyerabend dedicated *Against Method* "To Imre Lakatos: Friend and fellow-anarchist." He realized that Lakatos's argument inevitably leads to the conclusion that science is not rational (as Lakatos defines *rational*), and hence labeled Lakatos's philosophy "anarchism in disguise" (1978a, p. 181).

Certainly, Feyerabend (1978a, p. 179) attacks the critical rationalists, and Popper in particular.[39]

> [W]herever we look, whatever examples we consider, we see that the principles of critical rationalism (take falsifications seriously; increase content; avoid *ad hoc* hypotheses; 'be honest'—whatever *that* means; and so on) and, *a fortiori,* the principles of logical empiricism (be precise; base your theories on measurements; avoid vague and unstable ideas; and so on) give an inadequate account of the past development of science and are liable to hinder science in the future. They give an inadequate account of science because science is much more 'sloppy' and 'irrational' than its methodological image. And, they are liable to hinder it, because the attempt to make science more 'rational' and more precise is bound to wipe it out, as we have seen.

And in his section in *Science in a Free Society* entitled "Life at the LSE?" (1978b, p. 217), Feyerabend concludes: "Considering that there are many people engaged in these interesting activities we must admit that methodology is still very much alive, even at the LSE; but it is not the kind of life a reasonable person would want to live." Yet, there are various hints throughout Feyerabend's work that indicate that he is really just a critical rationalist who adheres unfalteringly to his critical creed. (Unlike Feyerabend, I do not

conflate critical rationalism and falsificationism.) For instance, consider
Feyerabend's explanation (1978b, pp. 184–85) of his dedication of *Against
Method* to Imre Lakatos.[40]

> When AM [*Against Method*] was ready for print Imre Lakatos and I dis-
> cussed various possibilities for a dedication. I considered: 'To Imre Lakatos,
> friend and fellow *rationalist*'—an ironical allusion to Lakatos' often voiced
> suspicion that I was a rationalist at heart and would recoil in horror if every-
> one became an anarchist (he was right). Next I considered dedicating the
> book to three alluring ladies who had almost prevented its completion.
> Lakatos approved for he knew two of them. Then I suggested 'To Imre
> Lakatos, friend, and fellow *anarchist*'. Lakatos said he was 'flattered' pro-
> vided the comma I had put after 'friend' was removed (it wasn't).

(The comma has not disappeared in the 1978 edition of *Against Method*.) In
addition, Feyerabend (1978a, pp. 32–33), in his typical ironic style, describes
his task. "An anarchist is like an undercover agent who plays the game of
Reason in order to undercut the authority of Reason (Truth, Honesty, Justice,
and so on.)[4]" A careful reading shows Feyerabend (1978b, p. 185) is against
rationalists who do not play by their own rules: "Although our new intellectuals extol the virtue of a rational debate they only rarely conform to its rules.
For example, they don't read what they criticize and their understanding of
arguments is of the most primitive kind." Feyerabend complains that critics
do not understand his argument, that they distort it so it is no longer his. But
irrationalists have no love of argument (that is, the process of reasoning).
Feyerabend is really a critical rationalist who shows that the "orthodox" rationalists take themselves so seriously that they no longer adhere to their own
values. And the thrust of his argument is that critical rationalists are forced to
conclude that their own "programmes"—that is, falsificationism and MSRP—
show that science is not and should not be a fully rational (= governed by
rules) enterprise. The values of critical rationalism have been labeled conservative; Feyerabend is for stricter adherence to them than are the critical
rationalists whom he attacks. Admittedly, Feyerabend's style is radical (see
the next section); the content, despite its form, does not qualify him to be a
darling of the Left.[41]

Dadasophia or Dadasophistry?

Webster's Ninth New Collegiate Dictionary defines *Dada* in the following way:
"a movement in art and literature based on deliberate irrationality and nega-
tion of traditional artistic values." And there does seem to be an intentional
irrational element involved with Feyerabend's work. Indeed, Feyerabend
proudly asserts that his work is inconsistent in certain places. In reply to his
critics he (1978b, p. 191) queries: "What is wrong with inconsistencies?" He
argues quite rightly that contradictions can be valid, that they are at the heart
of dialectic philosophy. Yet, he (1981b, p. xiv) also argues: "The reader will
notice that some articles defend ideas which are attacked in others. This

reflects my belief (which seems to have been held by Protagoras) that good arguments can be found for the opposite sides of any issue." Playing devil's advocate with one's own work—pointing out the opposing arguments or weaknesses in one's own argument—is to be welcomed as extremely fair play. But defending both sides without point of reference causes confusion for the reader. Not only does this technique obfuscate the message, it makes Feyerabend immune to criticism: he is right no matter what.[42]

Feyerabend tends to use extreme terminology to describe moderate ideas. Lenk (1982, p. 8) characterizes "epistemological anarchism" as a poor choice of words, and something that Feyerabend is aware of:[43]

> Has Feyerabend gone too far with his epistemological anarchism? The real anarchist—if he's really serious—would drown in his own swamp of absolute "un-rule-iness." Being against rules and dogmatism can sometimes also be dogmatic. Complete extremes are dogmatic. And "anything goes" is indeed an extreme. (Feyerabend knows that himself.)

For the same reason, one cannot take literally Feyerabend's slogan "citizens' initiatives instead of philosophy." Feyerabend wants to reform philosophy, not destroy it.[44]

Lenk (1982, p. 8) mentions that Feyerabend tends to confuse methodology with method. He also seems to use *proliferation* and *pluralism* as if they were interchangeable. Feyerabend also uses *relativism* in opposition to *rationalism*. Both Popper and Lakatos argue passionately against relativism, the view that one idea or tradition is as good as the other. Feyerabend's relativism is represented as the interplay of tolerance and democracy: all ideas and traditions should flourish so long as they allow others to coexist. Feyerabend (1981a, p. 30) describes the advantages of democratic relativism:

> A second argument in favour of a democratic relativism is closely connected with Mill's argument for proliferation. A society that contains many traditions side by side has much better means of judging each single tradition than a monistic society. It enhances both the equality of the traditions and the maturity of its citizens.

Hence, Feyerabend uses *relativism* in an unusual sense and, like anarchism and anything goes, as a part of his dadaism (which is meant to shock the critical rationalists).

The dadaist irrationalism can certainly be an extremely effective art form in theater and film. Yet, in Feyerabend's case it detracts from his message. And there is wisdom in his message. For instance, his chief theme in *Against Method,* torn free of Dada, is quite reasonable (Feyerabend, 1978b, p. 98):

> . . . there is no 'scientific method'; there is no single procedure, or set of rules that underlies every piece of research and guarantees that it is 'scientific' and, therefore, trustworthy. Every project, every theory, every procedure has to be judged on its merits and by standards adapted to the processes with which it deals. The idea of a universal and stable *method* that is an unchanging measure of adequacy and even the idea of a universal and stable *rationality* is

as unrealistic as the idea of a universal and stable measuring instrument that measures any magnitude, no matter what the circumstances.

Once the Dada is extracted from the substance, one finds many less than radical, rather sensible themes—for example, methodologies and methods have their limits; higher education no longer teaches students to think or be original (not even students of critical rationalism); a well-informed person should be acquainted with the supporters' and opponents' arguments; science is still (wrongly) taught as if it were an ahistorical phenomenon; part of being critical is keeping an open mind; the line between science and non-science does not exist; professionals of all sorts are not so intelligent that they cannot err, and they should be subject to criticism and held responsible for their actions just as laymen are; *homo academicus* takes him- or herself far too seriously.[45]

This is no charge that Feyerabend's dadaism is not creative and entertaining; it certainly is. It is just that Dada—the use of parody, pointed and exaggerated commentary, irony, and so on—can be fully appreciated only as a *visual* and *audio* form. In written form the signals that tell us this is a joke, exaggeration, or the like are not present. Thus, every reader of *Against Method* and the accompanying works may go away with a completely different understanding of Feyerabend. This is fine for certain types of plays or films but not for philosophical works. (Even if Feyerabend asserts that *Against Method* is a letter to Lakatos, he still wanted Lakatos to understand it.) When one cannot decipher jokes from substance of the text, there is risk of extreme interpretations, and, as mentioned, these extremes are mistakes. There seems to be some indication that Feyerabend himself has reached this conclusion because in his preface to the 1983 German edition of *Against Method* (1983, p. 12), he writes that he dedicates to Lakatos an edition that is "unfortunately less ironic" than the past editions of the book.[46]

Feyerabend exposes the hypocrisy and failings of his contemporaries—what is often the object of the fine arts. Perhaps Feyerabend could pretend to be Arthur Miller and compose plays, Ernst Mach when doing philosophy of science, so that his own work may be better appreciated. Feyerabend could think of this as cultivating "artistic schizophrenia." (Or perhaps Feyerabend should consider law because it supposedly combines both the theatrical and analytical.)[47]

I want to conclude with the reflections of two philosophers. Lenk (1982, p. 4) recounts Feyerabend's reception at the Free University of Berlin:

> What did Feyerabend say as he arrived from the airport late at the Free University of Berlin? Climbing up to the rostrum as guest professor (sleeves rolled up, and at a time less critical of ideologies than critically ideological) and unleashing a storm of applause after the dean, a professor of religion, had welcomed him with the words: "Habemus papem!" [Here's the pope!], he replied: "In the philosophy of science there are no popes—and if there should ever be one, he must be overthrown."

Herbert Feigl (1981d, p. 89), an eminent member of the Vienna Circle, has the following impression:

> Immediately, during my first conversation with Feyerabend, I recognized his competence and brilliance. He is, perhaps, the most unorthodox philosopher of science I have ever known. We have often discussed our differences publicly. Although the audiences usually side with my more conservative views, it may well be that Feyerabend is right, and I am wrong.

William Bartley: Pancritical Rationalism

William Bartley (1934–) focuses on the rationality of beliefs. His contribution has been called (Koertge, 1974, p. 75) a "more practical theory for the social scientist or historian to use." The standard approach to the rationality of beliefs is inductivist and justificationist. The rationality of X's beliefs is hence related to the evidence on hand (and not to an evaluation of X's goals):

> Most western philosophies—philosophies of science as much as philosophies of religion—sponsor *justificationist metacontents of true belief.* They are concerned with how to justify, verify, confirm, make firmer, strengthen, validate, make certain, show to be certain, make acceptable, probilify, cause to survive, *defend,* particular contents and positions. Most such philosophies— *again,* philosophies of science as much as philosophies of religion—end in commitment and in identification. (Bartley, 1982a, p. 128)

Bartley attacks two problems: Popper's solution to the demarcation issue and the question of how to be rational about rationality. This latter problem has come to be known as the *tu quoque* or boomerang argument. Briefly, the dilemma arises from the fact that the traditional solution to the problem of rational belief—to justify belief—leads to infinite regress as one authority is replaced by another. The grounding of a belief in a reason, R_1, demands justification of R_1; each new R_i in turns needs to be justified. Bartley (1987c, p. 212) argues that justification is not the answer.

> Rather, we locate rationality in *criticism.* A rationalist is, for us, one who holds *all* his positions—including standards, goals, decisions, criteria, authorities, and *especially* his own fundamental framework or way of life—open to criticism. He withholds nothing from examination and review.

The result is a more viable form of critical rationalism, which Bartley calls "pancritical rationalism" (originally called "comprehensively critical rationalism"). Essentially, he (1984b, p. 207) asks us to consider the following:

> How can our intellectual life and institutions, our traditions, and even our etiquette, sensibility, manners and customs, and behavior patterns, be arranged so as to expose our beliefs, conjectures, ideologies, policies, positions, programs, sources of ideas, traditions, and the like, to optimum criticism, so as at once to counteract and eliminate as much intellectual error as possible, and also so as to contribute to and insure the fertility of the intellec-

tual econiche: to create an environment in which not only negative criticism but also the positive creation of ideas, and the development of rationality, are truly inspired.

There is a bit of confusion involved with the relationship of pancritical rationalism to Popper's critical rationalism and with the discussion of the criterion of demarcation. Bartley first conceived of pancritical rationalism as a criticism of Popper's theory; now he considers his work to be compatible with and an extension of Popper's.[48] Bartley (1982a, p. 201) does claim to have a criterion of demarcation:[49]

> The development of Popper's thought, and the generalization and application of his ideas outside science, have then rendered his discussion of demarcation obsolete. Popper suggested to the positivists that the problem lies not in demarcation of the meaningful from the meaningless, but in the demarcation of the scientific from the nonscientific. In fact, the problem lies not simply in the demarcation of the scientific from the non-scientific, but in the demarcation—contextually and metacontextually—of the critical from the uncritical or pseudo-critical.

Yet, it is obvious that Bartley cannot be talking about demarcation in the sense of clear separation (or in the sense in which the positivists or Popper use it). And he (1984b, p. 206) does admit this:

> What, then, is the criterion of demarcation between a good idea and a bad one?
> There is none. There are, of course, certain qualities that are highly desirable in theories, and whose absence signals danger. These include testability and high empirical content. But these are not *criteria:* their presence is not required, and a theory lacking in them may turn out to be excellent. There are some objectionable characteristics in theories, and these include inconsistency and incoherency. But their contraries are not criteria of goodness: consistency and coherency are desired, but they do not, in and of themselves, make a theory a good one.

Watkins, Post, and others have attacked Bartley's pancritical rationalism on related grounds that (1) an individual cannot hold all of his beliefs open to criticism (for example, I cannot criticize the fact that I am older than three), and (2) it is impossible to criticize pancritical rationalism.[50] The latter point can be summed up as follows (Bartley, 1987b, p. 320):

> Take the following two claims:
> (A) All positions are open to criticism.
> (B) A is open to criticism.
> Since (B) is implied by (A), any criticism of (B) will constitute a criticism of (A), and thus show that (A) is open to criticism. Assuming that a criticism of (B) argues that (B) is false, we may argue: if (B) is false, then (A) is false; but an argument showing (A) to be false (and thus criticizing it) shows (B) to be true. Thus, if (B) is false, then (B) is true. Any attempt to criticize (B)

demonstrates (B); thus (B) is uncriticizable, and (A) is false. And hence, so Post would contend, my position is refuted.

Bartley and others have responded that the critics misrepresent his position. Rationality is not a property of sentences, states Bartley, but an attitude. The demarcation criterion is not criticizability but vulnerability to criticism. Pancritical rationalism does not turn on a concept of falsity; Bartley (1987b, p. 335) does not assume "that *for a statement to be criticizable is for it to be possibly false.*"

Pancritical rationalism, then, amounts to an assertion of the following: a theory is held rationally if (1) it contains no criticism-deflecting devices at time t, (2) its holder does not maintain it in the face of cogent criticism, and thus its holder makes every attempt to expose himself or herself to arguments against the theory, and (3) points (1) and (2) are subject to this principle as well. Bartley (1982a, pp. 159–60) acknowledges that there is a relativistic element involved:

> When one position is subjected to criticism, others have to be taken for granted—including those with which the criticism is being carried out. The latter are used in criticism not because they are justified or beyond criticism, but because they are *at present* unproblematical. These are, *in that sense alone,* beyond criticism. We stop criticizing—temporarily—not when we reach uncriticizable authorities, but when we reach positions against which we can find no criticisms. If criticisms of these are raised later, the critical process thus continues. This is another way of saying that there is no theoretical limit to criticizability—and to rationality.

And thus:

> Pancritical rationalism is hence compatible with one kind of relativism. The survival of a position is relative to its success in weathering serious criticism. A position that survives at one time may be refuted later. This kind of relativism—due to the fact that we are not gods, are ignorant, lack imagination, are pervasively fallible—is harmless. It is an example of the way that learning proceeds by trial and error—by guessing and trying to criticize guesses: the making and destroying of theories is part of the evolutionary process.

Pancritical rationalism appears to be the crystallization of common sense. It has been taken over by Karl Popper as a generalization of his philosophy. Thus, the early Popperian position has undergone a major transformation. Popper now embraces a (nonnoxious) relativistic position. The mature position gains rationality: Popper's rationalism no longer "rests on an irrational faith in the attitude of reasonableness" (1972a, p. 357). At the same time one must admit that falsification in the sense of empirical refutation recedes well into the background.[51] The sense of demarcation remaining is the idea of the exclusion from science of theories that have "*built-in* devices for *avoiding* or *deflecting* critical arguments—empirical or otherwise" (Bartley, 1982a, p. 196). Perhaps understated is the fact that this attitude, given a democratic society, is inextricably tied to education.[52]

Notes

1. Popper's doctoral thesis was entitled "On the Problem of Method in the Psychology of Thinking" (Popper's translation of his *Zur Methodenfrage der Denkpsychologie,* appearing in his autobiography (1974a, pp. 61, 165 n. 97). Popper's two *Rigorosa* (the oral examinations leading to the Ph.D. in the German-speaking world) were in the history of music and in the philosophy and psychology of science. His testers were Karl Bühler (1879–1963) and Moritz Schlick, founder of the Vienna Circle. See Popper, 1974a, pp. 61ff. According to Bartley (1974, pp. 319–20, p. 331 n. 17), Popper's dissertation (unpublished) is a defense of Bühler's ideas and was aimed "against the associationist physicalist ideas of Schlick, which Popper vigorously attacked." Bartley (1974) has researched the influence of Austrian thinkers on modern philosophy, including the thought of Glöckel, Bühler, Wittgenstein, and Popper. Interestingly, Bartley (1974, pp. 321–22) comes to the following conclusion about Popper:

> When one views Popper's thought against this background, it is hard, surprising as it may seem to some, to find much of striking novelty in his philosophy. . . . Popper's attacks on the positivists may, then, be construed as direct *applications* of the attacks already mounted by Koffka and Bühler on the association psychologists. Even some of Popper's constructive ideas, including the emphasis on testability in connection with the hypothetico-deductive method, may be found in the work of his teachers: in particular, in that of Heinrich Gomperz. Popper's views acquire their distinctive form and emphasis from the fact that they were elaborated in dialogue with the logical positivists; but they acquire no originality from this circumstance.

Stuhlhofer (1986) discusses another forerunner of Popper: August Weismann, a zoologist.

2. I argued in note 8, chapter 3, that Fleck and Popper's intellectual evolution had elements in common: both men published an antipositivist work in the 1930s—works practically ignored in Europe. Yet, that note needs to be qualified with the following comments: (1) Popper's *Logik der Forschung* was read by Vienna Circle members and positively reviewed by Hempel and Carnap. It, however, never had the impact of the English version, *The Logic of Scientific Discovery,* which appeared in 1959. In contrast, Fleck was actually ignored. (2) Whereas both philosophers are against some tenets of logical positivism, they still remain worlds apart because Popper believes science is rational and Fleck views science as professional indoctrination.

3. Popper (1974a, p. 169 n. 165) notes in his autobiography that he missed a chance to emigrate to the United States:

> At the Copenhagen Congress—a congress for scientific philosphy—a very charming American gentleman took great interest in me. He said that he was the representative of the Rockefeller Foundation and gave me his card: "Warren Weaver, the European of the Rockefeller Foundation" [sic]. This meant nothing to me; I had never heard about the foundations and their work. (Apparently I was very naive.) It was only years later that I realized that if I had understood the meaning of this encounter it might have led to my going to America instead of to New Zealand.

It is amazing that Popper did not eventually emigrate to the United States. The Rockefeller Foundation especially recruited the marginal-utility economists and the

Vienna Circle logical positivists because, according to Krohn (1985 and 1987), the foundation was looking for right-wing support to combat the New Dealers. Krohn (1985, p. 332) describes the situation in the following way:

> Im Gegensatz zu der erzwungenen fluchtartigen Emigration der gesamten Gruppe der Reformökonomen fanden die Österreicher einen leichteren Übergang ins amerikanische Exil. Fast alle jüngeren Vertreter dieser Richtung hatten schon seit Ende der zwanziger Jahre intensive Kontakte zur anglo-amerikanischen Wissenschaft. In jenen Jahren war von der Rockefeller Foundation ein spezielles sozialwissenschaftliche Fellowship-Programm entwickelt worden, das insbesondere die Grenznutzentheoretiker wie auch die Wiener Philosophen des logischen Positivismus zu jahrelangen Amerika-Aufenthalten einlud, weil man sich von deren konservativer ideologischer Position eine wirksame Unterstützung versprach, um den neuen Herausforderungen der 'linken' Gesellschaftstheorien, des Keynesianismus und des künftigen New Deal, offensiv begegnen zu können. Nicht von ungefähr hatten Schumpeter und Hayek daher bereits 1932 Rufe nach Harvard und London erhalten. Andere Österreicher fanden nach Ablauf ihrer Fellowships bruchlos Anstellung in den berühmtesten Universitäten.

See also Coser (1984).

4. The full passage from Kraft (1974, pp. 185–86) indicates how closely the Vienna Circle members and Popper interacted.

> Popper never belonged to the Vienna Circle, never took part in its meetings, and yet cannot be thought of as outside of it. Already in my 1950 article dealing with the Vienna Circle I found it necessary to refer to him repeatedly. On the other hand, Popper's work cannot be genetically understood without reference to the Vienna Circle. As Popper stands in a close, inextricable relationship with the development of the Vienna Circle, so the Circle was also of essential significance for his own development. Thus Popper himself says of Carnap's *Logical Syntax of Language* that "the book . . . marks the beginning of a revolution in my own philosophical thinking." And in his latest book, *Conjectures and Refutations* (1963), Popper proves his relationship with the Vienna Circle by mentioning it several times and producing valuable insights into it, especially in Chapter 11, "Demarcation Between Science and Metaphysics" (pp. 253ff.). Popper was in personal contact with only a few members of the Vienna Circle, especially with Carnap and Feigl, but also with Waismann, Menger, Gödel, and with myself. In 1928 or 1929 he participated in Carnap's seminar. In 1931 or 1932 his first book, which has remained unpublished, was read and discussed by several members of the Vienna Circle. In 1932 Popper spent his summer vacation with Carnap and Feigl in the Ötz Valley in Tyrol, an opportunity filled with philosophical discussions, both "long and fascinating." Popper's relationship with the Vienna Circle is shown by the fact that in Paris in 1935 and in Copenhagen in 1936 he participated and discussed in Congresses sponsored by the movement which started from the Vienna Circle. At the first meeting Neurath even called him "the official opponent" of the Circle. This relationship with the Vienna Circle becomes also clear from such details as that he took over Waismann's concept of the logical scope or that he uses a waving fog to illustrate a world without structure, as done by Zilsel, a member of the Vienna Circle.

Popper's direct contact with the Vienna Circle lasted only till 1936 or 1937, when he went to the University of Christchurch, New Zealand, as a Senior Lecturer. From then on he did not refer to the Vienna Circle for two decades, with the exception of a few criticial remarks on Wittgenstein and Schlick in his work, *The Open Society and Its Enemies*. But, how closely he remained attached to the Vienna Circle and how important it still appeared to him to come to terms with it is shown by the fact that he renewed his relations with it in later years. He was concerned especially with its most important representative, Carnap, even before his contribution to the volume, *The Philosophy of Rudolf Carnap* (1964) in the "Library of Living Philosphers." In his essay, "Degree of Confirmation," he already gave a basic critique of Carnap's inductive logic, and in *Conjectures and Refutations* he mentions the Vienna Circle again and again.

Popper (1974b, p. 974) found Kraft's contribution to the Schilpp volume fair: "Victor Kraft in his contribution gives a crisp intellectual history of me, seen from the point of view of a leading member and historian of the Vienna Circle. I enjoyed his contribution greatly, although I sometimes see things, necessarily, in a different light."

Popper represents a middle point between logical positivism and the new historical philosophy of science. Some of the chief differences between Popper and the Vienna Circle are listed below.

Popper	*Vienna Circle*
1. Science starts with a problem.	1. Science starts with observation.
2. Popper radically rejects induction.	2. Assume Hume's proof of the logical invalidity of induction but still find induction useful.
3. Against verifiability.	3. For verifiability.
4. Advocates demarcation between pseudoscience and science.	4. Advocate demarcation between meaningful and meaningless statements.
5. Perceives his work as philosophy; critical of traditional philosophy.	5. Most members reject philosophy.

Popper and the logical positivists shared an empirical and antimetaphysical stance, a concern for language, and a belief that demarcation rests on empirical testability. As logical positivism drifts into logical empiricism, it becomes difficult to establish major differences between the two.

5. Explains Carnap (1966, p. 12):

When I was young and part of the Vienna Circle, some of my early publications were written as a reaction to the philosophical climate of German idealism. As a consequence, these publications and those by others in the Vienna Circle were filled with prohibitory statements similar to the one I have just discussed (i.e., science should not ask 'Why?' but 'How?'). These prohibitions must be understood in reference to the historical situation in which we found ourselves. Today, especially in the United States, we seldom make such prohibitions. The kind of opponents we have are of a different nature, and the nature of one's opponents often determines the way in which one's views are expressed.

6. Popper defines *rationalism* in *The Open Society and Its Enemies* (1966b, p. 225) as "an attitude of readiness to listen to critical arguments and to learn from experience." He says further:

> It is fundamentally an attitude of admitting that '*I may be wrong and you may be right, and by an effort, we may get nearer to the truth*'. It is an attitude which does not lightly give up hope that by such means as argument and careful observation, people may reach some kind of agreement on many problems of importance; and that, even where their demands and their interests clash, it is often possible to argue about the various demands and proposals, and to reach—perhaps by arbitration—a compromise which, because of its equity, is acceptable to most, if not to all. In short, the rationalist attitude, or, as I may perhaps label it, the 'attitude of reasonableness', is very similar to the scientific attitude, to the belief that in the search for truth we need cooperation, and that, with the help of argument, we can in time attain something like objectivity.

This definition of *rationality* is, however, not the classical (seventeenth-century) version that is used in juxtaposition to empiricism. It is quite compatible with empiricism.

7. Kraft's (1974, pp. 200–201) full summary is reproduced below.

Summary. Popper's relationship to the Vienna Circle, if we now summarize it, was of a critical nature. But it was not a criticism from an incommensurable point of view and there was no permanent feud between them; but they agreed with each other again and again. There was no unbridgeable opposition; rather a common basis. Both faced the same problems: the foundations of empirical knowledge and the criterion of science, and in addition to that they had the same basic attitude: empiricism. But the way they tackled these problems in the first instance was different. The approach of the Vienna Circle was determined—apart from views of Mach and Russell—above all by Wittgenstein's *Tractatus*. Popper had formed a view of his own. But his philosophical development proceeded in contact and in disputation with the Vienna Circle, even though this does not stand out as strongly as does Popper's impact on the Vienna Circle. This impact was achieved not so much by Popper's critique of the views of the Vienna Circle as by the presentation of and confrontation with his own results. By the insight into the justification of those results the views within the Vienna Circle underwent a considerable change. Popper replaced Wittgenstein in his influence on the Vienna Circle. In the later period he was its opponent. It must be attributed to this influence that a rapid and productive development took place within the Vienna Circle by which a new, fruitful movement in epistemology was introduced. The Vienna Circle owes Popper gratitude for an essential contribution to this development outside of its own forces. Through this influence an agreement was reached again and again between the Vienna Circle and Popper so that the initial disagreements between them finally disappeared to a great extent; yet not by the assimilation of Popper to the Vienna Circle but partly through acceptance of Popper's insights, and partly also through the independent development within the Vienna Circle. The meaning-criterion of scientific knowledge was given up and Popper's criterion of testability was taken over; the hypothetical, conventionalistic and physicalistic character of the test sentences was reached by both sides; philosophy was recognized at least as logic of science, as theory of knowledge. But the agreement was reached with only one wing of the Circle, the one led by Carnap and Neurath. Schlick, on the other hand, moved less far from the original basis; he stuck to his assertion of the nonhypothetical, indubitable validity of observation and admitted philosophy only as the clarification of concepts. But with Carnap also differences remained; he did not only acknowledge falsification as valid but continued to maintain verification in its weaker form of confirmation. Reichenbach erected a quite different structure of knowledge against Popper's; he advocated inductivism and completely negated Popper's interpretion. But Reichenbach did not belong to the Vienna Circle.

8. Popper (1972b, p. 30) writes:

The theory to be developed in the following pages stands directly opposed to all attempts to operate with the ideas of inductive logic. It might be described as the theory of *the deductive method of testing,* or as the view that a hypothesis can only be empirically *tested*—and only *after* it has been advanced.

9. For instance, chapter 1, §1 of Popper's *Logic of Scientific Discovery* is entitled "The Problem of Induction." Agassi's *Towards an Historigraphy of Science* discusses induction throughout; the first chapter is called "The Inductivist Philosophy." Chalmer's elementary philosophy of science textbook, *What Is This Thing Called Science?*, dedicates its first two chapters to the problem of induction.

10. The proof for the reader interested in technicalities can be found in *The Logic of Scientific Discovery,* Appendix *viii, "Zero Probability and the Fine-Structure of Probability and of Content." Newton-Smith (1981, pp. 49–50) provides an intuitive, less rigorous proof, which is reproduced in full below:

To support this contention Popper considers the universal statement 'If there is an A at location x at time t it will be a B'. This generalization entails an infinite number of instances of the form: at x at t either there is no A or there is an A which is a B. Suppose in the absence of any other information we set the probability of each particular statement of this form being true as $\frac{1}{2}$. The probability of any two of them being true will then be $\frac{1}{2} \cdot \frac{1}{2} = \frac{1}{4}$. The probability of any n of them being true will be $(\frac{1}{2})^n$. The limit of $(\frac{1}{2})^n$ as n goes to infinity is zero. Thus the probability that they are all true is zero and the prior probability of the generalizations being true is consequently zero. This argument is not affected if we assign any probability other than $\frac{1}{2}$ to each instance so long as we do not assign the value of 1. Clearly we would not wish to do that, as this would imply certain knowledge in the absence of any information of the truth of each instance of the generalization.

If with Popper and others (i.e. Mary Hesse) we set the prior probability of universal generalizations as zero, and if we use standard probability theory, no amount of evidence will raise the probability of the generalizations. To illustrate this, consider Bayes theorem which is standardly used to assess the extent to which new evidence raises the probability of a generalization h. Writing p(h/e) for the probability of h given evidence e and p(e/h) for the probability of e given h, one form of Bayes's theorem states:

$$p(h/e) = \frac{p(h) \cdot p(e/h)}{p(e)}$$

Given that the evidence is an instance of the generalization, p(e/h) is 1 and assuming we are absolutely confident in the truth of e, p(e) = 1. However, given p(h) = 0, p(h/e) = 0. Thus no matter how many instances of the generalization we observe, the probability of the generalization will remain where it started, namely at zero. This is a genuine problem which has received much attention in the literature on probability and confirmation. Some have argued with Hesse that universal generalizations do indeed have a prior probability of zero, and that in order to learn from experience we should utilize generalizations which by being restricted to have a finite scope can be assigned a non-zero probability. Others have sought to find principles justifying a non-zero probability to universal generalizations in the absence of all evidence notwithstanding the above argument.

Hintikka and others have in fact developed systems attributing a non-zero inductive probability to universal statements. Popper (1974a, p. 117) criticizes these systems as too restrictive.

11. Popper has not always been consistent in his discussions of the demarcation criterion. Writes Bartley (1982a, pp. 189–90): "In detail, his theory of demarcation is confusing: in his development, statement, and application of it, from 1932 to the present, Popper has been inconsistent." I have taken pains to portray fairly the philosophy of each thinker, and have therefore drawn more heavily from more recent publications in accordance with the principle that this is the more mature philosophy. Thus, in Popper's case I have drawn heavily from his intellectual autobiography and his "Replies to Critics," both in the Schilpp volume (Popper, 1974a, 1974b), even though his philosophy there is not always in accordance with that of his past works. (Weinheimer, too, has also relied heavily on Popper's latest works in his recent book on Popper.)

Obviously, Bartley (a former student of Popper's) and Popper have had an altercation about the interpretation of the demarcation criterion, the role of criticism, and rationality in Popper's work. See *Problems in the Philosophy of Science,* edited by Lakatos and Musgrave, 1968. Because Popper has felt himself misunderstood, I want to include most of his discussion of "the Popper legend" (1974b, pp. 963–65):

> (1) I have always been a *metaphysical realist.* Consequently, I upheld the view that, though some metaphysicians did perhaps talk nonsense, as did also some antimetaphysicians, the meaningfulness of some metaphysical ideas (such as realism or atomism) can be shown by their historical influence on the growth of scientific theories. Thus I always opposed those who declared that all metaphysics is meaningless pseudotalk, and I especially opposed the attempts of the Vienna Circle positivists who tried to back their views by developing a criterion of meaningfulness or literal significance.
>
> (2) I criticized in general as a form of dogmatism the enterprise of trying to set up such a criterion of meaning, and I criticized in particular as "too narrow and too wide" the proposed verifiability criterion of meaning; for it excluded unintentionally the theories of science and included as meaningful, just as unintentionally, some typical existential statements of metaphysics.
>
> (3) I pointed out that this whole enterprise was an attempt to solve a pseudoproblem (an attempt to kill rather than to recognize metaphysics), and that this pseudoproblem had usurped the logical place belonging to a serious and real problem, whose significance had impressed itself on me years before I ever heard about the Vienna Circle. I mean the problem of demarcating

between empirical statements or statements of the empirical sciences on the one hand, and nonempirical statements on the other; examples of which are the statements of a pseudoscience like astrology, and also logical and metaphysical statements. This problem I called the problem of demarcation between science and nonscience, and I explained how it was the real problem hidden behind the positivists' (pseudo-)"problem of meaningfulness" or "significance". I showed that the positivists were convinced a priori that metaphysical talk was meaningless, and they therefore assumed *uncritically* that the problem of demarcation between science and metaphysics was to be solved by the formulation of a criterion of meaningfulness.

(4) I also suggested a solution of the problem of demarcation (but of course *not* of the pseudoproblem of meaning or significance), that is, the criterion of testability or refutability or falsifiability. And I argued that, whereas the criterion of meaning suggested by the positivists, that is, verifiability, leads to paradoxical consequences and to a wrong demarcation, my criterion of falsifiability—if taken as a criterion *not* of meaning, but of demarcation—has a number of fruitful consequences; more especially, it prepares the way for a theory of testability and content, and for a solution of the problem of induction. (I also pointed out that, if falsifiability were substituted for verifiability as a criterion of *meaning,* rather than of demarcation, it would lead, like verifiability, to absurd consequences.

Now the Popper legend affirms the opposite of every one of these points.

(1') The legend has it that Popper was (and perhaps still is) a positivist and perhaps also a member of the Vienna Circle.

(2') Accordingly, says the legend, Popper was also in favour of a criterion of meaningfulness or literal significance, in order to exclude metaphysics as meaningless.

(3') The legend is completely blind to the (for me) vitally important distinction between the problem of finding a criterion of meaning and that of finding a criterion of demarcation.

(4') The legend has it that Popper undertook a kind of rescue operation for the meaning criterion of "conclusive verifiability" (which somebody had shown to be too narrow and too wide) by introducing in its stead, as a new meaning criterion, falsifiability. Thus we can read in a recent and in many respects excellent paper on the subject by an American philosopher of the highest standing: "The difficulty presented (for) the principle of. . .verifiability. . .Prof. Karl Popper tried to circumvent (here there is a footnote reading '*Logik der Forschung* 1935') by a new criterion, namely: that a statement counts as meaningful if it is in principle *falsifiable.*" And it is further often pointed out that this leads to the new difficulty that purely existential statements like "Atoms exist" or "Unicorns exist" (though they are, after all, in principle verifiable) become meaningless, according to my legendary proposal; which, and I would hardly disagree, establishes the absurdity of this alleged proposal of mine.

The legend, in all its points, is clearly implicit in the quoted paper, which is fairly recent (1967). It is still implicit in the latest book (1970) on the positivists' problem of meaning to have come to my notice; here the mistake is hardly excusable because the author refers several times to my "Demarcation" paper in Schilpp's *Carnap* volume (1955(g)), (1964(b)); also (1963(a)), Chap. 11) where I defended at length, in section 2, my view that among the meaningful theories of a language, "(there) will be well-testable theories, hardly testable theories, and non-testable theories. Those which are non-testable. . .may be described as metaphysical". (Schilpp's *Carnap* volume, p. 187; and *C.&R.*, p. 257.) And a little later on the same page I wrote: "beginning with my first publication on this subject," (there is here a reference to (1933(a)); that is (1959(a)), pp. 312–14) "I stressed the fact that it would be *inadequate* to draw the line of demarcation. . .so as to exclude metaphysics as nonsensical from a meaningful language." One can hardly be more explicit.

It seems to me that the legend is widespread, although it is difficult to explain why a doctrine so obviously mistaken and indeed silly, should be so frequently referred to. The explanation is, no doubt, that the legend seems to have the highest authority to back it: it can be read into A. J. Ayer's *Language, Truth and Logic* (although not all of it is explicitly there, except for the identification of my refutability criterion with a criterion of meaning); it can be read into some remarks by Rudolf Carnap; it can be read into some remarks by C. G. Hempel (the hints or reservations added by Hempel turned out to be insufficient to prevent misinterpretation); and it is almost unambiguously stated by Jørgen Jørgensen in his authoritative historical narrative in the *International Encyclopedia of Unified Science*.

I shall say more about Carnap's and Hempel's remarks in the next section. Here I will only point out that we can learn from this not to put too much weight on the testimony even of learned professional witnesses: not a trace of Aristophanean intention to caricature my views can be found in the reports of any of the authorities mentioned.

Bartley (1976a, 1977, 1982a) began a five-part critical study of the work of Popper and a review of the Schilpp volumes in Popper's honor, published in *Philosophia* (Israel). Only three installments have appeared to date. Parts 4 and 5, according to Bartley, were to deal with Popper's social philosophy and philosophy of history and will never appear. Bartley is currently writing a three-volume biography of Popper in which these contributions will be discussed in detail.

12. Consider the following passage from Popper's autobiography (1974a, p. 31):

Early during this period I developed further my ideas about the *demarcation between scientific theories* (like Einstein's) *and pseudoscientific theories* (like Marx's, Freud's, and Adler's). It became clear to me that what made a theory, or a statement, scientific was its power to rule out, or exclude, the occurrence of some possible events—to prohibit, or forbid, the occurrence of these events—*the more a theory forbids, the more it tells us. . . .* I was, however, much concerned with the problem of *dogmatic thinking and its relation to critical thinking*. What especially interested me was the idea that dogmatic thinking, which I regarded as prescientific, was a stage that was needed if critical thinking was to be possible. Critical thinking must have before it something to criticize, and this, I thought, must be the result of dogmatic thinking.

One can clearly see the easy tie Popper makes between social and natural phenomena. His reflections on dogma are important for two reasons. First, he and Kuhn share an interest in dogma and its relationship to science. (See Kuhn, 1963.) The last sentence in the passage above, however, does not make good sense: anything that is not perfect, and thus not simply something dogmatic, can be criticized. Thus, dogmatism is no precondition for critical thinking.

13. Consider, in addition, this passage from Popper's autobiography (1974a, p. 29): "Thus I arrived, by the end of 1919, at the conclusion that the scientific attitude was the critical attitude, which did not look for verifications but for crucial tests; tests which could *refute* the theory tested, though they could never establish it." Popper (1974b, p. 977) is fully aware that criticism itself should be subject to criticism and has its limitations as well:

> 3. What we must avoid, like great artists, is the bad taste of a finicky scholasticism—getting tied up in little assertions or minor criticisms for the sake of criticism.
> 4. Criticism, as we shall see more clearly later, is the lifeblood of all rational thought. But we should criticize any theory always in its best and strongest form; if possible, repairing tacitly all its minor mistakes and concentrating on the great, leading, and simplifying ideas. Otherwise we shall be led into the swamp of scholasticism—of clever questions and answers which have a tendency of multiplying endlessly; a swamp from which there is no escape once we slipped in; a swamp over which the paralysing vapours of the publication explosion hold an eternal sway.
> To put it briefly: never lose sight of the main idea. Never get involved in little side issues if they can be avoided or solved in a simple straightforward way.

Finally, Popper (1974a, p. 92) describes the breadth of the concept:

> In *The Open Society* I stressed that the critical method, though it will use tests whenever possible, and preferably practical tests, can be generalized into what I described as the critical or rational attitude. I argued that one of the best senses of "reason" and "reasonableness" was openness to criticism—readiness to be criticized, and eagerness to criticize oneself; and I tried to argue that this critical attitude of reasonableness should be extended as far as possible. I suggested that the demand that we extend the critical attitude as far as possible might be called "critical rationalism," a suggestion which was later endorsed by Adrienne Koch, and by Hans Albert.
> Implicit in this attitude is the realization that we shall always have to live in an imperfect society. This is so not only because even very good people are very imperfect; nor is it because, obviously, we often make mistakes because we do not know enough; even more important than either of these reasons is the fact that there always exist irresolvable clashes of values: there are many moral problems which are insoluble because moral principles conflict. . . .
> One of the main arguments of *The Open Society* is directed *against moral relativism*. The fact that moral values or principles may clash does not invalidate them. Moral values or principles may be discovered, and even invented. They may be relevant to a certain situation, and irrelevant to other situations. They may be accessible to some people and inaccessible to others. But all this is quite distinct from relativism; that is, from the doctrine that any set of values can be defended.

14. It was the task of Popper's *Logik der Forschung* (1935) to lay out the theory of empirical falsification. But philosophers' ideas evolve, and hence it is advisable to read

Popper's *Logic of Scientific Discovery, Objective Knowledge, Conjectures and Refutations,* his autobiography (in volume 1 of the Schilpp volumes), his "Replies to My Critics" in volume 2 of the Schilpp work, and the three-volume *Postscript to The Logic of Scientific Discovery: Realism and the Aim of Science* (1), *The Open Universe: An Argument for Indeterminism* (2), and *Quantum Theory and the Schism in Physics* (3).

15. Although I am not qualified to evaluate the histories of (natural) science, I can say that there seems to be agreement in the literature that there are several historical accounts of science that conflict with the Popperian account. The two classic works are Feyerabend (1978a) and Kuhn (1970c). Others would add Duhem (1962).

16. Recall (from note 13) that Popper advocates criticism of theories in their "best and strongest form." Following his advice leads us directly to Newton-Smith's conclusions.

A bibliographical reference to Popper's philosophy follows: *Primary:* Popper (all works); Barlety (all works); Berkson and Wettersten (1982); H. I. Brown (1977, pp. 67–77); E. Harris (1972); Newton-Smith (1981, chap. 3); Schilpp (1974, 2 vols.); Spinner (1982); Weinheimer (1986). *Secondary:* Agassi (1975, 1967); Braun (1977); Chalmers (1973); Grünbaum (1976); Lakatos (1974a); Magee (1985); Mortimore and Maund (1976); Mulkay and Gilbert (1981); Musgrave 1973; Passmore (1960); Radnitzky (1976); Schilpp (1963); Stove (1982).

Ignored here is Popper's philosophy of social science.

17. For biographical references on Lakatos, see Worrall (1976) and Gethmann (1984c).

Most of Lakatos's works have been collected as his philosophical papers in two volumes, Lakatos (1978a, 1978b), edited by Worrall and Currie.

Lakatos did build up his own school of philosophy by the time of his death. Worrall and Zahar are two of his more prominent students.

18. In footnote 4, p. 122, Lakatos bemoans this confusion:

Cf. also *above,* p. 109, footnote 2. (*Added in press:*) Possibly it would be better in future to abandon these terms altogether, just as we have abandoned terms like 'inductive (or experimental) proof'. Then we may call (naive) 'refutations' anomalies, and (sophistically) 'falsified' theories 'superseded' ones. Our 'ordinary' language is impregnated not only by 'inductivist' but also by falsificationist dogmatism. A reform is overdue.

Therre are two more important passages wherein Lakatos discusses falsificationism that need to be included:

But when should a particular theory, or a whole research programme, be rejected? I claim, only if there is a better one to replace it. Thus I separate Popperian 'falsification' and 'rejection', the conflation of which turned out to be the main weakness of his 'naive falsificationism'. *One learns not by accepting or rejecting one single theory but by comparing one research programme with another for theoretical, empirical and heuristic progress.* (Lakatos, 1974b, p. 320)

The final passage contains a more technical definition of falsification (Lakatos, 1970, p. 116):

For the sophisticated falsificationist a scientific theory T is *falsified* if and only if another theory T' has been proposed with the following characteristics: (1)

T′ has excess empirical content over T: that is, it predicts *novel* facts, that is, facts improbable in the light of, or even forbidden, by T; (2) T′ explains the previous success of T, that is, all the unrefuted content of T is included (within the limits of observational error) in the content of T′; and (3) some of the excess content of T′ is corroborated.

19. The particular passage in Lakatos (1970, p. 177):

The dogmatic attitude in science—which would explain its stable periods—was described by Kuhn as a prime feature of 'normal science.' But Kuhn's conceptual framework for dealing with continuity in science is socio-psychological: mine is normative. I look at continuity in science through 'Popperian spectacles'. Where Kuhn sees 'paradigms', I *also* see rational 'research programmes'.

Lakatos's discussion of *normative* appears to be as suspect as Popper's reference to a "whiff of induction." Lakatos (1971a, p. 123 n. 2) defines *normative* in the following way:

This is an all-important shift in the problem of normative philosophy of science. The term 'normative' no longer means rules for arriving at solutions, but merely directions for the appraisal of solutions already there. Thus *methodology* is separated from *heuristics,* rather as value judgements are from ought statements. (I owe this analogy to John Watkins.)

20. Two excellent sources on the Duhem thesis are the classic work by Duhem (1962) and Harding (1976a).

21. Those interested in learning more about the Duhem thesis will find reading the literature easier if they recognize that the *modus tollens* is often represented as follows:

$$((H{\rightarrow}O) \cdot \sim O){\rightarrow} \sim H.$$

The above is a compact form of saying this: If from a given hypothesis H we predict a certain observation 0, and if that prediction turns out to be false, i.e., if ~0, then the hypothesis from which the prediction was deduced would be refuted, that is, ~H (Harding, 1976b, p. xxi). Duhem argues that the *modus tollens* rarely ever represents the structure of scientific argument because a scientist's argument is generally based on a locus of theories and assumptions. So conceived, the single hypothesis wins automatic immunity against falsification (rejection due to conclusive proof of falsity) because it cannot be isolated as the problem.

22. Many philosophers agree with Lakatos that there are two versions of the Duhem thesis: a stronger version, often referred to as the Quine thesis, and a weaker form, often designated as the Duhem thesis. Nonetheless, there is not universal agreement on this, or on the exact definition of either version. See Harding (1976a).

23. Koertge (1978) claims to have solved the Duhem problem.

24. See Feyerabend (1979a), Musgrave (1976), and Schramm (1974).

25. Newton-Smith (1981, p. 91) makes exactly that point: "It is in fact pointless to attempt to articulate a principle delimiting the scientific from the non-scientific. What matters is that we have a viable conception of what makes a theory a good one."

26. My translation appears in the text. The original German text (Koertge, 1979, p. 69) follows:

Jedoch sollten LAKATOS zufolge alle wissenschaftlichen Theorien so behandelt werden, als seien sie unfalsifizierbare Deutungsrahmen oder metaphysische Prinzipien. Ironischerweise hat ausgerechnet LAKATOS, obwohl er wenig Respekt vor der Sozialwissenschaft hatte, POPPER's Analyse der Sozialwissenschaft in ein normatives methodologisches Paradigma für die physikalischen Wissenschaften verwandelt! Aufgrund des DUHEMschen Problems— wenn Experiment und theoretisches System sich widersprechen, welcher Teil des Systems wird dann aufgegeben?—schloß LAKATOS, daß wissenschaftliche Theorien niemals falsifizierbar seien. Demnach verschwand die Abgrenzung zwischen Wissenschaft und Metaphysik.

It may be helpful to provide an overview of the differences between the philosophies of Popper and Lakatos.

		Popper	*Lakatos*
1.	Goal of science	increasing verisimilitude	increasing verisimilitude
2.	Unit of appraisal	theory	SRP
3.	Crucial experiments	important	unimportant
4.	When theory and observation conflict	falsify and/or reject theory	modify auxiliary belt
5.	The better theory	highest verisimilitude and corroboration	
6.	Criticism	central role in science	limited role
7.	Duhem thesis	not accepted	accepted as true

27. The passage from Bartley (1976b, p. 37) appearing immediately before that in the text reads:

> In the summer of 1965 we had an extraordinary and widely discussed quarrel, after which we did not speak. . . . There was a time, both before and after our quarrel, when I regarded Lakatos as the most immoral man I had ever met. I later came to think this judgement naive. Lakatos merely talked openly and appreciatively—with a certain connoisseurship—of the sort of behavior which is widespread and almost universally *covert*. I can now appreciate the merits of his practice: although I often saw Lakatos lie when it suited his purposes, he was never a hypocrite. He was remarkably without *self*-deception and quite without cant. In this regard he was morally my superior and light years ahead of some of our friends and colleagues at the London School of Economics. I remember vividly the scolding he gave me in the autumn of 1964: "Beel, you moralize too much. It doesn't get you anywhere. I used to do that in Hungary, and I ended up in prison."

28. Lakatos is not the only member of the Popper school to come under fire for less than scholarly historical representation. Popper himself has been charged with misrepresentation of the views of Plato (de Vries, 1952) and others. Agassi's published dissertation, *Towards an Historiography of Science* (1963), received scathing reviews from historians. For example, Smeaton (1962, p. 125) has the following to say about Agassi's historical accuracy:

> A number of well-known historians of science are condemned by Agassi on the ground that they have uncritically used information culled from earlier

writers. One of the very few writers to be praised by him is Thomas Thomson, who is described as a 'great historian of chemistry' (p. 8) and a 'great chemist and historian of chemistry' (p. 10); his *History of Chemistry* (2 vols., 1830–31) is an 'excellent history of chemistry' (p. 28). Now, the excellence of Thomson's book was partly due to his great ability as a transcriber, but Agassi seems to be unaware of this. Thomson's fifteen pages on Guyton de Morveau, for example, were, with the exception of one paragraph, copied without acknowledgement and without the formality of quotation marks from an earlier account by A. B. Granville. Attention was drawn to this failing of Thomson's by J. R. Partington and D. McKie in a paper to which Agassi refers (p. 105, note 118), but which he has obviously read with insufficient care.

Finlay (1962, pp. 128–29) assesses Agassi's philosphical message in the following way:

> . . . it does seem fairly evident that he is here interested only incidentally in the problem of what does or should count as history. Rather, his primary concern is merely to establish that conjectures meeting the requirements of his criteria, if approached with caution and intelligence, could often be useful to the historian; and he is mightily incensed because he believes many historians of science overlook this simple fact. What his arguments, such as they are, actually seem designed to show, therefore, is not that such conjectures should count as history, but they should count with historians—and historians of science, especially, should devote more effort than they presently do to the imaginative task of inventing interesting conjectures about the scientific work of the past. Agassi's intention, then, is to moralize; but as a result of this, and of his ardour, his analysis becomes wild and confused. In fact, his paper reads more like a revivalist sermon against the evil of following false gods than a critique of historial methods. Reading it, one suspects that Agassi does not really have any very well-substantiated positive views about history. Instead, he seems to have persuaded himself, evidently on philosophical grounds, that the writing of the history of science *must be* (not is) in a very sorry state: yet from this point he appears unable to make any appreciable advance. Perhaps, ironically, this is because he is so busy trying to vindicate his immoderate and excessively conjectural hypothesis.

29. Yet, Lakatos (1974c, p. 323 n. 60) seems to contradict this view in one of his footnotes: "There is nothing necessary about this process (of rational reconstruction of the history of science). I need not say that no rationality theory can or should explain *all* history of science as rational: even the greatest scientists make wrong steps and fail in their judgement."

30. Lakatos has a rather unusual concept of honesty. (See note 27.) Lakatos sometimes seems to equate honesty with rationality, dishonesty with irrationality. See Lakatos (1971a, pp. 104–5n.) and Hall (1976, p. 157). A bibliographical reference on Lakatos follows: Agassi (1986a, 1976, 1979); Andersson (1986); Bartley (1976b); Chalmers (1982, chap. 7); Cohen, Feyerabend, and Wartofsky (1976); Feigl (1971); Feyerabend (1970a, 1975b); Gähde (forthcoming); Gethmann (1984c); Hacking (1981a); Richard Hall (1971); Koertge (1971, 1976, 1979); Kuhn (1971); Lakatos (all works); Lakatos and Musgrave (1968, 1970); McMullin (1976a); Musgrave (1976); Newton-Smith (1981, chap. 4); Radnitsky and Andersson (1978); Sakar (1980); Schagrin (1973); Schramm (1974); Stove (1982); Worrall (1976).

31. Feyerabend (1978b, pp. 149–50), in his reply to Gellner's review of his work, explains:

> Now, first of all, what little bird has told Professor Gellner that I intended to write a scholarly treatise? In my dedication I make it clear that my book was conceived as a *letter* (p. 7) to Lakatos and that its style would be that of a letter. (Also, I am not a scholar, and I have no wish to be a scholar.)

The footnote (14), which appears on page 150, gives us the following information: "Gellner says history and philosophy of science are 'area(s) of (my) professional expertise'. They are not, as every historian and philosopher of science will be most happy to confirm. Besides—how would *he* know?"

32. See Broad (1979) for further biographical information; Feyerabend (1978b, pp. 107–22) for an intellectual autobiography of sorts.

33. Feyerabend qualifies his terminology in footnote 12, p. 21, of *Against Method* (1978a):

> When choosing the term 'anarchism' for my enterprise I simply followed general usage. However anarchism, as it has been practised in the past and as it is being practised today by an ever increasing number of people has features I am not prepared to support. It cares little for human lives and human happiness (except for the lives and the happiness of those who belong to some special group); and it contains precisely the kind of Puritanical dedication and seriousness which I detest. (There are some exquisite exceptions such as Cohn-Bendit, but they are in the minority.) It is for these reasons that I now prefer to use the term *Dadiasm*. A Dadaist would not hurt a fly—let alone a human being. A Dadaist is utterly unimpressed by any serious enterprise and he smells a rat whenever people stop smiling and assume that attitude and those faciial expressions which indicate that something important is about to be said. A Dadaist is convinced that a worthwhile life will arise only when we start taking things *lightly* and when we remove from our speech the profound but already putrid meanings it has accumulated over the centuries ('search for truth'; 'defence of justice'; 'passionate concern'; etc., etc.) A Dadaist is prepared to initiate joyful experiments even in those domains where change and experimentation seem to be out of the question (example: the basic functions of language). I hope that having read the pamphlet the reader will remember me as a flippant Dadaist and *not* as a serious anarchist. cf. footnote 4 of chapter 2.

And in footnote 4 of chapter 2, which appears on p. 33, Feyerabend adds:

> 'Dada', says Hans Richter in *Dada: Art and Anti-Art,* 'not only had no programme, it was against all programmes.' This does not exclude the skilful defence of programmes to show the chimerical character of any defence, however 'rational'. Cf. also Chapter 16, text to footnotes, 21, 22, 23. (In the same way an actor or a playwright could produce all the outer manifestations of 'deep love' in order to debunk the idea of 'deep love' itself. Example: Pirandello) These remarks, I hope, will alleviate Miss Koertge's fear that I intend to start just another movement, the slogans 'proliferate' or 'anything goes' replacing the slogans of falsificationism or inductivism or research-programmism.

34. See note 31.

35. Consider the note in Chalmers's preface to the second edition of his elementary philosophy of science text (1982, p. xiii):

> My friends Terry Blake and Denise Russell have convinced me that there is more of importance in the writings of Paul Feyerabend than I was previously prepared to admit. I have given him more attention in this new edition and have tried to separate the wheat from the chaff, the anti-methodism from the dadaism. . . .

As Chalmers (1982, p. 144) notes, the reviews and discussions of Feyerabend's work are "not of a high standard." The only works on Feyerabend that I would recommend are Chalmers (1982, chap. 12), Newton-Smith (1981, chap. 6), and the most entertaining Duerr anthologies (1980–1981), which have appeared only in German to date.

36. Feyerabend's works should be read in that order because with each new edition or translation he generally adds improvements and revisions.

37. The German passage in *Erkenntnis für freie Menschen* (1979, pp. 251–52), which diverges somewhat from the English in *Science in a Free Society,* appears below:

> Die Autoren des Wiener Kreises und die frühen kritischen Rationalisten, die die Wissenschaft verzerrten und den Ruin der Philosophie herbeiführten, gehörten einer Generation an, die noch eine gewisse physikalische Bildung besaß. Auch begannen sie eine neue "Forschungsrichtung"—sie haben nicht bloß ihre Vorgänger imitiert. Sie *erfanden* ihre Irrtümer, sie mußten für sie *kämpfen* und brauchten daher ein gewisses Minimum an *Intelligenz.* Sie sahen auch bald ein, daß die Wissenschaften komplexer sind, als ihre eigenen kindlichen Modelle und sie bemühten sich deshalb sehr, die Modelle plausibel zu machen. Sie waren Pionere, wenn auch nur Pionere philosophischer Einfalt. Hinsichtlich der Wissenschaftstheoretiker, die heute unserer Universitäten bevölkern, ist die Lage anders. Sie haben ihre Philosophie nicht erfunden, sie haben sie einfach von ihren philosophischen Vorfahren übernommen. Sie haben weder das Talent noch die Neigung, die Grundlagen dieser Philosophie zu untersuchen. Statt kühner Denker, die bereit sind, fruchtbare aber unplausible Ideen gegen eine Überzahl von Gegnern zu verteidigen, haben wir ängstliche akademische nagetiere, die ihre Unsicherheit hinter einer finsteren Verteidigung des Status quo verbergen. Diese Verteidigung hat nun das epizyklische Stadium erreicht: man untersucht Details, flickt hier und dort an Sprüngen, Löchern, rostigen Stellen herum, kann aber nur leere Phrasen zur Begründung des ganzen Verfahrens anführen. Auch haben viele Wissenschaftshistoriker von heute größeres Interesse an der Logik und an den Sozialwissenschaften. Ihr Bild der Physik beziehen sie von Laien wie Popper, Stegmüller odor Radnitzky. Es ist zuzugeben, daß einige Wissenschaften (Teile der Physik, große Teile der Sozialwissenschaften) sich heute auch in einem Stadium der Stagnation befinden. Sie sind in der Tat dem öden Bild sehr ähnlich, das sich die Philosophen von den Wissenschaften machen. So wird Mangel an Talent auf der einen Seite durch Mangel an Bildung auf der anderen unterstützt—und das alles auf Kosten der Steuerzahler. Es ist hohe Zeit, daß sich die freien Bürger um diese Phänomene in ihrer Umgebung kümmern und das nutzlose, analphabetische *und teure* Gerede der Philosophen durch ihre eigenen konkreten Entschlüsse ersetzen!

38. The analogy appears in Feyerabend (1978b, pp. 139–40 n. 8):

> Having seen hundreds of unsmiling and narrowminded young people moving grimly along the Path of Reason (critical reason, dogmatic reason—that does not make any difference) I ask myself what kind of culture it is that has eulogies, prizes, respect for such killing of *souls* while turning with standard- ized revulsion from the killing of bodies. Is not the soul more important than the body? Should not the same or an even greater punishment be extended to our 'teachers' and our 'intellectual leaders' than is now extended to individual and collective murderers? Should not guilty teachers be found out with the same vigour one applies to the hunting of Nazi octogenarians? Are not the so- called 'leaders of mankind'—men such as Christ, Buddha, St. Augustine, Luther, Marx, some of our greatest criminals (it is different with Erasmus, or Lessing, or Heine). All these questions are pushed aside with the facile 'of course' that freezes standard reactions instead of making us *think*.

39. In another passage Feyerabend (1981b, pp. 141–42) attacks the critical ratio- nalists' position on proliferation:

> It is depressing and a sign of the historical illiteracy of most modern philoso- phers (including modern philosophers of science) to see that they are either unaware of the importance of proliferation or ascribe dessicated versions of it to authors far removed from Mill's humanity, simplicity and perceptiveness. Popperians even confront us with the amusing spectacle of fights for a priority none of the participants deserves. Thus Imre Lakatos grants Popper that he has introduced proliferation as an 'external catalyst' of progress but reserves for himself the discovery that 'alternatives are not merely catalysts, which can later be removed from the rational reconstruction' but *'necessary parts'* of the falsifying process; while some Germans hail Helmut Spinner as the real inven- tor of proliferation. The true story is of course that the only contribution these gentlemen made was to bowdlerize Mill's great principle.

Footnote 4 on page 142 tells us:

> Does it matter who the inventor was? Well, it seems to matter to the Pop- perians. Besides, different authors have different background philosophies which put a different complexion on their (alleged) discoveries. For Mill proliferation is an instrument of understanding and social reform. For the Popperians it is a clever trick within a narrowly technical philosophy.

40. Feyerabend's foreword to the 1983 edition of *Wider den Methodenzwang* discusses the making of *Against Method* in greater detail:

> Im Jahre 1970 zog mich Imre Lakatos, einer der besten Freunde, die ich je besessen habe, zur Seite und sagte mir: "Paul", sagte er, "du hast so komische Ideen. Warum schreibst du sie nicht nieder, ich schreibe eine Antwort, wir publizieren die Sache und haben einen Heidenspaß." Der Vorschlag gefiel mir und ich machte mich an die Arbeit. Das Manuskript meines Teils des geplanten Buches war im Jahre 1972 beendet und ich schickte es nach London. Dort verschwand es auf geheimnisvolle Weise. Imre Lakatos, der dramatische Gesten liebte, verständigte die Interpol, und in der Tat, die Interpol fand mein Manuskript und schickte es an mich

zurück. Ich las es noch einmal und schrieb es zum großen Teil um. Im Februar des Jahres 1974, nur einige Wochen nachdem ich meine Revision beendet hatte, starb Imre Lakatos. Ich habe dann meinen Teil ohne seine Antwort publiziert.

Diese Entstehungsgeschichte erklärt die Form des Buches. Das ist kein systematischer Traktat, es ist ein Brief an einen Freund und geht dabei auf die Eigentümlichkeiten des Addressatan ein. Zum Beispiel: Imre Lakatos war ein Rationalist, also spielt der Rationalismus im Buch eine große Rolle. Imre Lakatos bewunderte auch Popper, darum kommt Popper viel öfter vor, als es seiner "objektiven Bedeutung" entspricht. Imre Lakatos nannte mich, in etwas scherzhafter Weise, einen Anarchisten, darum stelle ich mich selber als einen Anarchisten vor. Imre Lakatos war ein Freund der Ironie, und darum machte ich von der Ironie häufigen Gebrauch. Zum Beispiel ist das Ende von Kapital I ganz ironisch gemeint; denn *anything goes* ist nicht *mein* Grundsatz— ich glaube nicht, daß man "Grudsätze" unabhängig von konkreten Forschungs-problemen aufstellen und diskutieren kann, und solche Grundsätze ändern sich von einem Fall zum anderen—, sondern der erschreckte Ausruf eines Rationalisten, der sich die von mir zusammengetragene Evidenz etwas genauer ansieht. Bei der Lektüre der vielen ernsthaften und gründlichen Kritiken, die mir nach Publikation der englischen Fassung ins Haus flatterten, dachte ich oft mit Wehmut an meine Diskussionen mit Lakatos: wie hätten wir beide doch gelacht, wäre es uns vergönnt gewesen, diese Kritiken zusammen zu studieren.

Die neue Fassung unterscheidet sich von der des Jahres 1975 durch Kürzungen, stilistische Änderungen, Einschübe und drei völlig neue Kapitel. Die politischen Anwendungen sind jetzt zur Gänze in *Erkenntnis für freie Menschen* (Suhrkampf 1980) zu finden. Weiteres (und älteres Material) zu den Problemen findet der Leser in den beiden Bänden meiner *Philosophical Papers,* Cambridge 1981, sowie in den (etwas anders gestalteten) *Aus-gewählten Schriften,* 2 Bde., Vieweg 1980, 1981. Anwendungen auf die Kunst sind zu finden in *Wissenschaft als Kunst,* 1983. Noch einmal betone ich, daß die im Buch vorgetragenen Auffassungen nicht neu sind—für Physiker wie Mach, Boltzmann, Einstein, Bohr waren sie eine Selbstverständlichkeit. Aber die Ideen dieser großen Denker wurden von den Nagetieren des Wie-ner Kreises und den sie wieder benagenden kritisch-rationalistischen Nage-tieren bis zur Unkenntlichkeit entstellt. Imre Lakatos war einer der wenigen Denker, die diese Diskrepanz bemerkten und durch die Entwicklung einer sehr viel komplexeren Rationalitätstheorie beseitigen wollte. Ich glaube nicht, daß ihm das gelungen ist. Aber die Anstrengungen war der Mühe wert und hat zu vielen interessanten Ergebnissen geführt. Ihm widme ich daher auch diese, leider schon etwas weniger ironische, Fassung meines verein-samten Teils unserer gemeinsamen Arbeit.

41. It is not difficult to imagine that Feyerabend would be pleased as punch to be anybody's darling. Nonethless, Feyerabend cannot be considered radical just because his argument leads him to the conclusion that *science* is sometimes an irrational enter-prise. Feyerabend did, after all, reach that conclusion via the process of reasoning.

Crittenden (1983, p. 215) does not share my conviction that Feyerabend is truly a rationalist:

The teaching of science distorts by massive simplification and the invention of facts; it blunts imagination and the development of critical abilities; and is in short a systematic indoctrination in the manner of religious instruction at its worst. Objections along these lines could be seen as a tribute to reason. In that case, Feyerabend would be advocating more and better rationality, not less of it. That might be what he has in mind at certain times; but that interpretation would be a misconception of his overall position.

I do not share his conclusion. Feyerabend is an advocate of keeping an open mind, and he shows why this is rational. My own view, however, hangs on my interpretation of Feyerabend's Dada. See the next section. There is also a sense in which Feyerabend's work and the philosophy of science are turning on the definition of *rationalism*. Feyerabend often uses *rational* to mean sensible, but when used to describe the Popper school, he uses *rational* to mean rule-bound. Hence, the resulting confusion.

42. On contradictions in Feyerabend's work, there is one passage that is particularly problematic. Feyerabend claims to be a humanitarian, yet he (1978b, p. 27) writes, "*Traditions are neither good nor bad, they simply are.* 'Objectively speaking' i.e. independently of participation in a tradition there is not much to choose between humanitarianism and anti-semitism." This is a noxious form of relativism, which seemingly clashes with the rest of his theory: anti-Semitism does not allow some traditions to coexist. Thus, unless I missed something, Feyerabend's enlightened relativism fails.

43. Because there was no official translation available, I undertook the translation of Lenk (1982, p. 8):

Schoß Feyerabend mit seinem Total-anarchismus der Erkenntnis über jedes dieser Fernziele hinaus? Der Total-anarchist, machte er wirklich ernst, ertränke im eigenen Sumpf der absoluten Regellosigkeit. Regellosigkeit und Antidogmatismus können manchmal auch dogmatisch wirken. Totale Extreme sind dogmatisch. Und "Alles geht" ist eben ein Extrem (Feyerabend weiß das selbst).

In another passage Lenk (1982, pp. 6–7) complains about Feyerabend's extreme formulations:

Heuristisch ist Feyerabend zweifellos im Recht, obgleich man die pauschale Schlagwortformulierung des "Alles geht", diese ungeschützte, überdehnte Formulierung seines anarchistischen Erkenntnisansatzes für die Beurteilung natürlich nicht so wörtlich nehmen kann. Feyerabend selbst wehrt sich später gegen den Vorwurf, er habe mit dieser "scherzhaften" Superregel und mit "zwei, drei Beispielen" schon "das Ende der Rationalität" verkündet: "Es ist schon wahr", sagt er, "daß zwei oder drei irrationale Schwalben noch keinen irrationalen Sommer machen, aber sie entfernen doch Regeln und Maßstäbe, die in rationalistischen Gebetsbüchern an prominenter Stelle auftreten" (1979, 17).

44. Lenk (1982, p. 10) concludes:

Feyerabend oder Feierabend für die Erkenntnistheorie? Implizit wählte Feyerabend doch wohl—entgegen seinen Worten—das erstere. Oder sollte er selbst schon als Philosoph, der er ist und bleibt, die Feierabendglocke für

Feyerabends Philosophie eingeläutet haben—in selbstironisierend-dialektischer Selbstaufhebung sozusagen? Feierabend für Feyerabend? Glücklicherweise ist es noch nicht so weit.

In the German collection of Feyerabend's papers (1978c, p. 3) Feyerabend indicates not only that he is not giving up philosophy but that he does not even intend to withdraw from publication the old things he no longer cares for or agrees with:

Warum publiziere ich Aufsätze, die mir heute fehlerhaft erscheinen? Weil meine Fehler besser produziert sind, als die der Rationalisten; weil sie noch immer wirksame Gegenmittel bieten gegen *philosophische* Vorurteile; weil viele Leute die Fehler für Einsichten halten und daher an ihnen interessiert sind; und weil sich die Möglichkeit ergibt, aus jugendlichen Irrtümern im Alter finanziell zu profitieren.

My translation of the above passage follows:

Why do I publish papers that I consider to be wrong today? Because my mistakes are better produced than those of the rationalists; because they still offer effective antidotes to certain *philosophical* prejudices; because many people consider the mistakes to be insights and therefore are interested in them; and because it is possible for me to benefit financially in my old age from my youthful mistakes.

45. Feyerabend may be pleased to learn that scientists have discovered that humor is linked positively to intelligence.
46. See the last sentence of the foreword cited in note 40.
47. This is no plea for Feyerabend to relinquish his lively style. Feyerabend (1978b, p. 150) advocates a lively style for the following reasons:

Secondly, why should scholarly books be dry, impersonal, lacking in frivolity? The great writers of the 18th century, Hume, Dr. Johnson, Voltaire, Lessing, Diderot who introduced new ideas, new standards, new ways of expressing thought and feelings wrote a [sic] lively and vigorous style, they called a spade a spade, a fool a fool and an imposter an imposter.

48. Any differences that might have existed between Bartley and Popper in the past (they did break off contact for an extended period of time over differences in interpretation of Popper's work [see note 11]) now seem to be behind them. Bartley (1987c, p. 211) explains:

My approach is based on, interprets, partially corrects, and generalizes Popper's approach. Before stating it briefly, I would like to note briefly the corrections to Popper's position, since otherwise one might become sidetracked in textual exegesis.
In some of Popper's early works, there are occasional passages which might lead one to count him as a limited rationalist, or even as a fideist. These appear in *Die beiden Grundprobleme der Erkenntnistheorie,* in *The Logic of Scientific Discovery,* and also in the first three editions of *The Open Society and Its Enemies.* Thus in *The Open Society* (Chapter 24), Popper urges an "irrational faith in reason" by which we "bind" ourselves to reason. In *The Logic of Scientific Discovery* (Chapter 5), a "decisionism" emerges briefly in

his discussion of the acceptance of basic statements; and in *Die beiden Grundprobleme,* a fideism of sorts appears in passing in his remarks about the selection of aims and goals, and in his acceptance of "Kant's idea of the primiacy of practical reason."

In my view, these early fideistic remarks are relatively unimportant; they play no really significant role in Popper's early thought and none at all in his later thought, but are superfluous remnants, out of line with the main thrust and intent of his methodology, empty baggage carried over from the dominant tradition. They may be dropped without loss, as Popper himself has done, with considerable improvement in consistency, clarity, and generality as a whole. Thus, in 1960, when I proposed a contrast between justificationist and nonjustificationist theories of criticism as a generalization of his distinction between verification and falsification, he dropped the remaining fideism from his approach, and adopted instead the approach that I am about to describe. Our contrast between justificationist and nonjustificationist accounts was introduced at that time. The alternative appproach, which Popper continues to call "critical rationalism", and which I prefer to call "comprehensively critical" or "pancritical" rationality, is then an attempt to overcome the problem of the limits of rationality by generalizing and correcting Popper's original approach.

49. Bartley (1982a, p. 126) defines a "metacontext" as a "context of a context." He elaborates (1982a, p. 127):

A metacontext differs from a context as understood here. A metacontext has to do with how and why contexts are held, subjectively *and* objectively. While there are endless positions and thousands of contexts, there are comparatively few metacontexts. Theory of rationality and ecology of rationality are metacontextual: they are theory about how and why to hold contexts and positions; and they depend in part on goals: e.g., is it one's goal to justify or defend a particular position? Or to attain a more adequate representation?

50. Several criticisms of pancritical rationalism and Bartley's response are collected in Radnitzky and Bartley (1987). Bartley (1987b, p. 313 n. 1) provides a rather comprehensive listing of Watkins's and Post's publications criticizing pancritical rationalism. See Bartley (1987b, p. 314 n. 3) for critical reviews of Watkins's review of pancritical rationalism.

Weinheimer (1986, pp. 110–17) uncritically, and I believe unrightly, accepts Watkins's critique of Bartley, without investigating the replies of Bartley and others.

51. Bartley (1968, p. 43) makes this point:

By contrast to Popper, I believe that his falsifiability criterion of demarcation is relatively unimportant, at least for purposes of evaluation and criticism. And I shall argue that if the problem of the demarcation between science and nonscience is taken in Popper's sense, the problem of demarcation is an unimportant problem.

All members of the Popper school are not in complete agreement. Discussions of falsification (that is, empirical refutation) continue. Consider for instance Musgrave's conclusion in his "Falsification and Its Critics" (1973, p. 406): "In short, *empiricial falsifications are ONE source, though not of course the ONLY source, of problems whose solution requires theoretical innovation.* I take this to be the basic thesis of falsificationism."

52. Popper touches on education; Bartley, however, does actually discuss its significance. See Bartley (1969b, 1970, 1974).
Hans Albert's work on critical rationalism complements Bartley's and needs to be mentioned. See Albert (1964, 1965, 1980). Albert discusses critical rationalism in the context of economics. Albert, it is said, is the most important representative of critical rationalism in Germany. Spinner (1982, pp. 59ff.) discusses the German reception of Popper's philosophy. He characterizes it as (1) not specifically narrowed to physics, (2) narrower in doctrine and wider in application, (3) done almost exclusively by social scientists, (4) more oriented toward applications and practice, (5) marked by its striving to win a place beside Anglo-American philosophy. Spinner (1982, p. 60) remarks that Popper is rarely represented in German philosophy departments:

> Denn die akademischen *Träger* der deutschen Rezeption sind fast ausnahmslos von Haus aus *Sozialwissenschaftler,* denen es vor allem darum geht, die Lehren des Kritischen Rationalismus—als Philosophie, Epistemologie und Methodologie; als Wissenschaftstheorie und Sozialphilosophie—für ihre Studienfächer nutzbar zu machen. Daran hat sich bis heute nicht viel geändert. An den bundesrepublikanischen Universitäten ist der Kritische Rationalismus erstmals zunächst an sozialwissenschaftlichen Fakultäten "institutionalisiert" worden und mit deren Fächern in engere Berührung gekommen, die sich im deutschen Sprachgebiet als bester Resonanzraum erwiesen. So hat POPPERS Denken in den Sozialwissenschaften—vor allem in der Ökonomie und Soziologie, inzwischen aber auch der Pädagogik, Psychologie und einigen Nachbarwissenschaften—seine weitaus stärkste Bastion gewonnen und zumeist auch halten können, sei es in "formaler" Anwendung als *Wissenschaftslehre der Sozialwissenschaften* (vgl. die Übersicht in ALBERT 1967), sei es in "materialer" Anwendung als inhaltliche *Sozial-, Polit-, und Geschichtsphilosophie.* Nicht als "reine" oder naturwissenschaftlich vorprogrammierte Wissenschaftstheorie, die POPPERS Methodologie ursprünglich ausschließlich gewesen und später überwiegend geblieben ist, wurde der Kritische Rationalismus im deutschen Sprachgebiet rezipiert, sondern als sozialwissenschaftlich ausgerichtete Wissenschaftstheorie oder gleich als angewandte Sozialphilosophie, -methodologie und -technologie.

And further (Spinner, 1982, p. 61):

> In den traditionell geprägten philosophischen Fakultäten hat POPPERS Philosophie dagegen bis heute nur ausnahmsweise Fuß fassen können, wobei insbesondere, nicht ohne Vorbehalte hinsichtlich ihres früher oder jetzt abweichenden Standpunktes, ERNST TOPITSCH (Heidelberg/Graz), HANS LENK (Karlsruhe) und GERARD RADNITZKY (Bochum/Trier) genannt werden könnten.

5

Whither the History of Science?

History and the Philosophy of Science

Not all philosophers of science have been interested in historical aspects of science. The current emphasis on the history of science and on theories as complex structural wholes is in part due to the work of the historian of science Thomas Kuhn. But as the historian of science I. Bernard Cohen (1977, p. 309) indicates, Joseph Agassi, Gerd Buchdahl, Joseph Cark, Paul Feyerabend, Adolf Grünbaum, Mary Hesse, N. R. Hanson, Ernan McMullin, Imre Lakatos, Ernest Nagel, Dudley Shapere, Karl Popper, Hans Reichenbach, Stephen Toulmin, and others are philosophers of science who are very concerned with aspects of history of science. To the contrary, I. B. Cohen (1977, p. 310) mentions that Carnap and Quine are notable for their lack of interest in history.[1]

The history of science as a discipline is relatively new. Arnold Thackray (1985, p. 17) points out that it began to emerge as a profession in the United States only after World War II.

> Until that time, the history of science in this country [the United States] was largely a part-time, amateur pursuit, one that was largely celebratory in nature. During the 1920s and 1930s, most of those few people who wrote about the history of science, if asked to discourse about scientific progress would probably have agreed that the sciences' internal criteria for gauging progress were adequate and that scientific progress continued unabated— particularly in the United States.

So in the fifties, when the history of science finally emerged as a discipline, hardly any Anglo-American universities had full-time teaching positions dedicated to the history of science.

> North America could boast perhaps five professional historians of science in 1950, twenty-five in 1960, and probably a hundred and twenty-five by the time this article appears in print [in 1970]. The fifties was thus the crucial decade for defining standards, agreeing on methods, enrolling students, and

creating a discipline. It was also the decade of the H bomb, the Cold War,
Senator Joseph McCarthy, loyalty oaths, militant anticommunism, and the
"silent generation" of students. There were therefore unusually complex
political, ideological, social, and professional factors at work in the shaping of
this new discipline. (Thackray, 1970, p. 117)

Paralleling the trends in the philosophy of science, the range of interest of
the few existing historians of science was narrow. Toulmin (1977a, p. 151)
describes the early history of the discipline as follows:

So, while consciously refraining from any disciplinary collaboration—which
would have been against both their principles—historians and philosophers of
science in the United States of the 1950s went parallel ways, under the influence of their common commitment to an empiricist conception of "scientific
method." If the philosopher's task was to establish the formal organon of
science, the task for the historian was to produce "rational reconstructions"
of the scientific achievements of the past.[2]

Thackray (1985, p. 13) sets the first phase in the discipline's history as running
from 1950 to 1970. The contemporary phase is marked by a shift in emphasis
to the social dimension of science. Although the resistance to the development of the history of science waned somewhat from the sixties on,[3] "most
philosophers of science have a schizophrenic attitude toward the history of
science" (J. Brown, 1980, p. 236). Indeed, the philosophy of science literature
abounds with articles devoted to examining how closely and in what ways the
two fields are interrelated.[4] Some philosophers remain unconvinced that the
relationship is a particularly intimate one.[5] Others believe the two fields are
inextricable; and many suggest that the distinction between the two should
not, and cannot, be drawn.[6] There seems to be some indication that the
debate has reached a median position, with both sides in agreement that "the
philosopher of science must appeal in *some* way to actual scientific practice
and to the historical development of actual science if his analysis is to have any
content" (Merrill, 1980, p. 223), and that a history of science that does not pay
attention to philosophy is flawed: Kant's maxim.

A closer examination of the "soul-searching" of philosophers of science
reveals that one side of the issue seems to be settled: "Do philosophers need
history?" has been answered with a resounding yes. Science is, after all, their
subject matter. And, according to I. B. Cohen (1977, p. 349): "Philosophers
use history to provide an empirical base for their statements, or at least to find
examples in the real world of science (as it has actually been practiced) which
may illustrate a thesis of their own or confute a thesis of their opponents."

On the other hand, it seems to me that the reverse question, "Do historians (especially those dealing with intellectual history) need to understand
philosophy?" has not been adequately explored. Yet, I believe that the answer
must be clear. Before the historian writes, does she or he consider "What
were the arguments?" or "How do I interpret the arguments?" (that is, were
they right or wrong, good or bad, and so on?). I do not know how one can
avoid a conclusion that good intellectual history begs for an answer to both

questions. Any kind of history calls for explanation, and that in itself demands an understanding of argument. Likewise, what is science, progress, and the like cannot be answered without a prior notion of what good science, progress, and the like are.

Stephen Toulmin

Because Frederick Suppe (1977b, p. 127) places the beginning of the historical movement in the philosophy of science in the publication of Stephen Toulmin's *The Philosophy of Science* in 1953, I want to begin with Toulmin. Stephen E. Toulmin (1922–) was born in London, England, and received a Ph.D. in philosophy at Oxford. He has written extensively in the philosophy of science, the history of scientific ideas, and epistemology. Currently, he is professor of social thought and philosophy at the University of Chicago.

In *Foresight and Understanding* Toulmin (1961, p. 99) argues that the task of science is not prediction, as widely believed, but "understanding—a desire to make the course of Nature not just predictable but intelligible—and this has meant looking for rational patterns of connections in terms of which we can make sense of the flux of events." Prediction, a tool of science, "an application of science rather than the kernel of science itself" (Toulmin, 1961, p. 36) is "all very well; but we must make sense of what we predict" (1961, p. 115). Science provides explanation through theories, which designate regularities and explain deviations from the expected. A theory presents an "ideal of natural order," that is, a principle of regularity or a paradigm, specifying a "natural course of events" that scientists regard as self-explanatory or natural.[7]

Theories consist of laws, hypotheses, and ideals of natural order, which are all formulated in terms of everyday language, explains Toulmin. The meanings of all terms (for example, from the natural sciences: *light, ray, mass,* and so on) undergo a "language shift" when integrated into technical language; hence, meanings of concepts are theory-dependent. The concepts change and develop through the history of science. The ideals of natural order are "preconceived notions," for

> it is always *we* who frame the questions. And the questions we ask inevitably depend on prior theoretical considerations. We are concerned, not with prejudiced belief, but rather with preformed concepts; and, to understand the logic of science, we must recognize that 'preconceptions' of this kind are both inevitable and proper—if suitably tentative and subject to reshaping in the light of our experience. If we fail to recognize the conceptions for what they are, we shall not appreciate the true character of our scientific ideas, nor the intellectual problems which face our predecessors, through whose labours our own ideas were gradually formed. (Toulmin, 1961, p. 101)

Because of the changing and developing nature of the ideals of natural order, Toulmin (1961, p. 109) cautions that the body of scientific thought and practice needs to be regarded as evolutionary. It is well worth quoting him at length.

> We need, accordingly, to see scientific thought and practice as a developing body of ideas and techniques. These ideas and methods, and even the controlling aims of science itself, are continually evolving, in a changing intellectual and social environment. To study in an effective and lifelike manner either the History of Scientific Ideas or the Logic and Methods of Science, we must take this evolutionary process seriously. Otherwise, we shall be in danger, as historians, of concerning ourselves too much with particular discoveries or doctrines or persons, with anticipations and anecdotes. And, as philosophers, we may end by replacing the living science which is our object of study by a formal and frozen abstraction, forgetting to show how the results of these formal enquiries bear on the intellectual and practical business in which working scientists are engaged. A purely chronological history of science and a purely formal philosophy of science thus have the same deficiency: each of them neglects to place the scientific ideas which are in question into their intellectual environment, so as to show what, in that particular context, gave these ideas and investigations their merit.

Having posed the question "What patterns of thought and reasoning give scientific understanding?" and having answered it with his notion of changing and developing ideals of natural order, Toulmin (1961, p. 99) then explores which "factors determine which of two rival theories or explanations yield greater understanding." Toulmin (1961, p. 110) rejects a single criterion of merit, and likens the process of competing theories to one of the Darwinian survival of the fittest:

> The common task which accordingly faces historians and philosophers of science has parallels elsewhere—in Darwinian biology. In the evolution of scientific ideas, as in the evolution of species, change results from the selective perpetuation of variants.

The Darwinian struggle between theorists encourages progress in science "only if men apply their intellects critically to the problems which arise in their times, in light of the evidence and the ideas which are then open to consideration" (Toulmin, 1961, p. 115). Concluding, he (1961, p. 115) states:

> The mainspring of science is the conviction that by honest, imaginative enquiry we can build up a system of ideas about Nature which has some legitimate claim to 'reality'. That being so, we can never make less than a threefold demand of science; its explanatory techniques must be not only (in Copernicus' words) 'consistent with the numerical records'; they must also be acceptable—for the time being, at any rate—as 'absolute' and 'pleasing to the mind'.

Toulmin's volume 1 of *Human Understanding* is a "reappraisal of our working ideas about rationality" (1972b, p. vii). The other two volumes of the planned trilogy have not appeared, and considering the time lag, may not appear. Rationality, asserts Toulmin (1972b, p. vii) "is concerned far more

directly with matters of function and adaption . . . than with formal considerations." The chief argument running through this volume is that

> in science and philosophy alike, an exclusive preoccupation with logical sys-
> tematicity has been destructive of both historical understanding and rational
> criticism. Men demonstrate their rationality, not by ordering their concepts
> and beliefs in tidy formal structures, but by their preparedness to respond to
> novel situations with open minds—acknowledging the shortcomings of their
> former procedures and moving beyond them. (Toulmin, 1972b, pp. vii–viii)

Toulmin (1972b, p. 259) states that "the search for a permanent and universal demarcation criterion, between 'scientific' and 'non-scientific' considerations, appears in vain." He (1972b, p. 484) proposes "an ecological approach
to the problem of scientific rationality," which has the advantage of "escap[ing]
the difficulties that beset Kuhn, Lakatos, and Popper alike." (But both Kuhn
and Popper currently favor an evolutionary epistemological approach.[8])

Without the two companion volumes, it is difficult to assess Toulmin's
philosophy of science. His account of science in volume 1 is highly descriptive,
and as Suppe (1977b, p. 678) remarks, it is "exciting, insightful, fascinating,
and vague."

Norwood Russell Hanson

Norwood Russell Hanson's *Pattern of Discovery* (1958) was published four
years before Kuhn's first edition of *The Structure of Scientific Revolutions*. It
could be said that Kuhn is intellectually indebted to Hanson to a certain
extent, for Hanson developed the doctrine of theory-ladenness of observation, a correction of the received view's assumption of a neutral observation
language.[9]

Norwood Russell Hanson (1924–1967),

> a Yale philosophy professor, onetime Marine fighter pilot and full-time indi-
> vidualist, whose own philosophy of life was that "it is very short and should
> be lived to the hilt," a proposition he assiduously followed by buying himself
> a 500-m.p.h. brute of a war-surplus F-8-F Bearcat, in which he buzzed the
> Yale Bowl and roared aloft in fantastic aerobatics, sometimes before the
> enthralled crowds at air shows, more often just for the pure, unadulterated
> hell of it[10]

died when his Bearcat crashed into a hillside as he was on his way to a lecture
at Cornell University in April 1967, cutting short a promising career. Hanson
was educated at Chicago and Columbia universities before completing his
graduate studies at Oxford. He taught for a while at Cambridge University,
then returned to the United States to teach at Indiana University, where he
founded the first interdisciplinary department of history and philosophy of
science in the United States. Hanson published two books while alive: *Patterns of Discovery* (1958) and *The Concept of the Positron* (1963). His first

textbook, *Perception and Discovery,* and several other collections of his essays were published posthumously.

In *Patterns of Discovery* Hanson investigates the procedure and nature of scientific discovery and develops a logic of discovery based on "retroductive reasoning." His task is twofold: Hanson seeks to discredit the received view's assumption of a neutral observation language and its preoccupation with "the context of justification" rather than the more interesting "context of discovery."

> More philosophers must venture into these unexplored regions in which the logical issues are often hidden by the specialist work of historians, psychologists, and the scientists themselves. We must attend as much to how scientific hypotheses are caught, as to how they are cooked. (Hanson, 1958a, p. 1089)

Why doesn't the logic of discovery lie in the domain of psychology or history? The key is "retroducive reasoning," which is tied to Hanson's theory of ladenness of facts, observation, and concepts.

All scientific inquiry, explains Hanson, is theory-laden. Johannes Kepler regarded the sun as fixed with the earth moving around it; Tycho Brahe considered the earth fixed with the sun revolving around it. Seeing, Hanson notes, depends upon the viewer's knowledge and experience; observation is theory-laden; it is an epistemic undertaking. Hanson provides numerous examples of visual doubles in which the same sense datum is perceived but interpreted differently. (Kuhn also makes use of these visual doubles, in particular the perspex cube and bird-antelope figure.) Hanson explains that a child, layman, and scientist may all see the same physical object, yet interpret it differently. Concepts are not just derived from sense experience; many factors predetermine our abilities to perceive, recognize, understand. Seeing is the key to discovery in the sense that one "sees in familiar objects what no one else has seen before" (Hanson, 1958b, p. 30).

Hanson's logic of discovery is based on the division between reasons for accepting a hypothesis and reasons for suggesting the hypothesis in the first place. Reasons for the acceptance of the hypothesis are reasons for believing it true: the logic of justification. Reasons for formulating the hypothesis (the logic of discovery) are reasons that make the hypothesis "a plausible conjecture" (Hanson, 1958a, p. 1074). Hanson's goal is to show that the logic of discovery and logic of justification are logically distinct. He bemoans the fact that the logic of discovery is an aspect of scientific inquiry neglected by philosophers: they either overlook the insight and genius bound up with the discovery process, or deny that logic and reason play a role in discovery. Hanson insists that discovery does not belong to the psychological or sociological realm but is more properly logical. Inductive reasoning is the basis for the logic of justification, retroductive (explanatory) reasoning the basis for the logic of discovery.

Hanson left unfinished work behind, and no one has taken up exactly where he left off.[11] His attempt to show that the two logics are distinct has come under fire. Achinstein (1977, p. 358) points out that reasons for suggesting a hypothesis in the first place can very well be reasons for accepting it. For example, that a hypothesis offers a plausible explanation of the data may be reason both for its suggestion and acceptance.

Hanson's major contribution may well be his founding of the first department of history and philosophy of science in the United States. The department in Indiana University (1960) was followed by the University of Pittsburgh in 1969, the University of Illinois in the mid-seventies, and by numerous institutions since then.

Notes

1. I. Bernard Cohen (1977, p. 309) also adds that the French have a long tradition of noteworthy histories produced by philosophers of science:

> The French have been particularly notable for historical studies made by philosophers: at once such names leap to mind as Gaston Bachelard, Léon Brunschvicg, Georges Canguilhem, Pierre Duhem, Arthur Hannequin, Alexandre Koyré, Hélène Metzger, Emile Meyerson, Gaston Milhaud, Abel Rey.

2. The term *rational reconstruction* truly belongs to the list of "weasel words." John Passmore (1967b, pp. 224–25) contrasts rational reconstruction with descriptive and prescriptive types of inquiry in his article "Philosophy" in the *Encyclopedia of Philosophy:*

> To employ the method of "rational reconstruction" is neither simply to prescribe nor simply to describe. A satisfactory reconstruction of explanation, for example, will almost certainly not apply to everything which has previously gone under that name. It may well reject, for example, explanations in terms of the guiding hand of Providence, not on the ground that they are false but because they are not the sort of thing which ought to be reckoned as an explanation; they do not enable us to account for the fact that things happen in one way rather than in another.
>
> So far, rational reconstruction is prescriptionistic, but not arbitrarily so. The rational reconstructor will try to show that there is no way of distinguishing between explaining and other forms of human activity (for example, reasserting), unless some of what have ordinarily been called explanations are excluded as pseudo explanations. Quite similarly, the test of whether a "rational reconstruction" of philosophy is a reasonable one is whether it provides us with a method of demarcating philosophy from other forms of inquiry, even if in order to do so it is obliged to exclude much that ordinarily passes as philosophy. The rational reconstructor is not so much prescribing as drawing attention to a difference. . . . Obviously, there can be disagreement about whether a reconstruction is rational. Some people will think it so intuitively obvious that Kierkegaard and Kant are both engaged in the same sort of activity that they will reject as prescriptionist any rational reconstruction which has the effect that Kierkegaard is not a philosopher; or they will think it so obvious that Marxism explains what happens in history that they will reject out of hand any theory of explanation which denies that the materialist interpretation of history can serve as an explanation. Controversy on this

point is all the more difficult because words like explanation, philosophy, science, and knowledge are used eulogistically.

This is from the 1967 edition of the *Encyclopedia of Philosophy,* which is outdated in the case of this entry. It is, then, not surprising that I do not find the passage above particularly enlightening. This is no isolated case: most philosophers do not even offer a definition of *rational reconstruction.* Consequently, the term is used differently by different authors. I think if that article were rewritten today, an updated definition of *rational reconstruction* would probably appear as "interpretation of the past based on assumptions of rationality, that is, that the scientist proceeded rationally."

   3. The resistance to history did not, however, cease to exist. See, for example, Brush's "Should the History of Science Be Rated X?" (1974). Brush's thesis (p. 1164) is reproduced below:

   I will examine arguments that young and impressionable students at the start of a scientific career should be shielded from the writings of contemporary science historians for reasons similar to the one mentioned above—namely, that these writings do violence to the professional ideal and public image of scientists as rational, open-minded investigators, proceeding methodically, grounded incontrovertibly in the outcome of controlled experiments, and seeking objectively for the truth, let the chips fall where they may.

These same "impressionable students" who are to be shielded from history are entrusted with the right to drink, drive, and so on, and more important, to vote in elections for the president of the United States.

   4. The fifth volume of the Minnesota Studies in the Philosophy of Science, edited by R. Stuewer and entitled *Historical and Philosophical Perspectives of Science* (1970), was dedicated to that end.

   5. Giere (1973, p. 296), for instance, concluded that the volume mentioned in the note above did not produce "good reasons for thinking that the union is particularly intimate."

   6. See Burian (1977), J. Brown (1980), Hanson (1971b), Kuhn (1977j), McMullin (1976b, 1970), Moulines (1983), Wartofsky (1976).

   7. Toulmin is not using *paradigm* in Kuhn's sense here.

   8. See Kuhn in Suppe (1977b, pp. 508–9) and Kuhn (1970c, pp. 172–73, the final and penultimate pages of his *Structure of Scientific Revolutions,* not counting the "Postscript"). Popper's latest publications appear in the Radnitzky and Bartley volume (1987) entitled *Evolutionary Epistemology, Theory of Rationality, and the Sociology of Knowledge.*

   9. Suppe (1977b, p. 152) supports this view.

   10. "Milestones," *Time* 89 (28 April 1967): 103.

   11. See Suppe (1977b, p. 635).

6

Conclusions

I have purposefully concluded with the role of the history of science because it has been serving as a thorny reminder to philosophers of science that their canons of rationality do not always correspond to actual practice. Although swinging too far to the opposite (sociological) extreme, the historical aspects have certainly brought to light many of the difficulties of the modern philosophy of science.[1] Those historically minded philosophers, remarks Kuhn (1977g, p. 121), "have at least raised problems that the philosophy of science is no longer likely to ignore. . . . There is as yet no developed and matured 'new philosophy' of science." The ideas of the historically oriented philosophers of science are still evolving. Hence, it is impossible at this juncture to assess how the school will fare.[2]

Because a historical approach has not replaced the logical empiricist program, the nature of contemporary philosophy of science is one of continuing transformation and lack of consensus on methodology. Disagreement between those who favor formal methods and those who advocate the historical approach marks the current state of the discipline. But not all philosophers of science find the conflict undesirable, as is well testified by Patrick Suppes's message (1979, p. 25) that there is an

> obvious intellectual tension between those who advocate historical methods as the primary approach in the philosophy of science and those who advocate formal methods. This tension in itself is a good thing. It generates both a proper spirit of criticism and a proper sense of perspective. Each group can tell the other about their weaknesses and the pursuit of philosophical matters can be undertaken at a deeper level. There is no worse fate for a developing theory or method than not to be confronted with opposing views that require a sharpening of concepts and a detailed development of arguments. On the other hand, there seems no reason not to find room in the philosophy of science for a vigorous pursuit of both historical and formal approaches.

Suppes (1979, p. 26) concludes on a still more emphatically pluralistic note that "the tyranny of any single approach or any single method, whether formal or historical, should be vanquished by a democracy of methods that will coalesce and separate in a continually changing pattern as old problems fade away and new ones arise." Nevertheless, it should be emphasized that Suppes's willingness to compromise history and formalism by integrating both represents no consensus within the profession. Suppes (1979, p. 16) acknowledges that one reason for formalist resistance to pluralism lies in the fact that "the present pluralistic and schematic view of the philosophy of science does have the danger of a lack of intellectual discipline." Clearly, the fear that there will be an encroachment of nonscientific elements into sciences if no rules exist to prevent it still persists.

Newton-Smith (1981, p. 232) notes that McMullin "has astutely remarked that the real difference between rival methodologies lies not in the advice they give to the scientist, but in the account they offer of why the methodology works as it does." And, in fact, there is truth to this. Most philosophers of science would agree that a theory should have scope for future developments, be internally consistent, support a successful extant theory, be able to correct its mistakes (a property called "smoothness"), be simple, and so on. All philosophers of science, I believe, would also agree that there is a correct institutional setting that allows science to flourish: assumed is a democratic society in which mutual criticism may be exercised, and so on.

In what way has the philosophy of science changed? Science is recognized as a highly complex process. The idea of method and science has evolved. The seventeenth-century view of scientific method as (1) observation and collection of facts, (2) experiment, and (3) derivation of laws and theories from those facts by a logical procedure is recognized as inadequate and naive. Science is not perceived as a strictly rational (rule-regulated) enterprise. One can talk about a sociology of knowledge, and its significance probably depends upon the individual discipline and the role that external factors play within that discipline. The goals of philosophers of science have become more modest; the urge to delimit or demarcate on the basis of structure or logical form is almost absent today. The measure of all things scientific is no longer taken to be the method of physics but depends on the aims of the individual discipline and how well those aims are fulfilled. Many of the old problems still preoccupy philosophers of science, although the focus has shifted to the way in which these factors contribute to the growth of knowledge.

One may begin to gain a sense that the philosophy of science has been of little assistance to scientists: rules or a methodology will not guarantee that efforts expended on research will yield scientific fruits. But science is not so simple that following a set of ready-made rules will guarantee instant success. Philosophers of science have demanded too much of science in the past; this is particularly clear when one looks at how what it means to be rational has evolved. Popper's early concept of *rational* meant striving to falsify hypotheses. *Rational* for Feyerabend means "rule-bound," and this rigid formulation allows him to be extremely critical of philosophy of science. For Lakatos, *rational* means adherence to the MSRP, and sometimes being honest. For Bartley and the latter Popper *rational* simply indicates a willingness to expose one's ideas to criticism. This less demanding, more realistic position gives scientific discovery the *Spielraum* necessary for growth and creativity. (It also indicates that methodologists have opened themselves to criticism and revision, and so methodology itself evolves.)

Does this absence of rules render the investigations of philosophers and historians of science meaningless? Quite to the contrary, their research yields provocative insights into the limits of knowledge, as well as fulfilling a nonsense-removal function. Philosophers of science from the Vienna Circle up to today have always been concerned with driving deception, sham argument, and even the ideology of science itself out of the sciences. It seems that the first place to eliminate nonsense, then, is to minimize external factors that thwart science. This cannot be accomplished by the imposition of scientific rules. Chalmers (1982, p. 169), in fact, elevates this nonsense-removal function to the most important purpose of the philosophy of sciences:

> In retrospect, I suggest the most important function of my investigation is to combat what might be called the *ideology of science* as it functions in our society. This ideology involves the use of the dubious concept of science and the equally dubious concept of truth that is often associated with it, usually in the defence of conservative positions. For instance, we find the kind of behaviourist psychology that encourages the treatment of people as machines and the extensive use of the results of I.Q. studies in our educational system defended in the name of science. Bodies of knowledge such as these are defended by claiming or implying that they have been acquired by means of the "scientific method" and, therefore, must have merit. It is not only the political right wing that uses the categories of science and scientific methods in this way. One frequently finds Marxism using them to defend the claim that historical materialism is a science. The general categories of science and scientific method are also used to rule out or suppress areas of study. For instance, Popper argues against Marxism and Adlerian psychology, on the grounds that they do not conform to his falsificationist methodology, whilst Lakatos appealed to his methodology of scientific research programmes to argue against Marxism, contemporary sociology, and other intellectual pollution!

In short, when one encounters arguments for the naive view of science as laws proven by observation and experiment, for falsification, and so on there should be immediate concern about possible deceptive intent.

Hinted at above is an additional ramification for the philosophy of science that should not go neglected: the influence of the philosophy of science itself on science is often overlooked. Alexandre Koyré (1961, p. 177) points out that

> it is, indeed, my contention that the role of this "philosophic background" (of conflicting theories) has always been of utmost importance, and that, in history, the influence of philosophy upon science has been as important as the influence—which everyone admits—of science upon philosophy.

The widespread influence of philosophy upon science warrants greater interest and participation in the philosophy of science from practitioners of all sciences.[3]

Notes

1. The more extreme outlook toward the historical group of philosophers of science has been expressed by Radnitzky (1976, p. 505):

> In *contemporary philosophy of science,* on the other hand, "psychologism" or "sociologism" has retained or more correctly *regained* a dominant position. The use of the history of science to challenge the philosophy of science has recently received a major impetus through the work of Polanyi, and above all, Kuhn, Feyerabend and Toulmin—to name only a few major figures.

2. Suppe (1977b, pp. 633–34) does not concur with this view. (What I call historically minded philosophers he refers to as *Weltanschauung* philosophers.)

> [T]he *Weltanschauungen* analyses are not widely viewed as serious contenders for a viable philosophy of science. Contemporary philosophy of science, although strongly influenced by these *Weltanschauungen* views, has gone beyond them and is heading in new directions. The *Weltanschauungen* views, in a word, today are *passé,* although some of their authors continue to develop them and they continue to be much discussed in the philosophical literature.

3. This critical survey could not be all-encompassing. One aim of this work has been to equip the reader with enough background information to tackle the literature. The reader interested in the details or in further reading should see the bibliography here, as well as Ian Hacking's bibliography and minihistory of the philosophy of science (1981b) and the bibliographies in Chalmers (1982), H. I. Brown (1977), Newton-Smith (1981), Stegmüller (1978, 1987; 3 vols. in German), and Suppe (1977b).

Also available are the *Scientific American* and *Science* series on the philosophy of science: Gardner (1976); Nagel and Newman (1956); Quine (1962); Salmon (1973); Shapere (1971); Tarski (1969).

For modifications of Kuhn's philosophy see Sneed (1971) and Stegmüller (1986) on German structuralism, and Laudan (1981).

Reichenbach (1938) coined the terms "context of discovery" and "context of justification" (see his 1938, §1).

Finally, D. Redman's *Economic Methodology: A Bibliography with References to Works in the Philosophy of Science (1860–1988)* (1989), treats in part 2 of the bibliography the philosophy of science in general, including the works of Popper, Kuhn, Lakatos, Feyerabend, and the German structuralists, as well as themes on holism, the Duhem thesis, and the role of the history of science. Many works are annotated and there is an introduction to unannotated sections. (Part I is devoted to sources on economic methodology, whether written by economists or philosophers.) The bibliography grew out of and is intended to be a companion piece to this work.

II

ECONOMICS
AND THE
PHILOSOPHY OF SCIENCE

7

Philosophy's Influence on Economics: Early Exchanges

Rewriting economics in view of one of the frameworks of the contemporary philosophy of science has become the vogue in economics in recent years.[1] Although much intellectual energy has been devoted to this end, the application of a philosophy of science framework to economics, as Coats (1983a, p. 312) has observed, has as yet failed to yield a "major conceptual or historiographical breakthrough."[2] At the same time, the "crisis" in economics (about which it has become so fashionable to speak since the 1970s) seems to have deepened, if one can measure its depth by the increase in literature devoted to methodology and by the wide range of economists engaged in this criticism.[3] There is indeed a plethora of literature amassing on the methodology of economics: since 1970 more than fifty books alone on economic methodology have been published (since 1975 at least forty-six books; since 1980 at least thirty-one books).[4] Growing discontent with the discipline, despite the increase in the exchange of knowledge between economics and philosophy, warrants a closer look at the relationship between economics and the philosophy of science. Why have economists paid attention to philosophers of science? Is the study of the philosophy of science a fruitful undertaking for economists? That is, does it improve theory, deepen our understanding of the economy, help to explain why economists do what they do? What is the relationship of methodology to "crises" in economics? Why have some economists put so much faith in certain philosophies of science? Should they continue to do so in the future?

The relationship between economics and the philosophy of science is not a new one but actually dates back to the emergence of economics as a discipline. Adam Smith, often credited as being the founder of the discipline, was and considered himself to be a philosopher.[5] He developed a system of natural philosophy[6] in accordance with the accepted philosophical methodology of the eighteenth century before engaging in his discourses on moral theory and economics.[7] His lectures on the development of astronomy and its relationship to the social sciences indicate the strong influence of the Newtonian system (despite the fact that Smith never read Newton[8]); Smith clearly planned to do for the social sciences what Newton had done for the natural sciences.[9] Not only did Adam Smith isolate economics as a discipline in its own right, he created the first link with the philosophy of science. That link can easily be criticized as scientism, that is, the imitation of the the physical sciences.[10] One cannot overlook the fact that Newton's system was a spectacular success: spin-offs had the advantage of association. Perhaps not surprisingly, Smith's economics was well received by many scientists and laymen alike, especially in the Anglo-Saxon world. But as fond as economists are of Adam Smith, the nexus with the philosophy of science must be regarded as misguided and overly optimistic about what economists can achieve; certainly, no social science could or can expect to predict with the accuracy with which Newtonian science could predict the positions of heavenly bodies in space.[11]

Perhaps the second major influence on economics from the philosophy of science was that of logical positivism/empiricism. Under its influence "political economy" evolved into "economics" and finally into its current form, "positive economics." The economists Mark Blaug (1978, p. 725) and Bruce Caldwell (1984a, p. 1) and the philosopher Daniel Hausman (1984b, p. 40) mark the union of logical positivism/empiricism and economics in the publication in 1938 of T. W. Hutchison's first major work on economic methodology, *The Significance and Basic Postulates of Economic Theory.*

The above needs to be qualified two ways. First, the word *positive,* as far as I know, did not orginally have any association whatsoever with logical positivism. Hence, Schumpeter (1954, p. 8 n. 4) could write in the 1940s: "The word 'positive' used in this connection has nothing whatsoever to do with philosophical positivism."[12] Although the origin of the usage of *positive* among economists is not clear, it was and is almost always juxtaposed with "normative" economics. John Neville Keynes (1852–1949) seems to be the first to use *positive* in this sense in his *Scope and Method of Political Economy* (1973 [1917], pp. 34–35):

> [A] *positive science* may be defined as a body of systematized knowledge concerning what is; a *normative* or *regulative science* as a body of systematized knowledge relating to criteria of what ought to be, and concerned therefore with the ideal as distinguished from the actual; an *art* as a system of rules for the attainment of a given end. The object of a positive science is the establishment of *uniformities,* of a normative science the determination of *ideals,* of an art the formulation of *precepts.*

As Hutchison makes clear in his work on positive economics (1964, pp. 23–50), the idea of a contrast between is and ought was not new at that time. John Neville Keynes's (1973, p. 34n) reservations about using the word *positive* are interesting, for the confusion was soon to be compounded by the binding of the tenets of logical positivism/empiricism to positive economics:[13]

> The use of the term *positive* to mark this kind of enquiry is not altogether satisfactory; for the same term is used by Cairnes and others in contrast to

hypothetical, which is not the antithesis here intended. It is difficult, however, to find any word that is quite free from ambiguity. *Theoretical* is in some respects a good term and may sometimes be conveniently used. In certain connexions, however, it is to be avoided, inasmuch as it may be understood to imply an antithesis with *actual,* as when theory and fact are contrasted; it may also suggest that the enquiries referred to have little or no bearing on practical questions, which is of course far from being the case. Professor Sidgwick in his *Methods of Ethics* employs the term *speculative;* but this term, even more than the term *theoretical,* suggests something very much in the air, something remote from the common events of every-day life. It seems best, therefore, not to use it in the present connexion.

Whereas T. W. Hutchison introduced economists to ideas of the logical positivists/empiricists, it was Milton Friedman who gave the appellation *positive* its positivistic tenor. Friedman, in his famous essay on positive economics (1953, pp. 4–5), ties empiricism to the is–ought dichotomy:[14]

> Positive economics is in principle independent of any particular ethical position or normative judgments. . . . Its task is to provide a system of generalizations that can be used to make correct predictions about the consequences of any change in circumstances. Its performance is to be judged by the precision, scope, and conformity with experience of the predictions it yields. In short, positive economics is, or can be, an "objective" science, in precisely the same sense as any of the physical sciences. Of course, the fact that economics deals with the interrelations of human beings and that the investigator is himself part of the subject matter being investigated in a more intimate sense than in the physical sciences, raises special difficulties in achieving objectivity at the same time that it provides the social scientist with a class of data not available to the physical scientist. But neither the one nor the other is, in my view, a fundamental distinction between the two groups of sciences.

Positive has since taken on multifarious meanings, relegating it to the burgeoning list of "weasel words" of the social sciences.[15] The is–ought distinction has not been spurned and in fact remains a fundamental part of most economics textbooks. For instance, in Lipsey's *Positive Economics* (1963, p. 4) we find, "Positive statements concern what *is* and normative statements concern what *ought to be.*" This distinction, of course, has long since come under fire.[16]

At any rate, when Hutchison's *The Significance and Basic Postulates of Economic Theory* was first published in 1938, economists' use of *positive* was unrelated to logical positivism/empiricism. This brings us to Hutchison's work and the second qualification. In this work, Hutchison adapts Popper's ideas of falsification to economics. Although Popper has views in common with the Vienna Circle, he also diverges significantly. Hence, Hutchison does not introduce economists to the ideas of pure positivism. All this does not change the fact that Hutchison's work represents a major turning point in economic methodology and the first link with twentieth-century philosophy of science.

Kötter (1980, p. 13) has nicely summed up Hutchison's goal in *The Significance and Basic Postulates of Economic Theory:*[17]

Hutchison recognized that the half-hearted inductive program of Mill and his successors did not allow scientific statements to be sharply demarcated/ distinguished from the nonscientific. Strongly influenced by Popper, Hutchison's primary objective was to demarcate scientific economic statements from the nonscientific. In a nutshell, Hutchison's position was the following. He distinguished between two classes of scientific statements: on the one hand *analytically true* statements or statements which are by definition logical, and on the other hand *synthetic* statements, which must at least in principle be falsifiable. Hutchison replaced Mill's weakly inductive line of argument by a radical falsificationism. In his opinion all nonanalytical economic statements—including the so-called "basic postulates" or axioms of the theory—must be directly empirically testable and therefore directly falsifiable.

Hutchison's message rings clear: economics as an empirical science has severe shortcomings. The premises or basic postulates of economic theory are tautologous and not at all self-evident. To lay a claim to scientific status, Hutchison demands that all propositions—even the fundamental assumptions of economics (for example, the maximization principle)—be empirically falsifiable, and thus he makes greater demands on science than does Popper (Kötter, 1980, p. 14).

Certainly, Hutchison's work is not without its problems. And one can readily perceive that it is a product of the times: it reflects the Vienna Circle–Popper zeal to create a program to end excesses in science. Writes Hutchison in 1938 (1965, pp. 10–11):

> The most sinister phenomena of recent decades for the true scientist, and indeed for Western civilisation as a whole, may be said to be the growth of Pseudo-Sciences no longer confined to the hole-and-corner cranks or passive popular superstitions, but organized in comprehensive militant and persecuting mass-creeds, attempting simply to justify crude prejudice and the lust of power. There is, however, one criterion by which the scientist can keep his results pure from the contamination of pseudo-science and there is one test with which he can always challenge the pseudo-scientist—a test which at once ensures precision and exposes the vague concepts and unsupported generalisations on which the pseudo-scientist always relies. As three scientists have insisted, who work in a field which is possibly in greater danger and in even greater need of a definite barrier against the rising tide of pseudo-science than is Economics: "The essence of science is the appeal to fact". This is an appeal which the scientist must always be ready and eager to see made, and where this appeal cannot conceivably be made there is no place for the scientist as such.

Hutchison has in the preface to the 1960 reprinting of his work (1965, p. xi) and elsewhere recanted his own excesses. His earlier insistence that there was no major difference between the social and natural sciences has been retracted as well:

> [M]y views have become considerably less "naturalist," or less naively or crudely so, than they were twenty years ago. Differences between the natural

and the social sciences seem more important and ineluctable than they did then. Indeed, though quite ready, for the most part, to accept and rely on Professor Popper's anti-naturalist thesis in *The Poverty of Historicism,* I would not always want to go so far as he seems to go in denying significance to the differences between the natural and social sciences. The much greater difficulty in securing adequate and convincing tests for statements and theories in human and social studies is, and it seems will always remain, a source of important differences. But these are differences of degree—though vital and consequential differences of degree—and not of principle. The doctrines of a fundamental discontinuity in epistemological criteria between the study of nature and the study of man do not seem acceptable.

Above all, the rejection of the demarcation criterion of intersubjective testability and falsifiability—as, for example, by Professor Knight—seems to leave subjective feelings of certainty, or what Professor Popper calls "psychologism," as the sole, hopelessly inadequate, basis for statements and theories in the social field, with no principle for marking them off or trusting them more rationally, as based on an intellectual discipline, than the utterances of the soothsayer, the ideologist and the crank. Moreover, a process essential, or invaluable, for diminishing and limiting the area of disagreement would disappear. (Hutchison, 1965, pp. xi–xii)

Noteworthy is that Hutchison does not abandon the demarcation criterion despite his admission of difficulties with testing. His and Popper's early work in methodology have evolved considerably. The mature form brings us to contemporary thought, which is the focus here.[18]

The postpositivists—Popper, Lakatos, Kuhn, and others—constitute the contemporary influence of the philosophy of science on economics. The philosophy of science has inspired a rewriting of the history of economics and has generated much controversy and a mass of literature, even prompting a few economists to learn the philosophy of science and a few philosophers to learn economics.[19] The modern relationship between the two disciplines has grown so confused, confusing, and involved that the following discussion focuses solely on more recent developments.

There are two final remarks before the discussion turns to Popper's philosophy and economics. Methodology, although enjoying a revival, is not a very popular theme among economists. Samuelson (1963, p. 231), in fact, has likened it to the work of the devil:[20]

It is more correct, albeit not very informative, to say that soft sciences spend time in talking about method because Satan finds tasks for idle hands to do. Nature does abhor a vacuum and hot air fills up more space than cold. When libertines lose the power to shock us, they take up moral pontification to bore us.

Those of us who blithely absorb the literature on methodology—and even dare to write about it—hope that it will sharpen the wits and drive out loose argument. Finally, because economists (notably Samuelson in the above passage) and even some philosophers of science forget, *methodology* and *method* should not be used interchangeably (that is, if one is interested in clear communication). For the sake of clarity, Machlup's classic discussion (1978y, pp. 54–55) on this follows:

> [A]lthough methodolgy is *about* methods, it is not *a* method, nor a set of methods, nor a description of methods. Instead, it provides arguments, perhaps rationalizations, which support various preferences entertained by the scientific community for certain rules of intellectual procedure, including those for forming concepts, building models, formulating hypotheses, and testing theories. Thus, investigators employing the *same method*—that is, taking the same steps in their research and analysis—may nevertheless *hold very different methodological positions.* Obversely, supporters of the *same methodological principles* may decide to *use very different methods* in their research and analysis if they differ in their judgments of the problem to be investigated, of the existing or assumed conditions, of the relevance of different factors, or of the availability or quality of recorded data. Thus, while we use a method, we never "use" methodology; and while we may describe a method, we cannot "describe" methodology. The confusion of methodology with method is, for a literate person, inexcusable.

Notes

1. Kuhn's and Lakatos's philosophies were met with enthusiasm by economists. The following is a partial bibliography:

Applications of Kuhnian Paradigms/Revolutions to Economics: Abele (1971); Aeppli (1980); Bliss (1986); Canterbery (1980); Canterbery and Burkhardt (1983); Chase (1983); Coats (1969); Dillard (1978); Eichner and Kregel (1975); D. Gordon (1965); Hennings (1986); Hutchison (1978); L. Johnson (1983) (1980); Katouzian (1980, chap. 4); Kunin and Weaver (1971); Meier and Mettler (1985); Pasinetti (1986);* Peabody (1971); Routh (1975); Stanfield (1979), (1974); Sweezy (1971);* Vroey (1975); Ward (1972); Worland (1972); Zinam (1978); Zweig (1971).* *With reservations:* Stigler (1969). *Against:* Baumberger (1977) and Rousseas (1973).

Applications of Lakatosian Research Programs to Economics: The Latsis volume, *Method and Appraisal in Economics* (1976a), was dedicated to Lakatos and economics. See therein Blaug (1976); Coats (1976); de Marchi (1976); Hicks (1976); Latsis (1976b); Leijonhufvud (1976). In addition, see Blaug (1983, 1980a, 1980b, 1978, pp. 713–24); E. Brown (1981); Cross (1982); Fisher (1986); Fulton (1984); Goodwin (1980); Hands (1985b, 1985); O'Brien (1983, 1976); Rosenberg (1986b); Weintraub (1987, 1979)

2. Goodwin (1980, p. 610) concurs with Coats:

> Over the past decade there have been numerous attempts to apply and adapt innovations in the history and philosophy of science to the history of economics. Success has been limited. The two writers whose ideas have been most influential are Thomas Kuhn and Imre Lakatos. . . . It seems high time that the historians of economics begin to propose their own models for the development of their subject.

*In the *Review of Radical Economics* Special Issue, "On Radical Paradigms in Economics" (1971).

3. The criticisms are authored by more than just an unrepresentative handful of economic unknowns or "rebels." Coats (1977, p. 10) briefly sketches the situation:

> It is impossible to list all the deficiencies of current economics noted by these critics, but the following sample will suffice to indicate their range and variety. It is said that: the empirical foundations of economics are inadequate (W. Leontief); there has been a scandalous waste of intellectual resources in the overdevelopment of mathematical economics and econometrics (F. A. Hahn); many of the economists' efforts are irrelevant (G.D.N. Worswick) and contribute little or nothing of value to the solution of major practical problems (O. Morgenstern); the profession is in a state of general confusion (P. J. Bauer and A. A. Walters); its members are suffering from a "collective hubris" (A. Leijonhufvud); an advanced training in economics inculcates a false sense of intellectual values and is therefore actively unhelpful (E. H. Phelps-Brown); and the profession generally maintains a perverse reward system (J. M. Blackman).

4. Since 1970 the following books on methodology have appeared: Arndt (1984, 1979); Balogh (1982); Baranzini and Scazzieri (1986a); Bensusan-Butt (1980); Blaug (1980a, 1980b); Boland (1982, 1986); Boulding (1970); Brunner (1979); Buchanan (1979); Caldwell (1982, 1984); Coats (1983c); de Marchi (1988); Dolan (1976b); Dyke (1981); Fisher (1986); Godelier (1972); Hahn and Hollis (1979); Hausman (1981b, 1984b); Hicks (1979a); Hollis and Nell (1975); Hutchison (1981, 1977b, 1981); Jochimsen and Knobel (1971); Katouzian (1980); Klant (1984a); Latsis (1976a); Machlup (1978m); Marr and Raj (1983); McCloskey (1985b); McKenzie (1983); Mini (1974); Mirowski (1986d); Myrdal (1973); O'Driscoll and Rizzo (1984); Pheby (1988); Pitt (1981); Rosenberg (1976a); Samuels (1980); Stegmüller et al. (1982); Stewart (1979); Thurow (1983); von Mises (1978); Ward (1972); Weintraub (1985); Wiles and Routh (1984); Wiseman (1983a)

The above list is meant to be exhaustive, but I have no doubt neglected someone.

5. Writes Hutchison (1978, pp. 4–5):

> For Adam Smith was, in fact, and undoubtedly always considered himself to be, a philosopher, in a highly comprehensive sense, not as interested in epistemology as Locke, Berkeley, and Hume, but penetrating much more deeply into social and legal philosophy and the psychology of ethics. Smith remained a philosopher from the beginning to the end of his life. He would never have regarded his work as a whole as primarily economic. He thought of economics, or political economy, as only one chapter, and not the most important chapter, in a broad study society and of human progress, which involved psychology and ethics (in social and individual terms), law, politics, and the development of the arts and sciences.

It is also good to remind oneself that the terms *philosophy* and *science* were indistinguishable in Smith's era. They were used interchangeably. Smith considered philosophy to be "the science of connecting principles of nature" (in Lindgren, 1967, p. 47).

On the issue of the founding of economics as a discipline, no doubt the Physiocrats should also be recognized as cofounders of the discipline.

6. This task was suggested to him by the French Physiocrats. See Thompson (1965, pp. 212–13). It was the French *philosophes*—d'Holbach, d'Alembert, Descartes, Montesquieu, Rousseau, Voltaire, Daubenton, and Buffon, among others—who first applied the mechanistic image to all of society. Helvétius, not Smith, first

applied the Newtonian concepts to economics. See D. Redman (1981) for a fuller discussion of Smith's philosophy of science.

7. The reference is to his *Theory of Moral Sentiments* with *The Wealth of Nations.*

8. Smith read popular versions of Newton, but not Newton in the original. Smith sets forth his philosophy of science in a group of works collected as *Essays on Philosophical Subjects,* and especially in the piece "The History of Astronomy." See Lindgren's *The Early Writings of Adam Smith* (1967, pp. 53–109).

9. There is no doubt that Smith was infatuated with Newton's system and intended to build a parallel system in the social realm. Smith (Lindgren, 1967, p. 108) writes in his "History of Astronomy":

> His [Sir Issac Newton's] system, however, now prevails over all opposition, and has advanced to the acquisition of the most universal empire that was ever established in philosophy. His principles, it must be acknowledged, have a degree of firmness and solidity that we should in vain look for in any other system. The most sceptical cannot avoid feeling this. They not only connect together most perfectly all the phaenomena of the Heavens, which had been observed before his time; but those also which the preserving industry and more perfect instruments of later Astronomers have made known to us have been either easily and immediately explained by the application of his principles, or have been explained in consequence of more laborious and accurate calculations from these principles, than had been instituted before. And even we, while we have been endeavoring to represent all philosophical systems as mere inventions of the imagination, to connect together the otherwise disjointed and discordant phaenomena of nature, have insensibly been drawn in, to make use of language expressing the connecting principles of this one, as if they were the real chains which Nature makes use of to bind together her several operations. Can we wonder then, that it should have gained the general and complete approbation of mankind, and that it should now be considered, not as an attempt to connect in the imagination the phaenomena of the Heavens, but *as the greatest discovery that ever was made by man, the discovery of an immense chain of the most important and sublime truths, all closely connected together, by one capital fact, of the reality of which we have daily experience.* (my emphasis)

Blaug (1980b, p. 57) evaluates Smith's methodology in the following way:

> Given the pivotal role of sympathy for other human beings in *The Theory of Moral Sentiments* and that of self-interested behavior in *The Wealth of Nations,* both of these books must be regarded as deliberate attempts by Smith to apply this Newtonian method first to ethics and then to economics. (Skinner, 1974, pp. 180–81)

10. Why can Smith's methodology be labeled scientism? Smith believed that the Newtonian method was the prototype scientific method and that economics should be patterned after it: scientism, by accepted definition. To be fair, Smith did recognize that patterning economics after Newtonian physics would bring with it difficulties.

11. Some historians of science suggest that Newton "sold" his work to the general public by incorporating a major role for the deity in his system. God is the mover of his universal machine. Similarly, one could say that Smith "sold" economics by tying economics to the highly successful Newtonian model. D. Redman (1981) develops the

view that Smith adopted the Newtonian analogy without fully understanding and foreseeing the consequences that it would bring (that is, with unquestioned optimism). It should not be forgotten that Smith was writing only sixty years after Newton had developed his system. Mirowski (1987b, p. 83) advances a much stronger thesis concerning not the classical but the neoclassical economists and their use of the physics' analogy in science:

> . . . examination of the origins of neoclassical theory reveals that its progenitors consciously and willfully appropriated the physical metaphor in order to render economics a "mathematical science" (Mirowski, 1984a, forthcoming). Jevons (1959b, p. 50), Walras (1960), Edgeworth (1881), and nearly every other early neoclassical economist admitted this fact.

12. Schumpeter (1954, p. 8) associates "positive" with "modern" or "empirical." But this is no allusion to logical positivism, which is mentioned only in passing in his *History of Economic Analysis*.

13. The reference Keynes gives is to Cairnes's (1965 [1888], pp. 60–61) use of *positive* in the passage below:

> In the first place, we may describe a science as "positive" or "hypothetical" with reference to the character of its premises. It is in this sense that we speak of Mathematics as a hypothetical science, its premises being arbitrary conceptions framed by the mind, which have nothing corresponding to them in the world of real existence; and it is in this sense that we distinguish it from the positive physical sciences, the premises of which are laid in the existing facts of nature. But "positive" and "hypothetical" may also be used with reference to the conclusions of a science; and in this sense all the physical sciences which have advanced so far as to admit of deductive reasoning must be considered hypothetical, in contradistinction to those less advanced sciences which, being still in the purely inductive stage, express in their conclusions merely observed and generalized facts.

14. Friedman's book *Essays in Positive Economics,* and especially his famous essay "The Methodology of Positive Economics" therein, popularized the usage of *positive economics*. It also generated considerable, as yet unabated, controversy and confusion, bringing with it a mass of literature. What follows is a partial bibliography of works on Friedman: Agassi (1971b); Archibald (1959); Bear and Orr (1967); Blaug (1980b); Boland (1979a), (1980); Caldwell (1980a); Coddington (1972, 1979); de Alessi (1965, 1971); Frazer and Boland (1983); Hamminga (1983); Hirsch (1980); Katouzian (1974); Klant (1974); Klappholz and Agassi (1959); McKenzie (1983, chap. 5); W. Mason (1980–1981); Melitz (1965); Musgrave (1981); Nagel (1963); Piron (1962); Pope and Pope (1972a, b); Rivett (1970, 1972); Rosenberg (1972); Rotwein (1959, 1980); Wong (1973)

15. See Machlup's "Weaselwords and Jargon" in his *Essays in Economic Semantics* (1967, pp. 73–96).

No one, it seems, has undertaken a study of the full history of the word *positive* in economics. There is, however, Machlup's "Positive and Normative Economics" (1978r, pp. 425–50), in which he treats the many meanings of *positive* since its having been fused with logical positivism and logical empiricism, and Hutchison's *'Positive' Economics and Policy Objectives* (1964, part I, pp. 23–119), in which he discusses "the positive-normative distinction in the history of economic thought."

16. A short history of the is–ought dichotomy in philosophy is provided in Appendix I.

Lipsey (1963, p. 5 n. 1) fully realizes that the normative–positive distinction is far too simple to be embraced without reservation, yet continues to use it in instruction for the following reason:

> Philosopher friends have persuaded me that, when pushed to its limits, the distinction between positive and normative becomes blurred, or else breaks down completely. I remain convinced, however, that, at this stage of the development of economics, the distinction is a necessary working rule the present abandonment of which would contribute to confusion rather than clarity.

In all fairness to Lipsey, he does warn the student not to turn the dichotomy into a dogmatic rule (in the first paragraph of the note partially cited above). Economists who are interested in seeing that students become problem solvers—and thus must first learn to recognize the existence of problems—may find Lipsey's decision to feed students easy answers objectionable. Like many moral issues, a clear solution to the is–ought problem may never be found. Not learning how to avoid logical error and to recognize how words function within a language is folly it would seem, for the is–ought problem has broad ramifications for policy issues. Lipsey does not discuss the is–ought problem in the fifth edition (1979) of his elementary economics text, *Positive Economics*.

The is–ought problem in economics is intertwined with the concept of "positive" economics and with the *Werturteilsproblem*. On this T. W. Hutchison and Hans Albert are the experts. See Hutchison's "From 'Dismal Science' to 'Positive Economics' " (1983), and his *'Positive' Economics and Policy Objectives* (1964), as well as Albert's "Das Werturteilsproblem im Lichte der logischen Analyse" (1956).

17. Kötter's (1980, p. 13) text in the original reads:

> Hutchison erkannte, daß das halbherzige induktionistische Programm von Mill und seinen Nochfolgern es nicht gestattete, einen wissenschaftlichen Satzbestand scharf von einem außerwissenschaftlichen zu trennen. Stark von Popper beeinflußt, ging es ihm in erster Linie darum, diese Demarkationslinie zwischen den wissenschaftlichen Aussagen der Ökonomie und nichtwissenschaftlichen Sätzen zu ziehen. Auf einen kurzen Nenner gebracht, sah seine Position folgendermaßen aus: Er unterscheidet zwei Klassen von wissenschaftlichen Sätzen. Zum einen *analytisch wahre* Sätze, d.h. Sätze, die aus Definitionen logisch gefolgert werden. Zum anderen *synthetische* Sätze, die zumindest prinzipiell falsifizierbar sein müssen. Hutchison ersetzte den schwach induktionistischen Ansatz durch einen *radikalen Falsifikationismus*. Seiner Meinung nach müssen nämlich *alle* nicht analytischen Sätze der Ökonomie, also auch die sog. "basic postulates" oder Axiome der Theorie unmittelbar empirisch überprüfbar und damit direkt falsifizierbar sein.

18. Before leaving behind Hutchison's early methodological work, a bibliographical reference is provided. Hutchison's *The Significance and Basic Postulates of Economic Theory* aroused a considerable amount of attention. See, for example, the exchange between Knight (1941) and Hutchison (1941), between Machlup (1978u) and Hutchison (1956), and between Melitz (1965) and Hutchison (1966). Knight (1940) reviewed the book. Other reviews include Stonier (1939), Shearer (1939), and Whitta-

ker (1940). Coats (1983c) provides an interesting discussion of the historical background of the book and contrasts with Hutchison's later writings.

19. The controversy has also raised the issue of whether or how much economists should pay attention to philosophers. See, for example, Hausman (1984a, b), as well as S. Gordon (1978, p. 728), who (I believe representing the profession in general) reaches the conclusion that

> the Age of Science in which we now live includes social science, the most advanced branch of which is economics. That mythical creature, the economist qua economist, need not pay much attention to philosophy, good or bad, but the philosopher of science had better pay attention to economics, good *and* bad.

20. Hahn (in E. Brown and Solow, 1983, pp. 31–32) gives us the following insight into Samuelson's attitude toward issues concerning the foundations of economics:

> Although Samuelson wrote (he said in jest) that "soft sciences spend time talking about method because Satan finds tasks for idle hands to do" (1963, II, Chap. 129, p. 1772), he seems to have been more conscious than many other economists of carrying out a methodological programme. This methodology is, of course, most clear and explicit in the *Foundations,* but it seems to me detectable in much of his subsequent work as well. In a nutshell, he brings economic theorising to a point at which it yields empirically falsifiable propositions.

Certainly, economic methodology does have a bad reputation. Two further examples illustrate this point. Daniel Hausman (1984a, p. 231), a philosopher, has the following to say about economists and methodology:

> Most methodological writing on economics is by economists. Although the bulk is produced by lesser members of the profession, almost all leading economists have at one time or another tried their hand at methodological reflection. The results are usually poor. If one read only their methodology, one would have a hard time understanding how Milton Friedman or Paul Samuelson could possibly win Nobel Prizes. It thus is less surprising that the economics profession professes such scorn for philosophizing than that its members spend so much of their time doing it.

Harrod (1983, p. 383) prefaced his "Scope and Method of Economics" with the following:

> In my choice of subject to-day, I fear that I have exposed myself to two serious charges: that of tedium and that of presumption. Speculations upon methodology are famous for platitude and prolixity. They offer the greatest opportunity for internecine strife: the claims of the contending factions are subject to no agreed check, and a victory, even if it could be established, is thought to yield no manifest benefit to the science itself. The barrenness of methodological conclusions is often a fitting complement to the weariness entailed by the process of reaching them.
>
> Exposed as a bore, the methodologist cannot take refuge behind a cloak of modesty. On the contrary, he stands forward ready by his own claim to give advice to all and sundry, to criticize the work of others, which, whether

valuable or not, at least attempts to be constructive; he sets himself up as the final interpretor of the past and dictator of future efforts.

And these examples abound. For all of this negativism about methodological considerations, it is noteworthy that a *considerable* number of notable economists' first works deal with methodology. See Appendix II, where economists whose *first* works were of a methodological nature are listed. (Samuelson's *Foundations* appears in the list.)

Methodology is here to stay. Perhaps natural scientists have less use for methodology because they can experiment and control their environment, but economists need methodology because the substance of their science is argument. Because philosophy is aimed at analyzing and improving argument, economics and philosophy form a natural partnership. Fritz Machlup (1978m, p. x), an occasional member of the Vienna Circle, always gave students and others interested in methodology this advice:

> I often tell my students that they should not publish any methodological notes, papers, or books until after they have done years of substantive research in their field and attained recognition for their mastery of its technical aspects. The danger of vacuous chatter is great if one engages in methodological discourse without previous work on substantive problems. On the other hand, even a lifetime of scientific research does not generate, let alone guarantee, comprehension of methodological problems. It also takes years of studying philosophy, not limited to just one philosophical school, but catholic in scope. No one should attempt to be a self-made methodologist.

8

Sir Karl Popper's Philosophy of the Social Sciences: A Disjointed Whole

Unlike many philosophers of science who generally concern themselves only with physics, Popper has aimed at influencing the social sciences. Whereas in the English-speaking world Popper is most often associated with his *Logic of Scientific Discovery,* where he first develops his falsification thesis, in the German-speaking world Popper is regarded as one of the world's most important philosophers of social science (Popper, 1987c, p. 54).

It is unfortunate that there seems to be no full treatment of Popper's philosophy of the social sciences, or of economics, his favorite social science.[1] If one wants to assess whether and to what extent economists have used and benefited from Popper's philosophy, one must inevitably first understand Popper's philosophy itself. This rests on interpretation, which is made especially difficult because Popper's philosophy has evolved considerably since the 1930s, leaving behind confusing and conflicting streams of thought. Moreover, Popper never wrote a systematic treatise on the methodology of the social sciences. His methodological *magnum opus* is his well-known *Logik der Forschung* (1934, 1935)/*The Logic of Scientific Discovery* (1959). That work has been extended by his *Conjectures and Refutations* (1963) and by a three-volume *Postscript to The Logic of Scientific Discovery,* edited by a former student of his, William Bartley: vol. 1, *Realism and the Aim of Science* (1983), vol. 2, *The Open Universe: An Argument for Indeterminism* (1982a, vol. 3, and *Quantum Theory and the Schism in Physics* (1982b). All of these works deal almost exclusively with the philosophy of natural science.

But Popper's philosophy of the social sciences cannot be understood properly without a knowledge of his philosophy of the natural sciences. There are in some instances clear ties between the two. For instance, much of Popper's philosophy is prohibitive, and in this respect there is a clear symmetry between his philosophy of the social and his philosophy of the natural sciences: the goal of science is the elimination of error, and similarly, the goal of society in Popper's view is the diminishment of suffering (= his "negative utilitarianism"). But whereas the scientific method is "revolutionary" because criticism sometimes induces major changes, social scientists are advised to recommend small adjustments to social policy. Finally, Popper clearly advocates a unity of method among all disciplines and a "naturalistic" view that rejects any telling difference between the social and natural sciences.

A systematic treatment of each work would fail in Popper's case because he keeps developing his ideas further and in divergent directions. Thus, the focus instead is on his messages and their development in his thought, for example, be like the natural sciences, reject historicism, Marxism, the Freudian philosophy, and relativism; be critical, use the rationality principle and situational logic, and so on. A consistent philosophy of the social sciences cannot be found in Popper's work to date. What can be established is that economics is treated apart from the other social sciences: it is given a place of honor. Popper's criticism of the method of the other social sciences is subsequently explored and his rationality principle is discussed. Then the influence of his philosophy of the natural sciences on economics is treated, and in particular, why falsification fails. In the final section conclusions are drawn about Popper's influence on economists, and in turn, economists' use of Popper's philosophy.

Economics: Queen of the Social Sciences

Popper (1960, p. 2) strengthens the conviction that the social sciences have somehow fallen behind the natural sciences and therefore should be characterized as "the less successful sciences."[2] In the opening page of *The Poverty of Historicism* (1960, p. 1) Popper writes:

> But with Galileo and Newton, physics became successful beyond expectation, far surpassing all other sciences; and since the time of Pasteur, the Galileo of biology, the biological sciences have been almost equally successful. *But the social sciences do not as yet seem to have found their Galileo.* (my emphasis)

Elsewhere Popper (1970, pp. 57–58) especially attacks sociology and psychology as "riddled with fashions, and with uncontrolled dogmas." Yet, according to Popper (1960, p. 60 n. 1), the one exception among the social sciences is economics, for "it must be admitted, however, that the success of mathematical economics shows that one social science at least has gone through its Newtonian revolution."

Why is Popper satisfied with economics? He adopts a "pronaturalistic" stance in *The Poverty of Historicism,* which means he "favour[s] the application of the methods of physics to the social sciences" (1960, p. 2). (Modern trends in economics have in fact paralleled developments in the natural sciences. Mirowski [1984, p. 377], among others, supports the argument that "neoclassical economic theory is bowdlerised nineteenth century physics.") *Naturalism,* as defined by Popper, smacks of scientism. Popper does in fact a most curious definition of *scientism* in *The Poverty of Historicism. Scientism,* explains Popper (1960, p. 105), is "a name for the imitation of *what certain people mistake* for the method and language of science." In other words, scientism in Popper's view seems to reduce to the use of a false method. This is question begging. But Popper's view has evolved. Consider, for instance, the strong stance he took against scientism in his *Postscript to The Logic of Scientific Discovery* (1983, p. 7):

> I dislike the attempt, made in fields outside the physical sciences, to ape the physical sciences, by practising their alleged 'methods'—measurement and 'induction from observation'. The doctrine that there is as much science in a subject as there is mathematics in it, or as much as there is measurement or 'precision' in it, rests upon a complete misunderstanding. On the contrary, the following maxim holds for all sciences: Never aim at more precision than is required by the problem at hand.

Popper does not doubt that economics can and should adopt the method of the natural sciences. He (1960, p. 61) declares that "every natural law can be expressed by asserting that *such and such a thing cannot happen;* that is to say, by a sentence in the form of the proverb: 'You can't carry water in a sieve'." Because the existence of real economic laws has been doubted, Popper (1960, p. 62) provides examples of some laws that can be formulated in economics.

> 'You cannot introduce agricultural tariffs and at the same time reduce the cost of living.'—'You cannot, in an industrial society, organize consumers' pressure groups as effectively as you can organize certain producers' pressure groups.'—'You cannot have a centrally planned society with a price system that fulfills the main functions of competitive prices.' —'You cannot have full employment without inflation.'

Yet these "laws" hardly correspond to anything seen in physics. They are of a general rather than specific nature. They lack the experimental potential of physical laws. The terminology lacks the exactitude of the counterparts in the natural sciences. These are not laws but trends. Popper (1960, pp. 115–16) knows this:

> But, it will be said, the existence of trends or tendencies in social change can hardly be questioned: every statistician can calculate such trends. Are these trends not comparable with Newton's law of inertia? The answer is: trends exist, or more precisely, the assumption of trends is often a useful statistical device. *But trends are not laws.* A statement asserting the existence of a trend is existential, not universal. (A universal law, on the other hand, does not assert existence; on the contrary: as was shown at the end of section 20, it asserts the impossibility of something or other.) And a statement asserting the existence of a trend at a certain time and place would be a singular historical statement, not a universal law. The practical significance of this logical situation is considerable: while we may base scientific predictions on laws, we cannot (as every cautious statistician knows) base them merely on

the existence of trends. A trend (we may again take population growth as an example) which has persisted for hundreds or even thousands of years may change within a decade, or even more rapidly than that.

It is important to point out that *laws and trends are radically different things*. There is little doubt that the habit of confusing trends with laws, together with the intuitive observation of trends (such as technical progress), inspired the central doctrines of evolutionism and historicism—the doctrines of the inexorable laws of biological evolution and of the irreversible laws of motion of society. And the same confusions and intuitions also inspired Comte's doctrine of laws of succession—a doctrine which is still very influential.

Despite a warning that laws and trends are "radically different," his footnote 1 (1960, p. 116) hopelessly muddles the distinction: "A law, however, may assert that under certain circumstances (initial conditions) certain trends will be found; moreover, after a trend has been so explained, it is possible to formulate a law corresponding to the trend; see also note 1 on p. 129, below." Popper wrongly conflates *law* to mean "law corresponding to a trend." Statistical laws, or laws corresponding to trends, are not universal, and thus prediction becomes very limited, as Popper notes in the passage cited above. Although economists have insisted on referring to their trends as laws, they should enclose laws in quotation marks if they want to remain realistic about economic theory and prediction. Sir John Hicks reminds us in a recent work (1979a, pp. 1–2) that economic "laws" are really trends:

> There are very few economic facts which we know with precision; most of the 'macro' magnitudes which figure so largely in economic discussions (Gross National Product, Fixed Capital Investment, Balance of Payments and so on—even Employment) are subject to errors, and (what is worse) to ambiguities, which are far in excess of those which in most natural sciences would be regarded as tolerable. There are few economic 'laws' which can be regarded as at all firmly based.

Hicks (1986, p. 100) succinctly summarizes the epistemological ramifications of the trendlike nature of economics:

> Economics, I have said elsewhere (Hicks, 1979) is 'on the edge of science and on the edge of history'. It is on the edge of science, because it can make use of scientific, or quasi-scientific, methods; but it is no more than on the edge, because the experiences that it analyses have so much that is non-repetitive about them. If a scientific theory is good, it is good now, and it would have been good a thousand years ago, if it had been available; but the aspects of economic life which we need to select in order to make useful theories can be different at different times (see Hicks, 1976, 1983). Economics is in time, and therefore in history, in a way that science is not.

It is, incidentally, a shame that Popper did not have at hand a copy of Oskar Morgenstern's *Wirtschaftsprognose: eine Untersuchung ihrer Voraussetzungen und Möglichkeiten* (Economic prediction: an investigation into its implications and possibilities; 1928), which unfortunately has not been translated, or a copy of his *Die Grenzen der Wirtschaftspolitik* (1934; translated as *The Limits of*

Economics). In the former work, Morgenstern (1928, pp. 96, 108, 112ff.) presents the argument that economic prediction is in principle impossible: forecasts are self-defeating as economic agents understand them and will hence form anticipations that render the original forecast invalid. Morgenstern originally overstated his case (see Marget, 1929), and he has since refined his ideas. Nonetheless, his work, despite its shortcomings, could have effectively counteracted Popper's almost unlimited faith in economic prediction, and it represents in itself a very significant contribution to economic methodology.[3]

Popper does not, however, believe that the social sciences can develop *exactly* like the natural sciences. In his "Prediction and Prophecy in the Social Sciences," written in 1948 (in Popper, 1972a, pp. 342–43) he tells us:

> The view that it is the task of the theoretical sciences to discover the unintended consequences of our actions brings these sciences very close to the experimental natural sciences. The analogy cannot here be developed in detail, but it may be remarked that both lead us to the formulation of practical technological rules stating *what we cannot do*.

He then provides the reader with several examples of social laws such as those appearing above (p. 183) and concludes (1972a, p. 343): "These examples may show the way in which the social sciences are practically important. They do not allow us to make historical prophecies, but they may give us an idea of what can, and what cannot, be done in the political field." Here Popper is advocating the formulation of social theories in a way such that they can be easily falsified, which means that he again erroneously assumes the existence of universal laws. This confusion between laws and trends and subsequently the application of falsification to the "laws" of social theory runs throughout Popper's work. He needs social laws because his empirical basis is the falsifiability of laws by "basic" empirical statements. He needs falsification to eliminate Marxism, psychologism, and so on. We should not forget that Popper's *Poverty of Historicism* was meant as an allusion to Marx's *Poverty of Philosophy*, which in turn alluded to Proudhon's *Philosophy of Poverty*. At any rate, it is perhaps not insignificant that "The Poverty of Economics" (Kuttner, 1985, and also Matuszewski, 1980) has now appeared.

As inconsistent as it may be, Popper does appreciate the problems that economics will encounter if it is to become more lawlike. Consider the following passage from *The Poverty of Historicism* (1960, pp. 142–43):

> But it cannot be doubted that there are some fundamental difficulties here. In physics, for example, the parameters of our equations can, in principle, be reduced to a small number of natural constants—a reduction which has been successfully carried out in many important cases. This is not so in economics; here the parameters are themselves in the most important cases quickly changing variables. *This clearly reduces the significance, interpretability, and testability of our measurements.* (my emphasis)

Popper does possess an excellent grasp of statistics; he (1960, pp. 137–38) professed, however, to having no knowledge of the social sciences when he wrote *Logik der Forschung* in the early 1930s.[4] Although the first edition of *The Poverty of Historicism* appeared in 1957 and the second in 1960, Popper had great difficulty finding a publisher, and hence the work was finished much earlier—around 1936 he tells us (1960, p. vii). Given his admitted ignorance of economics and the youthful state of the social sciences, perhaps his misgivings about extending the method of the natural sciences to the social sciences should have been taken more seriously by economists.

T. W. Hutchison has reached a similar conclusion. There seems to be little evidence of a Newtonian revolution in economics, although the analogy of Newtonian mechanics has been extensively adopted. Hutchison's (1977c, p. 40) insight into Newtonian revolutions in economics is particularly enlightening:

> Certainly one might find similar claims suggested by mathematical economists. But the mathematical 'revolution' in economics has been one mainly (or almost entirely) of *form, with very little or no empirical, testable, predictive content involved.* In accepting as 'Newtonian' a purely, or almost purely, formal, or notational, 'revolution', Sir Karl seems to have allowed himself to be taken in by over-optimistic propaganda. Not only has nothing genuinely describable as 'a Newtonian revolution' taken place in economics, it is reasonable to suggest that it is not probable that anything of the sort is going to occur in the foreseeable future.

In spite of all this, some economists have cited the overly optimistic Popper and ignored his reservations. As Hutchison (who was Popper's colleague at the London School of Economics for a time) (1977c, p. 58) advises,

> one must distinguish between the vitally important issues of Popper's general methodological principles, and his highly over-optimistic, but more-or-less incidental, comments regarding a 'Newtonian revolution' in economic theory, and the comparability of economic laws with those of physics. Such comments as these latter are bound to be highly popular with 'theoretical' economists and their influence could probably be shown to have been quite important in the fifties and early sixties.

In order to put Popper's excess enthusiasm for the economic method in perspective, it is necessary to understand that his animosity toward the other social sciences stems from his attack on Marxism, or historicism, and to consider his theories of situational logic and falsificationism.

The Polemical Element: The Case Against Historicism

What is the problem with the social sciences? Why are they less successful than the natural? According to Popper, historicism, a false method, is the reason for their lack of progress.[5] Economics, to the contrary, is the notable exception (Popper, 1960, p. 3). What does Popper mean by this word *historicism,* which he coined? According to Popper (1960, p. 3), historicism is "an approach to the social sciences which assumes that *historical prediction* is their principle aim, and which assumes that this aim is attainable by discovering the 'rhythms' or the 'patterns', the 'laws' or the 'trends' that underlie the evolution of history." Popper does distinguish between *historism,* explanation by reference to the historical period in which an idea and so on develop (that is, historical relativism), and *historicism* as defined above.[6] He also acknowledges that his usage of *historicism* is an extreme form.[7] Despite his efforts to clarify what he means, his usage of *historicism* remains problematic in many ways.

The Poverty of Historicism (which is the companion piece to Popper's *Open Society and Its Enemies*), as well as various other articles of Popper's are polemics against the methods of Marx, Plato, Hegel, Karl Mannheim, August Comte, J. S. Mill, and Arnold Toynbee. The polemical element of his works has been seriously criticized for not being scholarly, especially for inaccurately portraying views of these authors.[8] For example, the historicist position that Popper describes is not one represented by Marx,[9] Hegel, or any of the members of the German historical school.[10] As Robert Ackermann (1985, p. 172) stresses: "Popper clearly lays down a conception of methodology which is compatible with the features of historicism that he wishes to combat." *Historicism* is undoubtedly a *Kampfbegriff* (Lee and Beck, 1953–1954, p. 575), and a term closer in meaning to the negative usage of the German term *Historismus,* in which historicism "is identified with the abandonment of theory, particularly in economics and law" (Iggers, 1973, p. 457). In the German version of *The Poverty of Historicism, Das Elend des Historizismus,* Popper translates *historicism* as *Historizismus* rather than *Historismus,* adding to the confusion. If nothing else, it can safely be said that Popper have *historicism* a meaning "which has not been generally accepted" (Iggers, 1973, p. 457).

The overriding theme in *The Poverty of Historicism* is that individuals control their own destinies. There are no laws of history, no historical forces that determine people's futures. Large-scale, long-term prediction is prophecy and not science. Social sciences should thus be grounded in methodological individualism,

> the quite unassailable doctrine that we must try to understand all collective phenomena as due to the actions, interactions, aims, hopes, and thoughts of individual men, and as due to traditions created and preserved by individual men. (Popper, 1960, pp. 157–58)

Psychology, sociology, or history may not be the basis of a social science. Methodological individualism combined with the logic of the situation is "a logical method" (Popper, 1960, p. 158) and Popper's answer to the problem of holism.

Popper's attack on historicism loses its potency because historicism seems to be equated with holism, relativism, and the large-scale prophesying of the social sciences. Popper was reacting to scientists (including natural scientists) who, at that time, were attracted to Marxism because it claimed to use the scientific method of the natural sciences (Popper, 1972a, p. 337). Popper's purpose is to draw a distinction between the valid use of science and the pseudo social sciences, and thereby to expose Marxism as unscientific and dangerous. Passmore (1974, p. 47) mentions that Popper simply conflates too many issues to remain clear:

> At the same time it is quite vital, I am convinced, to distinguish the issues which Popper has conflated: to see that such questions as whether one can construct a physics-type social science or base such a science on methodological individualism are logically quite independent of the question whether historical prophecies can be rationally based on the discovery of historical laws. For the conflation of these issues has had two effects, equally bad. In some quarters it has led to the too-rapid rejection of anti-naturalistic accounts of the social sciences or anti-individualistic accounts of social institutions in the belief that to accept them is at once to be committed to evolutionary holism. And, on the other side, it has led to the neglect of quite crucial questions Popper has raised—especially, perhaps, about historical inquiry—out of sheer impatience with the convolutions of Popper's argument or anger at his cavalier attitude to the historical phenomenon of historicism. In short, Popper's work is an object lesson in the way in which the use of a label can obfuscate discussion.

Certainly, a strict application of Popper's methodological individualism would eradicate modern macroeconomic theory: the aggregation problem has not been (and may never be) solved, and hence macroeconomics has no solid microeconomics foundation. Robert Ackermann (1985, pp. 179–80) points out that neoclassical economic theory makes assumptions that Popper would almost assuredly consider to be holistic:

> What seems not to be noticed in the Popper camp is that neo-classical economic theory (as well as Marxist economic theory) sees economic relationships as penetrating the entire fabric of society. A price change, or a supply change, even in some relatively small market sector will instantaneously affect the entire society, that is all other prices, or an entire schedule of demands. Economic theory then drives us to the conclusion that the social effects of individual action may well be holistic, and this is independently of Marxism, and not obviously compatible with the presumption of piecemeal social engineering that small changes have small effects.

The meaning of historicism has not been discussed for purely pedantic reasons. Out of the confusion arises a message that Popper simply was not delivering but that probably supported the twentieth-century neoclassical economist's repugnance for history.[11] Popper's usage of *historicism* invites a divorce of science (especially social science) from history itself. This confusion may arise when the reader encounters, for example, the following passage: "Social science is nothing but history: this is the thesis [of historicism]" (Popper, 1960, p. 45), or the title of the twenty-fifth chapter of Popper's *The Open Society and Its Enemies* (1966b, p. 259): "Has History Any Meaning?" Since *historicism* has taken on a meaning of (moderate) historical relativism in most English-speaking countries, and especially in the United States (a definition more akin to Popper's *historism*), it does not take a great leap in logic for the casual reader of these works to conclude that history itself is being rejected.

That would be a gross misrepresentation of Popper, and folly as well, because Iggers (1973, p. 462) reminds us that "in a sense the modern outlook continues to be historicist, if by historicism is meant merely the recognition that all human ideals and institutions are subject to historical change." Popper (1960, p. 39) himself admits: "It would be ridiculous to deny the importance of history in this narrow sense [that is, nonhistoricist sense] as an empirical basis for social science." Popper did not refute the theory of historicism in its normal usage,[12] prove that history is not important for the social sciences, refute the position(s) of the German historical school, prove that the methods can be adapted easily to the social sciences. Even his attack on Marx has been challenged.[13] One component in the two works that has weathered time is Popper's attack on intellectual relativism.[14] Perhaps the most important insight in *The Poverty of Historicism* (p. 88) is to be found in the following passage:

> Scientific method in politics means that the great art of convincing ourselves that we have not made any mistakes, of ignoring them, of hiding them, and of blaming others for them, is replaced by the greater art of accepting the responsibility for them, of trying to learn from them, and of applying this knowledge so that we may avoid them in the future.

Popper (1974a, p. 91) has the following to say in his autobiography about the importance of these two works:

> *The Poverty* and *The Open Society* were my war effort. I thought that freedom might become a central problem again, especially under the renewed influence of Marxism and the idea of large-scale "planning" (or "dirigism"); and so these books were meant as a defence of freedom against totalitarian and authoritarian ideas, and as a warning against the dangers of historicist superstitions. Both books, and especially *The Open Society* (no doubt the more important one), may be described as books on the philosophy of politics.

The Poverty, nonetheless, is the more important work for economists and other social scientists—because of its confusions. It is in these two works that Popper first develops his thesis of situational logic, to which we now turn.

Rationality and Situational Logic

Discussions of rationality are concerned with either rational action or rational belief. Popper's theory of situational logic is a theory of rational action. Koertge (1974, p. 75) provides a helpful introduction to the topic:

> There are two major theories of rational action at present—decision theory and Popper's situational logic. Decision theory is a prescription for rational action: List the options open to you, estimate the utilities and probabilities of the various outcomes and choose the option which maximizes expected utility (the minimax variant of decision theory recommends minimizing the maximum loss.)

> Popper's situational logic is a descriptive theory with methodological impli-
> cations: If you want to understand X's action, find out what X's goals were
> and what X perceived his situation to be; X's action will then be seen to be
> one appropriate to that perceived situation.

This introduction seems to be rather forthright. Popper's development of situational analysis, however, creates exegetical problems that cannot be fully resolved because situational analysis (like historicism) represents too many things. It is a statistical generalization, an ideal law, a mathematical method, a rational (re)construction, the assumption that the human being is rational. It is false but objective, not psychological. The first discussion of situational logic appears in *The Poverty of Historicism;* it is more fully developed in his 25th to 27th "theses" listed in his article "The Logic of the Social Sciences" (1976a, pp. 87–104), in a French article entitled "La rationalité et la statut du principe de rationalité" (1967, pp. 142–59), and in his autobiography (1974, pp. 93–94).

Popper's most recent discussion of situational logic is the brief summary in his autobiography (1974a, pp. 93–94). There he expresses his admiration for the method of economics:

> [T]he method of situational analysis, which I first added to *The Poverty* in
> 1938, and later explained a little more fully in Chapter 14 of *The Open
> Society,* was developed from what I had previously called the "zero method".
> The main point here was an attempt *to generalize the method of economic
> theory (marginal utility theory) so as to become applicable to other theoretical
> social sciences.* In my later formulations, this method consists of constructing
> a *model of the social situation,* including especially the institutional situation,
> in which an agent is acting, in such a manner as to explain the rationality (the
> zero-character) of his action. Such models, then, are the testable hypotheses
> of the social sciences; and those models that are "singular", more especially,
> are the (in principle testable) singular hypotheses of history.

The "zero method" is set forth in the original discussion of situational logic (1960, p. 141):

> I refer to the possibility of adopting, in the social sciences, what may be called
> the method of logical or rational construction, or perhaps the 'zero method'.
> By this I mean the method of constructing a model on the assumption of
> complete rationality (and perhaps also on the assumption of the possession of
> complete information) on the part of all the individuals concerned, and of
> estimating the deviation of the actual behavior of people from the model
> behavior, using the latter as a kind of zero co-ordinate. An example of this
> method is the comparison between actual behavior (under the influence of,
> say, traditional prejudice, etc.) and model behavior to be expected on the
> basis of the 'pure logic of choice', as described by the equations of economics.

In the footnote to the first sentence in the above passage, Popper refers the reader to an article written by Jacob Marschak (1943). Marschak (1898–1977) is regarded as one of the first "pioneer(s) of econometrics" (Coser, 1984, p. 151). In the introduction to this particular article Marshak (1943, p. 40) cites

Schumpeter's conclusion to his founding contribution to *Econometrica*, "The Common Sense of Econometrics" (1933, p. 12):

> We should not indulge in high hopes of producing rapidly results of immediate use to economic policy or business practice. Our aims are first and last scientific. We do not stress the numerical aspect just because we think that it leads right up to the core of the burning questions of the day, but rather because we expect, from constant endeavor to cope with the difficulties of numerical work, a wholesome discipline, the suggestion of new points of view, and helps in building up the economic theory of the future. But we believe, of course, that indirectly the quantitative approach will be of great practical consequence. The only way to a position in which our science might give positive advice on a large scale to politicians and businessmen, leads through quantitative work. For as long as we are unable to put our arguments into figures, the voice of our science, although occasionally it may help to dispel gross errors, will never be heard by practical men. They are, by instinct, econometricians all of them, *in their distrust of anything not amenable to exact proof.* (my emphasis)

In Popper's footnote referring us to Marschak's article he directs us to "see the 'null hypothesis' " discussed by Marschak. This is the method whereby a statistical hypothesis is tested and accepted or rejected in favor of an alternative hypothesis—that is, statistical analysis completely consonant with Popper's falsification thesis. By a "Newtonian revolution" it appears that Popper was in fact referring to economics' adoption of mathematical and statistical techniques. And it is clear that Marschak and Popper influenced each other. An article written by Marschak two years earlier than the one cited above (that is, Marschak, 1943) closes on the following note (Marschak, 1941, p. 448): "I hope we can become 'social engineers' . . . I don't believe we are much good as prophets." The echo of Popper's *Poverty of Historicism* rings clear. How much Marschak was responsible for Popper's belief in an economic Newtonian revolution is unclear; the influence was probably considerable.[15]

In the article on situational logic of Popper's that appeared in French (1967, pp. 142–50), we learn that situational logic does not play the role of an empirical, explanatory theory and it is not a testable hypothesis, that is, it is metaphysical. "Ce principe ne joue pas le rôle d'une théorie empirique explicative, ou d'une hypothèse testable" (Popper, 1967, p. 144). Moreover, Popper compares his rationality principle to the laws of Newton: Popper poses the question "What will move this model of situational logic?" and replies that the principle of *rationality,* and *not* the laws of psychology, corresponds to Newton's laws:

> L'erreur habituellement commise sur ce point consiste à supposer que, dans le cas de la société humaine, un modèle social doit être animé par *l'animus* ou la *psyché* de l'homme, et qu'ici nous devons pour cette raison remplacer les lois de Newton soit par les lois de la psychologie humaine en général, soit peut-être par les lois de la psychologie individuelle s'appliquant aux caractères individuels qui interviennent comme acteurs dan la situation envisagées. (1967, pp. 143–44)

Implicit in Popper's discussion is a black-and-white choice between Newton-like laws of physics and laws of psychology, which is misleading and indefensible.

Popper's treatment of the rationality principle has stirred up controversy. He ties the rationality principle to the curious view that the social world is less complex than the natural. In his *Poverty of Historicism* (1960, pp. 140–41) he explains that

> there are good reasons, not only for the belief that social science is less complicated than physics, but also for the belief that concrete social situations are in general less complicated than concrete physical situations. For in most social situations, if not in all, there is an element of *rationality*. Admittedly, human beings hardly ever act quite rationally (i.e. as they would if they could make the optimal use of all available information for the attainment of whatever ends they may have), but they act, none the less, more or less rationally; and this makes it possible to construct comparatively simple models of their actions and inter-actions, and to use these models as approximations.

Yet, a belief that the social scientist starts out at an *advantage* over the natural scientist is a typically Austrian view, and one not peculiar to Popper. Excess claims for economics have been made in the past by many utility theorists, Austrian and British alike.[16] Despite the fact that Oskar Morgenstern (1902–1977), a third-generation Austrian and Popper's contemporary, takes a much more modest and realistic methodological stance, the Austrian optimism nontheless shines through. Morgenstern (with von Neumann, 1953, p. 4) assumes, in contrast to Popper, that the Newtonian revolution lies in the future:

> Next, the empirical background of economic science is definitely inadequate. Our knowledge of the relevant facts of economics is incomparably smaller than that commanded in physics at the time when the mathematization of that subject was achieved. Indeed, the decisive break which came in physics in the seventeenth century, specifically in the field of mechanics, was possible only because of previous developments in astronomy. It was backed by several millennia of systematic, scientific, astronomical observation, culminating in an observer of unparalleled caliber, Tycho de Brahe. Nothing of this sort has occurred in economic science. *It would have been absurd in physics to expect Kepler and Newton without Tycho,—and there is no reason to hope for an easier development in economics.* (my emphasis)

It is this view that the social world is less complex than the natural world that has led the Frankfurt school (of neo-Marxists) to call Popper a positivist.[17] In an attempt to clarify the differences between Popper and the Frankfurt school, a conference was held by the German Sociological Association in 1961 in Tübingen. The result was an ill-titled book, *Der Positivismusstreit in der deutschen Soziologie* (Adorno, 1976). In his contribution, Popper (1976a, pp. 87–122) enumerated twenty-seven "theses" on the "logic of the social sciences." The twenty-fifth and twenty-sixth theses are the heart of his theory of situational logic and, although lengthy, are worth reproducing in full:[18]

> The logical investigation of economics culminates in a result which can be applied to all social sciences. This result shows that there exists a *purely*

objective method in the social sciences which may well be called the method of *objective* understanding, or situational logic. A social science orientated towards objective understanding or situational logic can be developed independently of all subjective or psychological ideas. Its method consists in analysing the social *situation* of acting men sufficiently to explain the action with the help of the situation, without any further help from psychology. Objective understanding consists in realizing that the action was objectively *appropriate to the situation*. In other words, the situation is analysed far enough for the elements which initially appeared to be psychological (such as wishes, motives, memories, and associations) to be transformed into elements of the situation. The man with certain wishes therefore becomes a man whose situation may be characterized by the fact that he pursues certain objective *aims;* and a man with certain memories or associations becomes a man whose situation can be characterized by the fact that he is equipped objectively with certain theories or with certain information.

This enables us then to understand actions in an objective sense so that we can say: admittedly I have different aims and I hold different theories (from, say, Charlemagne): but had I been placed in his situation thus analysed—where the situation includes goals and knowledge—then I, and presumably you too, would have acted in a similar way to him. The method of situational analysis is certainly an individualistic method and yet it is certainly not a psychological one; for it excludes, in principle, all psychological elements and replaces them with objective situational elements. I usually call it the 'logic of the situation' or 'situational logic'. (1976a, pp. 102–3)

The twenty-sixth thesis reveals that situational logic is false due to simplification:

> The explanations of situational logic described here are rational, theoretical reconstructions. They are oversimplified and overschematized and consequently in general *false*. Nevertheless, they can possess a considerable truth content and they can, in the strictly logical sense, be good approximations to the truth, and better than certain other testable explanations. In this sense, the logical concept of approximation to the truth is indispensable for a social science using the method of situational analysis. Above all, however, situational analysis is rational, empirically criticizable, and capable of improvement. For we may, for instance, find a letter which shows that the knowledge at the disposal of Charlemagne was different from what we assumed in our analysis. By contrast, psychological or characterological hypotheses are hardly ever criticizable by rational arguments. (1976a, p. 103)

This brings the exegetical exercise to an end. Out of it we learn that situational analysis is the method of economics. It is objective, rational, empirically criticizable, criticizable by rational argument, individualistic, not psychological, capable of improvement. It is a rational (re)construction. It is metaphysical. It is false because it is a simplification (but all scientific theories simplify and thus are false). It is clear that Popper's theory of situational analysis is only very sketchily developed and is not particularly original, given that concepts of rationality have a long tradition in economics, being a

116   ECONOMICS AND THE PHILOSOPHY OF SCIENCE

standard feature of neoclassical economics. Koertge (1975, p. 457) draws the following conclusion:

> Thus, in addition to setting forth a model for the explanation of human activities, Popper has provided systematic methodological and heuristic advice. He has laid out a research program in the Lakatosian sense (1970). The fundamental methodological maxim for his research program (what Lakatos would call the *negative heuristic* which protects the *hard core*) might be formulated as follows: Try to explain all actions and beliefs in terms of situational analysis and the Rationality Principle. If a given action or belief appears to be irrational always blame your model of the agent's situation, *not* the Rationality Principle.

This advice has a familiar ring to the economist's ear. The rationality principle is the backbone of modern economic theory, yet economists have never been able to agree on the exact role it should play. Hence, a whole range of rationality concepts are used and are still being developed. The subject is far too broad to treat well in limited space. Tisdell (1975) provides a lucid introduction to rationality in economics and philosophy.[19]

There are several things to note about Popper's situational logic. It and his antihistoricism work hand in hand to build an economics on sound foundations (that is, not on Marxism, historicism, psychologism, and so on). Popper admits he knew little about the social sciences, and hence it is not surprising that he made assertions about the social sciences that are too demanding. First, he claims that the social world is less complex than the natural. Second, he associates the concept of a law with the form of Newton's laws and then implies that economics has the choice of either physicslike laws or psychological "laws." There is no reason to accept either claim as it stands. Finally, Popper grounds economics' "purely objective method" in "logic" and in the testability of its hypotheses. That then puts Popper's falsification thesis, our next topic, at the heart of the objectivity of economics.

Why Falsification in Economics Fails

> Popper, more than any other philosopher of science, has had an enormous influence on modern economics. It is not that many economists read Popper. Instead, they read Friedman but Friedman is simply Popper-with-a-twist applied to economics.

This assertion of Blaug's (1978, p. 714) contains a grain of truth: Milton Friedman does contend that he and his famous essay on economic methodology (1953) are Popperian. But there is little evidence that Friedman was influenced to any extent by Popper (Frazer and Boland, 1983, de Marchi, 1986). It is, however, true that many economists believe that economic theories can be falsified whether they have read Popper or not. Paul Samuelson (1983 [first edition, 1947], p. 4), for instance, has been writing for years (without reference to Popper) that by a *meaningful theorem* I mean simply a hypothesis about empirical data which could conceivably be refuted, if only under ideal conditions. A meaningful theorem may be false. It may be valid but of trivial importance. Its validity may be indeterminate, and practically difficult or impossible to determine. Thus, with existing data, it may be impossible to check upon the hypothesis that the demand for salt is of elasticity -1.0. But it is meaningful because under ideal circumstances an experiment could be devised whereby one could hope to refute the hypothesis.

More recently Silberberg (1978, p. 10) has asserted that "the paradigm of economics, therefore, in order to be useful, must consist of refutable propositions. Any other kind of statement is useless." Boland (1977b, p. 104) even admits that "I have heard of theory papers being rejected by editors because they did not show that they contributed to an *increase* in the testability of standard demand theory (since that supposedly is the primary criterion of progress—increased testability)."

Why do economists believe they can falsify their hypotheses and theories? They assume that this is what makes economics scientific. That is, economics' scientific status is alleged to hang on its methods. But not everyone agrees on which method is central to economics' scientific status. Boland (1977b, p. 94) has noticed: "Economics today is alleged to be "scientific" mainly because economic theorists now have accepted the need to express their ideas in terms of specific models. Mathematical models of economic theories supposedly have enabled us to put our economic theory to test." Canterbery (1980, p. 26), in contrast, draws a direct connection between the naive, seventeenth-century conception of scientific method and economics' scientific status:

> The scientific method requires: (1) a statement of the scientific problem to be solved, (2) one or more suppositions (hypotheses of what the solution might be), (3) tests of the hypotheses to find out which is correct, (4) a statement of the correct hypothesis, and (5) forecasts to predict outcomes. *Economics is a science because economists use the scientific method to study and explain economic phenomena in a systematic way.* (my emphasis)

This latter assertion is, of course, not tenable. (See part I.) The above passage appears in Canterbery's elementary text, *The Making of Economics,* between a discussion of Bacon, Newton, and Galileo on the one side, and Kuhn's paradigms on the other—hence ignoring two to three centuries of scientific and philosophic change. Blaug (1978, p. 697), as a final example, is certainly more up-to-date than Canterbery but problematic nonetheless:

> In short, economists have always regarded the core of their subject as 'science', in the modern sense of the word: the goal was to produce accurate and interesting predictions that were, in principle at least, capable of being empirically falsified. In practice, they frequently lost sight of this scientific objective, and the history of economics is certainly replete with tautological definitions and theories so formulated as to defy all efforts at falsification. But no economist writing on methodology, whether in the 19th or in the 20th century, has ever denied the relevance of the now widely accepted demarcation

rule of Popper: theories are 'scientific' if they are falsifiable in principle or in practice and not otherwise.

Why is it wrong for some economists to believe that economic theories can be falsified (that is, refuted)? In brief, for the same reasons that physical theories cannot be refuted. (See part I, chapter 4.) In addition, Boland (1977b, p. 93) and Papandreou (1958) have explained that the refutation of a specific model of a theory does not necessarily refute the theory represented by the model. Caldwell (1984c) offers various reasons that falsification fails: too many assumptions involved (many of which cannot be included in the model), some initial conditions are not testable, there is an absence of general laws, and other restrictive conditions.

Often mentioned as a reason that falsification fails in economics is that the empirical basis is so uncertain. This is by far a thornier problem in economics than in physics because some economic variables cannot be observed or controlled, for example, expectations. Morgenstern's *On the Accuracy of Economic Observations* (1963) is a classic on problems encountered with economic observations and data. The now-famous passage from Popper's *Logic of Scientific Discovery* (1972b, p. 111), repeated here for emphasis, describes the uncertain state of physical science:

> The empirical basis of objective science has thus nothing 'absolute' about it. Science does not rest upon solid bedrock. The bold structure of its theories rises, as it were, above a swamp. It is like a building erected on piles. The piles are driven down from above into the swamp, but not down to any natural or 'given' base; and if we stop driving the piles deeper, it is not because we have reached firm ground. We simply stop when we are satisfied that the piles are firm enough to carry the structure, at least for the time being.

This uncertain state disturbs economists perhaps more than Popper, despite the fact that mathematics also rests upon the acceptance of basic axioms and hence does not rest on "solid bedrock" either. "A house built on sand . . . many have said," writes one economist (Wiles, 1984, p. 305). Indeed, there is a quaggy quality involved with economists' views on falsification. Again, this inability to accept the uncertainty of our knowledge that can be found in the philosophy of science literature is reflected in economic philosophical discussion.

And so it has become commonplace for some economists to assert falsification is necessary while in the same breath admitting that it is impossible. Mark Blaug is a paradigm example. Blaug (1980b) first introduced economists to the philosophy of science with his interesting, ambitious work *The Methodology of Economics or How Economists Explain* in 1980. Blaug's volume is separated into three parts: first, an introduction to the philosophy of science; second, a history of economic methodology; third, an appraisal of neoclassical economics in view of the modern philosophy of science. In the first part Blaug arrives at the conclusion that falsification is not possible; in the third part he advocates more falsification in economics! Hands (1984a, pp. 122–23) comments in his review of Blaug's work:

The fundamental problem of Blaug's criticism of economic practice is that it either completely neglects, or at least is inconsistent with, everything he stated in his survey of philosophy of science. His principal criticism is that there is not enough 'falsification' or even 'falsifiability' in modern economics. In light of the previously discussed work of Kuhn, Feyerabend, Lakatos, et al., one is inclined to retort—*so what?* The unambiguous conclusion of recent philosophy of science, which Blaug so accurately surveys in Part I, is that *no science* does that which Blaug criticizes economics for not doing.

Caldwell (1984c) makes the same point, and Boland (1985) also presents an illuminating discussion. The confusion in Blaug's position (in Hutchison, 1984, p. 32) persists:

> We (Caldwell and I) also agree that economic theories must sooner or later be confronted with empirical evidence as the final arbiter of truth, but that empirical testing is so difficult and ambiguous that one cannot hope to find many examples in economics of theories being decisively knocked down by one or two refutations.

A bit further, we find Blaug (in Hutchison, 1984, pp. 32–33) once again making a case for falsification:

> I argue more or less vehemently in favour of 'falsificationism', defined as 'a methodological standpoint that regards theories and hypotheses as scientific if and only if their predictions are, at least in principle, empirically falsifiable'. My reasons for holding this view are partly epistemological (the only way we can know that a theory is true is to commit ourselves to predictions about events and although a confirming instance does not prove truth, a disconfirming instance proves falsity) and partly historical (scientific knowledge has progressed by refutations of existing theories and by the construction of new theories that resist refutations.) In addition, I claim that modern economists do in fact subscribe to the methodology of falsificationism: despite some differences of opinions, particularly about the direct testing of fundamental assumptions, mainstream economists refuse to take any economic theory seriously if it does not venture to make definite predictions about economic events, and they ultimately judge economic theories in terms of their success in making accurate predictions. I also allege, however, that economists fail consistently to practise what they preach: their working philosophy of science is accurately characterized as 'innocuous falsificationism'. In that sense, I am critical of what economists actually do as distinct from what they say they do.

There seems to be considerable confusion here: Blaug asserts that economists say their goal is falsification but that they do not in fact practice falsification, and so economists are to be reprimanded for doing something that Blaug admits is impossible anyway. In addition, Blaug (1978, p. 724) maintains that falsification serves another function in economics: "The criterion of falsifiability can separate propositions into positive and normative categories and thus tell us where to concentrate our empirical work." Blaug's latest statement (1985, p. 288) helps us to put the confusion into perspective:

What I am saying is, for all our admiration of Popper, let us not make a fetish of everything he said. Actually, he is full of holes and that makes him even greater in my view. He recognizes problems he cannot solve: for example, how to measure degree of confirmation and degree of verisimilitude on scientific theories, and, to return to our subject, when to regard the rationality principle in social science as sacrosanct and when to abandon it as untenable.

As we learned in part I, Newton-Smith complained that many philosophers of science do not "think themselves into" Popper's system. Here this phenomenon is reflected in economics. (There is a similar discussion on Blaug in Hands [1984] and Pheby [1988, p. 34].) We should also keep in mind that Popper's discussion of falsification has been inconsistent. I assume that Blaug, like Hutchison, is defending testing and the need for empirical evidence in economics. Pheby (1988, pp. 108–11) provides a string of inconsistent quotations from F. A. Hayek that shows that Hayek's views on Popperian falsification are equally inconsistent. And this loyalty to Popper is commonplace in economics. Consider Wiles's (1984a, p. 297) conclusion about the state of economics: "Our [economists'] particular sickness is opposition to falsification, i.e., disrespect for fact; and an almost religious reluctance to challenge existing paradigms."

It is tempting at this point to inquire what is behind this inability of economists to think themselves into Popper's system. That will be treated. First, however, this view that empirical evidence is the final arbiter of truth in economics needs to be investigated. This is the econometrics–mathematics connection. The transition to econometrics is easy, for the economist asserts only what she or he can justify empirically. Testing, then, is a main source of economic knowledge. Typical, for instance, is Bray's assertion (1977, p. 1) about econometrics: "Economic theory is a prototheory, which is not fully falsifiable, but which yields falsifiable results if appropriate econometric methods, or a method-theory is applied to it." Bray's thesis is somewhat consistent with the ideas of Popper and Marschak discussed above. In all cases they are trying to give economics an absolutely sound empirical basis—which must of course fail. When Popper was writing *The Poverty of Historicism,* econometrics was in its infancy, Marschak being one of its founders. Neither Popper nor Marschak could foresee all future problems with this new branch of economics and both were very optimistic about its prospects, as are often the founders of new fields. Thus, it seems worthwhile to inquire just how sound the empirical basis of economics is.

For many economists, prediction is the central task of economics. This point is made by Worswick (1972, p. 80) as follows: "The idea of economics as positive science makes predictability the test of its performance, the prediction of relationships in situations not previously observed, as well as the prediction of future events, which in some ways is the acid test." But in the past decade or so, a "certain disenchantment with econometrics" (Stewart, 1979, p. 209) has become quite noticeable. Leamer (1983, p. 36), in a very controversial article, recently implored economists to "take the con out of econometrics":

Economists have inherited from the physical sciences the myth that scientific inference is objective, and free of personal prejudice. This is utter nonsense. . . . The false idol of objectivity has done great damage to economic science. Theoretical econometricians have interpreted scientific objectivity to mean that an economist must identify exactly the variables in the model, the functional form, and the distribution of the errors. Given these assumptions, and given a data set, the econometric method produces an objective inference from a data set unencumbered by the subjective opinions of the researcher.

The philsopher of science Clark Glymour (1985, p. 290) reviewed Leamer's article, reaching the conclusion that

[i]t is easy for a professional philosopher who reads Leamer's essay "Let's Take the Con Out of Econometrics" to find a great deal in it that seems contentious, cavalier, or objectionable. . . . My guess is that the sorts of complaints philosophical readers are likely to make about Leamer's paper are more the result of style than substance. The substance is very important.

And the substance, in a nutshell: "Statistical tests don't inform us as to whether or not a model is *approximately* true. They don't permit us to compare false models to determine which is closer to the truth" (Glymour, 1985, p. 293).

Econometrics has not humorously been dubbed "economagic" without due cause. It is a very uncertain tool with many drawbacks. Keynes referred to it as "statistical alchemy" (in Hendry, 1980, p. 396). Hendry (1980, p. 396) lists the drawbacks of econometrics—first, those named by Keynes, which are still valid today, and then some additional concerns:

using an incomplete set of determining factors (omitted variables bias); building models with unobservable variables (such as expectations), estimated from badly measured data based on index numbers . . . ; being unable to separate the distinct effects of multi-collinear variables; assuming linear functional forms not knowing the appropriate dimensions of the regressors; misspecifying the dynamic reactions and lag lengths; incorrectly pre-filtering the data; invalidly inferring "causes" from correlations; predicting inaccurately (non-constant parameters); confusing statistical with economic "significance" of results and failing to relate economic theory to econometrics. (I cannot resist quoting Keynes again—"If the method cannot prove or disprove a qualitative theory and if it cannot give a quantitative guide to the future, is it worth while? For, assuredly, it is not a very lucid way of describing the past".) To Keynes' list of problems [i.e., the above], I would add stochastic misspecification, incorrect exogeneity assumptions (see Koopmans, 1950 and Engle *et al.,* 1979), inadequate sample sizes, aggregation, lack of structural identification and an ability to refer back uniquely from observed empirical results to any given initial theory.

In addition, Morgenstern's *On the Accuracy of Economic Observations* discusses not problems with the theory but problems with collecting and interpreting economic statistics.

Econometrics' unreliability is well known. Hendry (1980) shows in his

"Econometrics—Alchemy or Science?" that when attempting to explain inflation in the United Kingdom, he got a good fit by using "cumulative rainfall in
the UK" as the explaining variable. And Hendry (p. 395) concludes: "It is
meaningless to talk about 'confirming' theories when spurious results are so
easily obtained." The reference to confirmation gives us insight into the nature of the process. According to Popper, falsification occurs if just one (reliable) refuting instance or crucial experiment occurs. Let's turn for a moment
to the role Popper (1960, pp. 133–34) gives to testing in science:

> [A]ll tests can be interpreted as attempts to weed out false theories—to find
> the weak points of a theory in order to reject it if it is falsified by the test. This
> view is sometimes considered paradoxical; our aim, it is said, is to establish
> theories, not to eliminate false ones. But just because it is our aim to establish
> theories as well as we can, we must test them as severely as we can; that is, we
> must try to find fault with them, we must try to falsify them. Only if we
> cannot falsify them in spite of our best efforts can we say that they have stood
> up to severe tests. This is the reason why the discovery of instances which
> confirm a theory means very little if we have not tried, and failed, to discover
> refutations. For if we are uncritical we shall always find what we want: we
> shall look for, and find, confirmations, and we shall look away from, and not
> see, whatever might be dangerous to our pet theories. In this way it is only
> too easy to obtain what appears to be overwhelming evidence in favour of a
> theory which, if approached critically, would have been refuted. In order to
> make the method of selection by elimination work, and to ensure that only
> the fittest theories survive, their struggle for life must be made severe for
> them.

This, in outline, is the method of all sciences that are backed by experience. But this is not the way econometrics works; in econometrics the theory
is *drawn* out of the statistics, and not vice versa. In econometrics, the *relevant
theory* suggests a relationship, for example, inflation is explained by variable
x. In other words, economic testing is completely theory-laden.[20] The econometrician then uses regression analysis to try to find trends in and relationships among economic variables over time. Because the substance of economics is general *trends,* an econometrician would be elated to find only a few
instances that do not fit the trend. H. G. Johnson (1971, p. 23) characterizes
econometrics in this way: "[T]he "testing of hypotheses" is frequently a euphe-
mism for obtaining plausible numbers to provide ceremonial adequacy for a
theory chosen and defended on *a priori* grounds." Lipsey (1979, p. 45), unlike
most economists, admits that economic theories can neither be confirmed as
true nor refuted (falsified). Kuttner (1985, p. 78), in an excerpt from his
"Poverty of Economics," sums up the current state of econometrics as follows:

> By manipulating time lags the determined econometrician can "prove" al-
> most anything. Moreover, though many economists argue that the fair way to
> test a theory is to specify the hypothesis, and run the regression equations
> once, it is common practice to keep fiddling with the equations, manipulating
> lag times, lead times, and other variables, until the equations more or less
> confirm the hypothesis. Some correlations, of course, may be just coinci-

dence; other apparent correlations may disguise real causes that have been overlooked.

We come full circle with the commentary of Erich Streissler (1970, p. 53), a (Viennese) econometrician, who cites Popper's warning against *historicism* in order to criticize modern econometric methods:

> It is a commonplace of long standing that exact forecasts are impossible in economics. This has been most forcefully stressed by Sir Karl Popper. He once said:
>
>> Long-term prophecies can be derived from scientific conditional predictions only if they apply to systems which can be described as well-isolated, stationary and recurrent. These systems are very rare in nature; and modern society is surely not one of them.
>
> Strictly speaking the same impossibility theorem is true for short-term economic forecasts as well. It may sometimes be appropriate to refer to Popper's stern measuring rod and denounce the machinations of naive—or even dishonest—forecasters as an intellectual sham.

Streissler (1970, p. 73) refers to these short-term forecasts as no more than "inspired guesses." He adds, "If they are supplemented by other guesses, they may be quite useful. But it would be grievously wrong to consider them just because of their complicated and often costly apparatus as anything necessarily by far superior to rule-of-thumb guesses." It is clear that although econometrics does "back economics by experience," it does not act as a final arbiter of truth. There is no such thing. Niehans (1981, pp. 171–72) aptly sums up the situation concerning falsification and testability in economics:

> Economists have learnt to see their own activities through the methodological eyeglasses provided by KARL POPPER. Our doctrines are supposed to consist of hypotheses which could conceivably be falsified by empirical observation, but have survived strenuous efforts to do so. A the time the Econometric Society was founded, almost half a century ago, econometrics was expected to provide the techniques by which economic doctrines could be transformed into such empirical hypotheses. Controversies would then begin to be settled by the appeal to correlation coefficients and confidence intervals.
>
> In the meantime, econometrics has indeed helped to transform economics, bringing enormous progress. In the process it became part of scientific fashion or folklore to require that papers submitted to scholarly journals have 'testable implications' and lead to 'predictions'. Nevertheless it can hardly be denied that econometrics has not transformed economic doctrines into testable (and perhaps tested) hypotheses. In fact, few controversies were ever settled by econometric tests, and most economic doctrines continue to be purely logical propositions. Whatever belief we may have in, say, the KEYNESIAN system, the PHILLIPS curve, the quantity theory of money, purchasing-power parity, the welfare loss from cartels or free trade is only marginally based on formal econometric testing—if at all.
>
> We have to face it: By the standards of KARL POPPER economics is not an empirical science. I believe, however, that this points not so much to a shortcoming of economics, but of POPPER's conception of science. POPPER's

conception is, it seems to me, appropriate for fields like physics, where it is relatively easy to isolate processes which remain stationary over a long time.

Niehans errs in this final comment: falsification (as conclusive disproof warranting theory rejection) of theories in physics does not exist either. (See part I, chapter 4.) And what we need to face is the fact that Popper's "standards" for social science are so inconsistent that they create confusion and leave one wondering if he has a methodology of social science.

Here it is helpful to pause and reexamine the many *inconsistencies* in Popper's methodology before investigating the reasons that economists still cling to falsification. First, Popper asserts that the methods of physics can be applied successfully to economics (his thesis of naturalism). But he admits that there will be problems nonetheless with this application and goes on to develop a philosophy of the social sciences. Second, he asserts that laws are not trends. But then he asserts that trends can in fact be formulated as laws. Third, Popper asserts that social science has a "purely objective method," that is, situational logic. Yet, a close examination of Popper's situational logic—his "zero method," the construction of an economic model on the assumption of complete rationality yielding *testable hypotheses*—indicates that its objectivity rests on an empirical base. Still, Popper does not tell us he has provided science with a certain empirical foundation. He in fact stresses the opposite, that we can never know for certain. On the other hand, his empirical basis for science is falsification by "basic" empirical statements. Further, Popper does at times (especially in older writings) leave us with the impression that falsification is *final,* which would mean that the observation statements upon which falsification rests would have to be certain. Popper, however, does not believe in conclusive disproof: the decision to reject or accept observation statements is based on the scientific community's judgment or decision (that is, the "corroboration" of the basic observation statements)—in other words, on convention.

This gross inconsistency seems to arise in part from his being in a middle or transition position: Popper does represent a middle station between the logical positivists and empiricists and the historically minded philosophers of science, and his work reflects a tension from being pulled in both directions. He, like the logical positivists and logical empiricists, stresses testing and logic, clings to the notion that knowledge is founded in experience. Nonetheless, he knows there is no epistemological certainty, that the empirical basis of science is fallible, that logical relations hold between statements but not theories. These inconsistent features pervade his treatment of naturalism, laws, objectivity, logic, falsification, and demarcation—something that promises to make the task for economists who want to understand Popper more difficult. This confusion, rather than having been resolved or identified, appears to be mirrored in the transfer of his methodology, and its various versions, to economics.

Why, then, do economists—and good ones—still hang on to falsification (disproof) of theories? The most plausible explanation is linked with the striving to be objective. This confusion has three facets. The first problem: If we cannot falsify (disprove) a theory, then how do we eliminate the bad theories? Some economists obviously fear that economics will not be able to cleanse itself of nonsense unless economics retains some clean and easy guidelines (which, however, do not exist). "All epistemological certainty is self-fabricated and hence worthless for grasping reality," writes Albert (1980, p. 30): "Alle Sicherheiten in der Erkenntnis sind selbstfabriziert und damit für die Erfassung der Wirklichkeit wertlos."

A related reason for this confusing vein is that economists in general do not want to emphasize the limitations of econometrics because prediction is most often cited as the chief goal and purpose of economics. Dwelling upon the fact that economics is grounded in "inspired guesses" makes them visibly nervous.

The third factor is that the quantitative method accompanying prediction *looks* exact and objective. Popper does believe the statistical-mathematical method gives economics an objective foundation; certainly, Marschak believed it would too. Leijonhufvud (Craver and Leijonhufvud, 1987, p. 181) reminisces about Marschak's role in the birth of econometrics: "For the immigrants who had lived through the interwar period in Europe—and some, like Marschak, had fled first Lenin and then Hitler—this hope of building a *wertfrei* social science, immune to propaganda of every kind, gave motivating force to the econometric movement." And certainly mathematics and econometrics have a rightful place in economics, but they cannot immunize the discipline against nonsense.

Yet, when even the mathematicians themselves complain that these tools are being abused, it is time for reflection. Morgenstern (1976b, p. 447), one of the greatest advocates of the use of analytic techniques in economics, wrote in 1963:

> As far as the use of mathematics in economics is concerned, there is an abundance of formulas where such are not needed. They are frequently introduced, one fears, in order to show off. The more difficult the mathematical theorem, the more esoteric the name of the mathematician quoted, the better. Then one is "in." So it happens that statements are proved—laudable by itself and correctly done—by means of complicated reasoning and use of elaborate machinery, though they can also be proved by elementary means.

Leontief (1982) has expressed similar dissatisfaction with the discipline. And certainly, it is possible to define the limit to the use of mathematics in economics. Surpassing the limit, described by Herbert Simon (1960, p. 59) as "mathematician's aphasia," is a phenomenon that occurs when

> the victim abstracts the original problem until the mathematical or computational intractibilities have been removed (and all semblance of reality lost), solves the new simplified problem, and then pretends that this was the problem he wanted to solve all along. He hopes the [editorial] manager will be so dazzled by the beauty of the mathematical formulation that he will not remember that his practical operating problem has not been handled.

Simon is describing the decision-making process of the business; the analogue, nonetheless, clearly fits academia as well. A philosopher (Rosenberg, 1983, p. 311), examining economics from the outside, characterizes it as "a branch of mathematics somewhere on the intersection between pure and applied axiomatic systems." It is doubtful that even the most avid mathematical economists intended economics to be classified as a branch of mathematics.

Why has this occurred, and why is it allowed to persist? Mayer (1980, pp. 176–77) concludes that

> the stress on using advanced mathematical tools allows us to have a good conscience while ignoring some very elementary rules of good research procedure. Much of the published research consists of taking a new technique out for a walk rather than of really trying to solve a problem. And also, economics has become much too isolated from other social sciences, since being hard scientists we do not want to use either the results or tools of those who cannot claim our exalted status. (Cf. Leontief, 1971.)

Mayer's (1980, p. 176) solution to the prediction problem is an appeal to economists to use all tools available:

> [G]iven all the weaknesses of econometric techniques, we should be open-minded enough to accept that truth does not always wear the garb of equations, and is not always born inside a computer. Other ways of testing, such as appeals to qualitative economic history, should not be treated as archaic.

This appeal to return to teaching and using history is far more common than most economists would expect. "It is the mark of intellectual poverty to know only one's time and place," Boulding (1971a, p. 234) contends. Gerschenkron (1969, p. 1), certainly a supporter of quantitative developments in economics, defends a place for qualitative history:

> Let me begin by saying that in our time the history of economic doctrines has come upon evil days. To some extent, this may be a still lingering result of the Keynesian revolution. Revolutions in general are unkind to history. It takes them a long time before they discover that they themselves are but another thread in the seamless web. More important, probably, has been the influence of science, and primarily of physics. The Department of Physics at Harvard has completely eliminated history of physics from its curriculum; such history has been shifted to an independent History of Science Department. By contrast, in the Department of Political Science, history of political thought is still the daily bread of the discipline. Today's economics finds itself between those extremes, but certainly not in the middle. We are getting closer and closer to physics. I find this neglect of doctrinal history unfortunate for a number of reasons. From a pedagogical point of view, I feel, translation into modern terms of past contributions to economics is a very effective way to teach modern economics. Beyond that, history of economic analysis is not simply a history of intellectual progress. It is also a history of shifts of emphasis and interest, of to-ing and fro-ing, of abandoned and rediscovered problems, of parallel and disparate strains. As such it may well

be a continual source of inspiration for new departures and directions in economic theory.

Sir John Hicks (1979a, p. 3) discusses the inevitable bond between economics and history.

> Now the mere fact that the economist is so largely concerned with current affairs, the affairs of the present, gives him a particular responsibility with respect to time. It is a responsibility which is akin to that of the historian. What the past is to the historian, the present is to the economist. The work of each of them is in time, in historical time, as the work of natural scientists is not.
>
> Experimental science, in its nature, is out of historical time; it has to be irrelevant, for the significance of an experiment, at what *date* it is made, or repeated.

Popper (1960, p. 38) supports this conclusion as well: "A non-experimental observational basis for a science is, in a certain sense of the term, always 'historical' in character. That is so even with the observation basis of astronomy." Interestingly enough, Niehans (1981, p. 174) argues that the true test of economic theories is not related to econometrics: "Economic doctrines are usually tested, not by systematic methods, but by a DARWINian struggle for survival in the arena of history." In addition, Sir Henry Phelps Brown and Robert Solow attest the usefulness of history. Phelps Brown (1972, p. 2) takes the following view in his "Underdevelopment of Economics":

> It may even be that training in advanced economics is actively unhelpful. I find it a common experience that when graduates in economics first assume practical responsibilities they have something to unlearn. One lecturer in economics, latterly much concerned with international aid, has written me, "I find I've learnt a good deal in these last years—particularly how misleading most of my economic training has been. Apart from the basic tools of the trade, I find more and more that I draw on economic history rather than on anything in development theory." An academically distinguished economist who has also had long experience in government service has told me, "By far the best preparation for a useful career in economics after the university, is to go to an organisation working on practical problems, partly so as to understand how little use a great many of the academic gadgets are."

Robert Solow (1985, p. 329) argues in uncharacteristic MIT style:

> If economists set themselves the task of modeling particular contingent social circumstances, with some sensitivity to context, it seems to me that they would provide exactly the interpretive help an economic historian needs. That kind of model is directly applicable in organizing a historical narrative, the more so to the extent that the economist is conscious of the fact that different social contexts may call for different background assumptions and therefore for different models.

Yet, even one branch of economic historians can be faulted for quantitative excesses. The goal of cliometrics, according to Robert Fogel (in Wood-

128 ECONOMICS AND THE PHILOSOPHY OF SCIENCE

man, 1972, p. 323), is "to reconstruct American economic history on a sound quantitative basis." The new economic historians, from their own viewpoint (Woodman, 1972, p. 324), assert:

> The most advanced of the new historians are the cliometricians. Usually economists by training, they are adept in the use of modern statistics, aware of the value of the computer in handling data, and capable in the application of modern economic theory. Moreover, modern economic theory seems to be (and probably is) more scientific than other social science theories and therefore more safely borrowed for historical analysis.

An outsider, Keith Thomas (in Woodman, 1972, p. 325), praises the new economic history as well:

> In America the new economic history, less than ten years old, is already sweeping all before it. Resting upon an alliance between mathematically sophisticated tools of measurement and the construction of elaborate theoretical models, it promises a definitive solution to such problems as the economic efficiency of slavery or the contribution of the railways to American economic growth.

But if economics is scientific, it is not because of mathematical or statistical technique. No approach or technique provides a "definitive solution" to the multitude of problems involved with economic explanation. Since Woodman's article appeared in 1972, Fogel and Engerman's *Time on the Cross* (1974) has appeared. No work more clearly demonstrates how the employment of these tools fails to guarantee a scientific outcome and how a method can be abused. Their work is now used to teach economic students how *not* to do economic history.

We have reached the end of a rather lengthy digression on the relationship of falsification to quantification, econometrics, and economics' foundations. In summary, what are the reasons that some economists persist in invoking falsification when it is not defensible? I find three plausible reasons.[21] First, they are afraid that abandonment of a demarcation criterion will invite nonsense into the discipline. Second, they are unwilling to face the fact that prediction is limited and the foundations are not solid. Third, the mathematical and statistical shroud that economic theory wears makes economics *look* more objective. (This last phenomenon has, in turn, led to a spurning of history—something that is quite unnatural, given that one purpose of economics is to explain the past.) In addition, I might add that some economists use falsification to *defend* their theory; they assert that their theory is valid because it has not yet been falsified.

Recapitulating, there is no demarcation criterion for any science. The foundations of natural science, like social science, are not solid. Mathematics and statistics may allow economics to look like physics, but a method cannot guarantee objectivity or success. No matter how enthusiastic Popper or Marschak were about economics' "Newtonian revolution," the word prediction, as used in economics, commonly has a rather different meaning from what it has in experimental sciences. This is because our predictions are in time, in historical time, in a way that most scientific predictions are not. They are predictions about the future, the future from now, the moment of time in which we are living: predictions which, if we are wise, we make as conditional predictions—so that, for instance, they are changeable by changes in policy—but predictions about the future all the same. Yet though the predictions relate to the future, the evidence on which we base them comes from the past. (Hicks, 1986, p. 98)

There is no tried-and-sure mechanical rule for determining whether a theory is good or bad, no methodology that guarantees success. In the 1920s Keynes (in Hicks, 1983b, p. 375) wrote: "The Theory of Economics does not furnish a body of settled conclusions immediately applicable to policy. Likewise, methodology does not furnish a body of settled rules for the successful appraisal of theories."

Economists' professionalism, credibility, and objectivity rest upon exercising sound judgments, which in turn depends upon their being able to recognize what can and cannot be achieved in their discipline. Worswick (1972, p. 84) concludes: "The more the impression is allowed to persist that economics is an exact science, or if not already one, then with the aid of mathematical models and the computer is about to become one, the more damage will be done to the subject when it fails to live up to exaggerated expectations." The nonexistence of a demarcation criterion or a rule for theory choice does not mean that nonsense will pervade the discipline. Economic theories can be submitted to criticism. Economists can demand that economic ideas be expressed clearly and that logic, mathematics, and history be mastered so that errors that can be avoided are avoided. As disappointingly prosaic as the lesson from the philosophy of science may be, there are no shortcuts to knowledge or scholarship.[22]

The Neglected Messages: Clarity and Criticism as Objective Method

There are several messages won from Popper's search for a demarcation criterion and subsequent development of falsification that have been neglected by economists. First, his development of falsification was aimed at bringing clarity to the subject under examination. Clearly, if a hypothesis, idea, or theory is stated unambiguously, it is easier to criticize. Economists, however, have always been known for their lack of communication skills. "[I]t has certainly become an academic cliché that economists write as gracefully and felicitously as a hundred monkeys chained to broken typewriters" (Mirowski, 1987b, p. 68). Hutchison (1981, pp. 270–71) has not missed Popper's (1979, p. 44) prescription, which begins his discussion of economics below:

'Aiming at simplicity and lucidity is a moral duty of all intellectuals: lack of clarity is a sin' (1979, p. 44). This value commitment to clarity in communication is very much stressed here and one which is and has been, frequently disregarded and even systematically contravened, in and around the fringes of economics, where there is, and has been, much theoretical writing which seems designed to impress by sheer obscurity, complexity and 'profundity'. . . . One thinks especially of some work in the areas of management science, econometric and mathematical model-building, and Marxian theorizing.

But mainstream economics displays no immunity to this malaise.[23] Galbraith (1978b, p. 105), with his usual dry wit, discusses the pros and cons of such (lack of) style:

Complexity and obscurity have professional value—they are the academic equivalents of apprenticeship rules in the building trades. They exclude the outsiders, keep down the competition, preserve the image of a privileged or priestly class. The man who makes things clear is a scab. He is criticized less for his clarity than for his treachery.

Additionally, and especially in the social sciences, much unclear writing is based on unclear or incomplete thought. It is possible with safety to be technically obscure about something you haven't thought out. It is impossible to be wholly clear on something you do not understand. Clarity thus exposes flaws in the thought. The person who undertakes to make difficult matters clear is infringing on the sovereign right of numerous economists, sociologists, and political scientists to make bad writing the disguise for sloppy, imprecise, or incomplete thought. One can understand the resulting anger. Adam Smith, John Stuart Mill, John Maynard Keynes were writers of crystalline clarity most of the time. Marx had great moments, as in *The Communist Manifesto*. Economics owes very little, if anything, to the practitioners of scholarly obscurity.

Fritz Machlup (1967) has dedicated an entire book to cleaning up economic terminology. More will be said on clarity when Kuhn and his "paradigms" are discussed.

The second message that has been neglected is the role of criticism in science. Popper has always insisted that objectivity, through the role of criticism, is common to both the natural and social sciences. Popper (1976c, pp. 292–93) explains in *The Poverty of Historicism* that

what I combated, mainly, was Mannheim's belief that there was an essential difference with respect to objectivity between the social scientist and natural scientist, or between the study of society and the study of nature. The thesis I combated was that it was easy to be 'objective' in the natural sciences, while objectivity in the social sciences could be achieved, if at all, only by very select intellects: by the 'freely poised intelligence' which is only 'loosely anchored in social traditions'.

That a common objectivity exists between the natural and social sciences is the message of Popper's sixth thesis of his "Logic of the Social Sciences" (1976a, pp. 89–90). Lengthy to cite in its entirety, it is nonetheless quite illuminating as it sets forth his "main thesis" of the method for the social sciences.

(a) The method of the social sciences, like that of the natural sciences, consists in trying out tentative solutions to certain problems: the problems from which our investigations start, and those which turn up during the investigation.

Solutions are proposed and criticized. If a proposed solution is not open to pertinent criticism, then it is excluded as unscientific, although perhaps only temporarily.

(b) If the attempted solution is open to pertinent criticism, then we attempt to refute it; for all criticism consists of attempts at refutation.

(c) If an attempted solution is refuted through our criticism we make another attempt.

(d) If it withstands criticism, we accept it temporarily; and we accept it, above all, as worthy of being further discussed and criticized.

(e) Thus the method of science is one of tentative attempts to solve our problems; by conjectures which are controlled by severe criticism. It is a consciously critical development of the method of 'trial and error'.

(f) The so-called objectivity of science lies in the objectivity of the critical method. This means, above all, that no theory is beyond attack by criticism; and further, that the main instrument of logical criticism—the logical contradiction—is objective.

It is unclear here whether by "method of trial and error" Popper means falsification; certainly, a good argument could be made for this interpretation, which is, as has been discussed earlier, a cul-de-sac. Again we can see that Popper is torn between two positions, that of the logician's *modus tollens* and the approach of J. S. Mill, which hangs on free discussion. Popper's critical rationalist message with the falsification extracted harks back to Mill:

> To question all things; never to turn away from any difficulty; to accept no doctrine either from ourselves or from other people without a rigid scrutiny by negative criticism, letting no fallacy, or incoherence, or confusion of thought, slip by unperceived; above all, to insist upon having the meaning of a word clearly understood before using it, and the meaning of a proposition before assenting to it; these are the lessons we learn from the ancient dialecticians. (J. S. Mill, 1984 (1867), pp. 229–30)

In his twelfth and thirteenth theses Popper (1976a, pp. 95–96) discusses the institutional ramifications for critical rationalism:

> *Twelfth thesis:* What may be described as scientific objectivity is based solely upon a critical tradition which, despite resistance, often makes it possible to criticize a dominant dogma. To put it another way, the objectivity of science is not a matter of the individual scientists but rather the social result of their mutual criticism, of the friendly-hostile division of labour among scientists, of their cooperation and also of their competition. For this reason, it depends, in part, upon a number of social and political circumstances which make this criticism possible. *Thirteenth thesis:* The so-called sociology of knowledge which tries to explain the objectivity of science by the attitude of impersonal detachment of individual scientists, and a lack of objectivity in terms of the social habitat of the scientist, completely misses the following decisive point: the fact that objectivity rests solely upon pertinent mutual criticism. What the

sociology of knowledge misses is nothing less than the sociology of knowledge itself—the social aspect of scientific objectivity, and its theory. Objectivity can only be explained in terms of social ideas such as competition (both of individual scientists and of various schools); tradition (mainly the critical tradition); social institution (for instance, publication in various competing journals and through various competing publishers; discussion at congresses); the power of the state (its tolerance of free discussion).

Such minor details, for instance, the social or ideological habitat of the researcher, tend to be eliminated in the long run; although admittedly they always play a part in the short run. . . .

Popper essentially takes for granted the existence of a critical climate among scientists: the social habitat of the scientist is a "minor detail." In reality, there is no reason to assume that an academic or scientific environment that fosters progress through honesty and mutual criticism will exist even in a democratic state. What Popper "misses" is that education plays a crucial role in guaranteeing an atmosphere of mutual criticism: a critical attitude must be taught, learned, and practiced. (Perhaps Popper takes this for granted because philosophy is by nature training in argument.)

Popper does, however, seem to sense that a critical climate does not always flourish among social scientists. In his "Reason or Revolution" (1976c, English ed. only, p. 293), a postscript of sorts and a review of *The Positivist Dispute in German Sociology,* Popper concedes that "if there is more 'objectivity' in the natural sciences, then this is because there is a better tradition, and higher standards, of clarity and of rational criticism." He (1976c, p. 295) also makes the following distinction between the social and natural sciences: "There is here, at first sight, a difference between the social sciences and the natural sciences: in the so-called social sciences and in philosophy, the degeneration into impressive but more or less empty verbalism has gone further than in the natural sciences." But this is to admit that these "minor details" play less than a minor role and that they may not be taken for granted. Indeed, these latter points are the message of William Bartley's work, which Popper has embraced. (See part I, chapter 4.) Popper (1974a, p. 91) in fact distinguishes "scientific" fields as ones in which the critical attitude flourishes, and "prescientific" fields as those in which the critical attitude is partially or totally impeded:

In *Logik der Forschung* I tried to show that our knowledge grows through trial and error-elimination, and that the main difference between its prescientific and its scientific growth is that on the scientific level we consciously search for our errors: *the conscious adoption of the critical method* becomes the main instrument of growth.

And with this formulation it becomes clear that Popper has no demarcation criterion here but, rather, a philosophy of education.[24] Constructive criticism is the only route to progress because it allows us to learn: "[W]e make progress if, and only if, we are prepared to *learn from our mistakes:* to recognize our errors and to utilize them critically instead of persevering in them dogmatically" (Popper, 1960, p. 87). In the same vein, Popper (1966b, p. 376) writes:

> This hint, very simply, is that *we must search for our mistakes*—or in other words, that *we must try to criticize our theories.*
> Criticism, it seems, is the only way we have of detecting our mistakes, and of learning from them in a systematic way.

Here criticism is not equated with logical contradiction, a formal property of sentences, or with empiricism and testing. It does not reduce to falsificationism. It is a theory of learning. And objectivity is not construed to mean "cleansed of all predilections," for that is impossible. Rather, objectivity comes about by exposing our theories, ideas, and beliefs so that they may be criticized, corrected, and bettered. Popper's theory seen from this perspective does not imply that bad theories are *rejected;* bad theories instead *evolve* for the better through the process of subjection to criticism, revision, further subjection to criticism, and so on.

Returning to economics, Popper wrote his twenty-seven theses on the method of the social sciences in the sixties. The seventies opened the discussion of the "crisis" in economics with the British expression of dissatisfaction with the state of the discipline. In the eighties it has become fashionable among American economists to talk of a crisis. Had Popper written his theses later, he might have become conscious of the fact that economists were complaining that economics demands a strict consensus of its students and faculty. For example, Patinkin (in Hutchison, 1977c, p. 61) laments:

> What generates in me a great deal of skepticism about the state of our discipline is the high positive correlation between the policy views of a researcher (or, what is worse, of his thesis director), and his empirical findings. I will begin to believe in economics as a science when out of Yale there comes an empirical Ph.D. thesis demonstrating the supremacy of monetary policy in some historical episode—and out of Chicago, one demonstrating the supremacy of fiscal policy.

Popper did not really understand the social sciences, probably because he (1974a, p. 96) admits that "the social sciences never had for me the same attraction as the theoretical natural sciences." (Still, Popper's doctoral dissertation was in *psychology*.) As for economists and their renderings of Popper's philosophy of science, they have ignored his pure (that is, uncontaminated by his confused falsification theory) critical rationalism, which is actually a precondition for the growth and thriving of science.

Hutchison (1977c, p. 58) has summed up economists' relationship with the philosophy of science: "The kind of 'methodology' which many economists want and value is one which boosts up their prestige. . . ." Hutchison's conclusion is hard to dispel, and it is all the more substantiated upon closer examination of economists' interpretation of the works of Lakatos and Kuhn.

Notes

1. Popper tells us the following in his autobiography (1974a, p. 96): "In fact, the only theoretical social science which appealed to me was economics."

2. The title of this section is taken wholly from the introduction to Samuelson's elementary textbook *Economics* (now in its twelfth edition) and is meant to be a bit ironic. (Samuelson is playing upon physics' distinction as "queen of the sciences.")

3. I find no indication that Popper was acquainted with Morgenstern's works. Why Morgenstern's work on methodology has been ignored and in part never translated into English (although a Rockefeller scholarship made his 1928 volume possible) is unclear.

Morgenstern did modify his methodological position in *Wirtschaftsprognose* in his later works. See Morgenstern (1928, 1934, 1972a), and Morgenstern and Schwödiauer (1976). Also informative are Dacey (1981) and Marget (1929). Briefly, Morgenstern's theory of self-defeating prophecies has evolved into his theory of "theory absorption" (1972a, pp. 706–7):

> Nature does not care—so we assume—whether we penetrate her secrets and establish successful theories about her workings and apply these theories successfully in predictions. In the social sciences, the matter is more complicated and in the following fact lies one of the fundamental differences between these two types of theories: the kind of economic theory that is *known* to the participants in the economy has an effect on the economy itself, provided the participants can observe the economy, *i.e.,* determine its present state. The latter must to some degree always be possible since without some knowledge of this kind no participant (individual or firm) could be functioning, *i.e.,* there would be no economy at all.
>
> However, the distribution of the kind of theory available, and the degree of its acceptance, will differ from one case to the other. This in turn will affect the working of the economy. There is thus a 'back-coupling' or 'feedback' between the theory and the object of the theory, an interrelation which is definitely lacking in the natural sciences.
>
> Clearly, this makes the creation of theory most difficult and confounds its application. To cap the matter, the economic 'reality' is never whole, as is the physical universe: Economic phenomena are embedded in a political, legal, moral and ideological world from which influences come, also determining events. Thus the prediction of economic phenomena implies statements about their separability (or lack of it) from the other social world of which they are only a part. The whole social world is finally embedded in the physical (taken in its widest sense).
>
> In this area are the great methodological problems worthy of careful analysis. I believe that the study of the degree of 'theory absorption' by the members of the economy and the study of the above mentioned embedding relationship will make all of us more modest in judging how far we have penetrated into economic problems.

And finally, Morgenstern and Schwödiauer (1976, p. 228) argue that non-self-defeating predictions are impossible in certain market games. They demonstrate that

> the core is a game-theoretic solution concept—i.e, a definition of rational behavior—which is not immune against "theory absorption": It is an example

of a theory of rational action the knowledge of which (on the part of the actors) destroys its predictive validity.

For those unfamiliar with game theory, Schotter (1973) surveys the major game-theoretic contributions on the core and competitive equilibrium. Schotter and Schwödiauer (1980) provide a survey of game theory. Shubik (1982, 1984) offers an introductory text on game theory. See also Johansen (1983) and Mirowski (1986a) for discussions of the methodological ramifications of game theory. The last papers are interesting because few discussions of the methodological significance of game theory exist.

4. Popper (1960, pp. 137–38) explains: "I have every reason to believe that my interpretation of the methods of science was not influenced by any knowledge of the methods of the social sciences; for when I developed it first, I had only the natural sciences in mind, and I knew next to nothing about the social sciences."

5. Popper (1960, p. 3) tells us: "Since I am convinced that such historicist doctrines of method are at bottom responsible for the unsatisfactory state of the theoretical social sciences (other than economic theory), my presentation of these doctrines is certainly not unbiased."

6. The respective passage from *The Poverty of Historicism* (1960, p. 17) reads:

Such a view may suggest the possibility of analysing and explaining the differences between the various sociological doctrines and schools, by referring either to their connection with the predilections and interests prevailing in a particular historical period (an approach which has sometimes been called 'historism', and should not be confused with what I call 'historicism'), or to their connection with political or economic or class interests (an approach which has sometimes been called the '*sociology of knowledge*').

7. Popper (1960, pp. 3–4) writes:

But I have tried hard to make a case in favour of historicism in order to give point to my subsequent criticism. I have tried to present historicism as a well-considered and close-knit philosophy. And I have not hesitated to construct arguments in its support which have never, to my knowledge, been brought forward by historicists themselves. I hope that, in this way, I have succeeded in building up a position really worth attacking. In other words, I have tried to perfect a theory which has been put forward, but perhaps never in a fully developed form. This is why I have deliberately chosen the somewhat unfamiliar label 'historicism'. By introducing it I hope I shall avoid merely verbal quibbles: for nobody, I hope, will be tempted to question whether any of the arguments here discussed really or properly or essentially belong to historicism, or what the word 'historicism' really or properly or essentially means.

8. Popper has been berated for casting Plato as a historicist and for unduly emphasizing certain elements in Plato's writings in order to form it into the historicist mould. Similarly, Popper has been criticized for holding a rather extreme view of inductivism (Grünbaum, 1976). Popper (1966b, p. 394) admits to having made (un-scholarly) historical mistakes when writing about Hegel:

All of this was clearly indicated in my book; also the fact that I neither could nor wished to spend unlimited time upon deep researches into the history of a

philosopher whose works I abhor. As it was, I wrote about Hegel in a manner
which assumed that few would take him seriously.

Similar criticisms apply to Popper's treatment of Mannheim, Comte, and Mill. With
respect to Marx and Hegel and their implications for historicism, Iggers (1968, p. 289)
warns that the critique has been too sharp at times.

> Karl Popper was therefore not as unjustified in applying the term to Hegel
> and Marx as his critics, e.g., Meyerhoff, have maintained. Popper sharply
> distinguishes between "historicism," which "insists upon historical predic-
> tion," and "historism," which "analys(es) and explain(s) the differences be-
> tween the various sociological doctrines and schools, by referring either to
> their conection with the predilections and "historicism," which "analys(es)
> and explain(s) the differences between the various sociological doctrines and,
> by referring either to their connection with the predilections and interests
> prevailing in a particular historical period . . . or to their connection with
> political or economic or class interests," in his *The Poverty of Historicism* (p.
> 17), first written in the 1930's. It is, however, the latter meaning of historical
> relativism and diversity which has been attached most commonly to the term
> "historicism" in America.

9. The "humanist Marxists" complain that Marx was no historicist, that the
heart of his work is alienation. Bartley (1987a, p. 426) sums up:

> Western literature on Marx is vast, and there is little point in generalizing
> about it. Yet it seems that many leading western Marxists—such as those
> influenced by such thinkers as Ernst Bloch, Walter Benjamin, Erich Fromm,
> Karl Korsch, and Herbert Marcuse—now tend to accept an interpretation of
> Marx that is wholly hostage to Marx's "Paris Manuscripts," his early philo-
> sophical and economic writings of 1844.
>
> Marx, it is claimed, was in fact not an "historicist", not a believer in
> historical laws; and his work is thus undamaged by criticisms such as those of
> Popper. Of course the myth that Marx was an historicist was, it is conceded,
> not Popper's fault. Rather, "what everyone knows about Marx is very largely
> a construct of the elderly Engels"—a construct begun in Marx's own lifetime
> and thrust into prominence on his death, when, in his speech at Marx's
> graveside in Highgate Cemetary, Engels stated: "Just as Darwin discovered
> the law of development of organic nature, so Marx discovered the law of
> development of human history. . . . Marx also discovered the special law of
> motion governing the present-day capitalist mode of production and the bour-
> geois society that this mode of production has created. . . . Such was the man
> of science." This account of Engels was taken over by and strongly promoted
> in the second and third Communist Internationals, and became a widely
> accepted popular understanding of Marx.

Bartley argues against the interpretation of the humanist critics in this paper. He also
points out that where Marxism was once popular because it was "scientific," today it is
attractive because it is humanistic. Yet, if the humanists' interpretation is wrong, they
are no longer Marxists.

10. One must be fair to Popper and admit that there are usages of historicism that
can be identified with Hegel. But generally speaking, Popper's definition conflicts with

the historicism of the German historical school, in particular with its emphasis on individuality and diversity, and its negative stance toward generalizations and a search for laws in society. (The view of the members of the German historical school, however, is not unified.)

11. I cannot support this claim that Popper's attack on historicism has in some cases been interpreted as an attack on history itself. Since I formulated this thesis I found that Katouzian (1980, p. 85) also concludes that Popper's discussion of historicism has "provided a good excuse for orthodox social scientists to reject *all historical knowledge* as irrelevant to the study of society."

12. There is a mass of literature on historicism. What follows is a partial bibliography: Bartley (1987a); Brinkmann (1956); Donagan (1974); Eisermann (1956); Iggers (1968, 1973); Lee and Beck (1953–1954); Mandelbaum (1967); Nipperdey (1985, pp. 498–533); Passmore (1974); Popper (1960, 1966a, b, 1972a, chap. 16); Rothacker (1960); Suchting (1972); Urbach (1978); Wilkins (1978).

The journal *History and Theory* publishes a "Bibliography of Works in the Philosophy of History" that is quite useful. See, for example, Beiheft 23 (1978–1982); Beiheft 18 (1973–1977); Beiheft 13 (1969–1972); Beiheft 10 (1966–1968); Beiheft 3 (1958–1961); Beiheft 1 (1945–1957); Beiheft 12 (1500–1800).

13. See note 9.

14. An example of Popper's (1966b, pp. 387–88) attack on relativism follows:

It is a great step foward to learn to be self-critical; to learn to think that the other fellow may be right—more right than we ourselves. But there is a great danger involved in this: we may think that both, the other fellow and we ourselves, may be right. But this attitude, modest and self-critical as it may appear to us, is neither as modest nor as self-critical as we may be inclined to think; for it is more likely that both, we ourselves and the other fellow, are wrong. Thus self-criticism should not be an excuse for laziness and for the adoption of relativism. And as two wrongs do not make a right, two wrong parties to a dispute do not make two right parties.

Popper points out that although there is no general criterion of truth, this does not mean that competition between ideas and theories is arbitrary. Furthermore, mistakes are positive because they constitute advances in knowledge for we learn from our mistakes.

15. Obviously, Popper was familiar with Marschak's work, and vice versa, despite the fact that the path of communications was not open after both emigrated from Europe just before World War II broke out. Popper does not mention Marschak in his autobiography, but he does in *The Poverty* (with four other economists: R. Frisch, Hayek, Florence, and C. Menger), so it is difficult to assess Marschak's influence on Popper. Coser (1984, pp. 153–54) describes Marschak's life in the years 1920–1940:

In the late 1920s and early 1930s, Marschak began to work in the field that was later to be named econometrics. His early paper on the elasticity of demand (1931), together with the contemporary works of Jan Tinbergen, Wasily [*sic*] Leontief, and Ragnar Frisch, provided the foundation for the development of this new field. His subsequent work, except for the final period of his life was almost entirely within the tradition of econometrics. His youthful political and practical interests subsided, and he became increasingly aloof from socialist thinking and even from concern with specific economic policy.

Marschak left Germany soon after Hitler came to power and went to Oxford, where, through the offices of the economist Revers Opie, he was appointed a Chichele lecturer at All Souls, a position specially created for refugees. Two years later, he was appointed the first director of the newly created Oxford Institute of Statistics upon the recommendation of Opie and the famous economist Roy F. Harrod. Most of the Oxford economic community treated the [i]nstitute with reserve, partly because of its emphasis on econometrics and partly because it had largely been created with the support of American funds, mainly from the Rockefeller Foundation. Despite its initially cool reception by Oxford economists, the institute soon established itself as a major center of statistical and empirical analysis in economics. . . . Marschak spent the period from December 1938 to August 1939 in America as a Rockefeller Foundation Traveling Fellow. Anticipating the outbreak of World War II, he brought his family over, and in 1940 he was appointed professor of economics in the Graduate Faculty of the New School for Social Research, largely on the recommendation of Emil Lederer, who was one of the leading figures there.

Marschak's influence on econometrics was wide-ranging; see, for example, Tintner's (1953) article entitled "The Definition of Econometrics," which provides historical insight into the making of econometrics.

16. Writes Hutchison (1977c, p. 159 n. 46):

It may now be largely forgotten how long, powerful, and confident the *a priorist* tradition in economics was, coming down from Senior and Cairnes to Wieser and Mises, and maintaining that far from facing greater difficulties, *the economist started* with *great advantages compared with the natural scientist:* 'The economist starts with a knowledge of ultimate causes. He is already, at the outset of his enterprise, in the position which the physicist only attains after ages of laborious research.' (Cairnes.) Moreover: 'We can observe natural phenomena only from outside, but ourselves from within. . . . What a huge advantage for the natural scientist if the organic and inorganic world clearly informed him of its laws, and why should we neglect such assistance?' (Wieser.) No wonder economists have been confident in their policy pronouncements. . . .

The Austrians are often classified in the following way: *First generation:* Carl Menger (1840–1925); Eugen von Boehm-Bawerk (1851–1914); Friedrich von Wieser (1851–1926). *Second generation:* Ludwig von Mises (1881–1973); Joseph Schumpeter (1883–1950). *Third generation:* Gottfried von Haberler (1900–); Friedrich von Hayek (1899–); Fritz Machlup (1902–1983); Oskar Morgenstern (1902–1977).

17. Popper's *Logik der Forschung* aimed at criticizing the views of the logical positivists. Thus, labeling Popper a positivist is poor taxonomy. The Frankfurt school, however, made Popper out to be a positivist because Popper shares with the logical positivists the unity-of-science view that there are no differences of kind between the social and natural sciences. These two groups have a history of talking past one another. This is substantiated by the publication that grew out of the 1961 conference held by the German Sociological Association: *The Positivist Dispute in German Sociology* (1976; orig. in German, 1969). See Frisby (1972, pp. 105–19) for a history of the Popper–Adorno conference of 1961 and the differences between Popper and Adorno.

The Frankfurt school is quite interesting, its history being intimately related to the

German student unrest movement of the 1960s. Fallon (1980, p. 78) describes the movement in general:

> The "Frankfurt School" was identified principally with the writings of Theodor Adorno, Herbert Marcuse, and Jürgen Habermas. Adorno had taught at the University of Frankfurt until his dismissal by the Nazis and subsequent exile to the United States. He returned in 1949 and headed the philosophy institute at Frankfurt until his death in 1969. The social philosophy associated with Frankfurt began from the simple perspective that educational reform in an advanced capitalistic society is not possible without structural changes. From this premise was developed an elaborate humanistic set of social theories drawing heavily from Hegel, Marx, and Freud, and focusing on the university as a starting point of reform. These cerebral and elegantly drawn neo-Marxist conceptualizations captured the imagination of large numbers of German students in the late 1960s.

The decline of the school is described by Fallon (1980, pp. 89–90) in the following way:

> One vignette during the period of reaction was the demise of the "Frankfurt School." At the height of student unrest, between 1968–70, radical student leftists turned upon the founding father, Theodor Adorno, accusing him of staying intellectually above the real revolution and thus of being a bourgeois defender of the social status quo. Just before he died, Adorno issued a statement warning against attempts to achieve university reform "through Molotov cocktails and violence." Many students viewed his statement as a defense of purely theoretical work in the tradition of the university's abstract search for truth. Since by implication Adorno's work was never meant actually to be put into practice, as the students were now attempting to do, his statement had the effect of discrediting him with many of the students he had inspired. Threatened by leftist violence, his last set of final examinations at Frankfurt before his death had to be held under police surveillance. Herbert Marcuse had never returned to Frankfurt from the United States and remained in retirement in San Diego. Jürgen Habermas eventually left students and the university for a pure research position at a Max Planck Institute in a rural area of Bavaria. The Frankfurt School simply ceased to exist.

18. The 27th thesis (Popper, 1976a, pp. 103–4) also deals with situational logic and hence is reproduced below.

Twenty-seventh thesis: In general, situational logic assumes a physical world in which we act. This world contains, for example, physical resources which are at our disposal and about which we know something, and physical barriers about which we also know something (often not very much). Beyond this, situational logic must also assume a social world, populated by other people, about whose goals we know something (often not very much), and, furthermore, *social institutions*. These social institutions determine the peculiarly social character of our social environment. These social institutions consist of all the social realities of the social world, realities which to some extent correspond to the things of the physical world. A grocer's shop or a university institute or a police force or a state law are, in this sense, social institutions. Church, state, and marriage are also social institutions, as are certain coercive customs like, for instance, harakiri in Japan. But in European society suicide is not a social institution in the sense in which I use the term, and in which I assert that the category is of importance.

19. Problems and aspects of rationality in economics are discussed in Albert (1972a); Axelsson (1973); Baier (1977); Benn and Mortimore (1976); Caldwell (1982, pp. 146–64); Godelier (1972); Hollis and Nell (1975); Kristol (1980); Machan (1981); Machlup (1978f, g, l, x); McKenzie (1979); Mongin (1984); Mortimore and Maund (1976); Sen (1987); Simon (1976, 1983); Watts (1981).

20. The positivist (logical positivist and logical empiricist) view of theories is that they receive meaning from experience. The Feigl diagram (Feigl, 1970, p. 6) is a paradigm representation of the "orthodox" view of theories and is thus reproduced below. By this view a theory is composed of theoretical, or primitive, concepts (○) connected by postulates. Other concepts (△) are defined in terms of the primitive concepts and are given meaning by correspondence rules (broken lines) connecting them with empirical concepts (□) that refer directly to observation. "[T]here is an 'upward seepage' of meaning from the observational terms to the theoretical concepts" (Feigl, 1970, p. 7). This means that the entire theoretical system gains meaning from experience, that is, observation statements are *not* theory-laden. This sharp distinction between observation and theory is indefensible. (See discussion in H. I. Brown, 1977, pp. 47–48; I am indebted to Professor Brown for bringing the diagram to my attention.)

In economics, it seems fair to say, the model makes more sense if you turn it on its head: the relevant *theory,* not experience, suggests a relationship. The theory is then modified and tested. Observation plays a limited role in the conception and discovery of economic theories. Brown (1977, p. 90) comments that in physics as well, theories impose meaning and organization on experience, and thus an inverted diagram is more realistic.

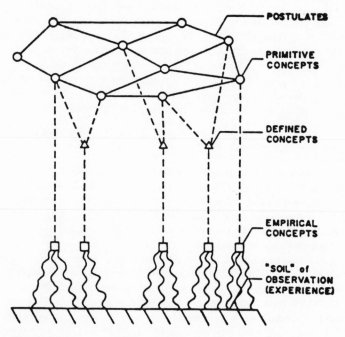

FIGURE 8-1 The Feigl Diagram (Feigl, 1970, p. 6)

21. Boland (1985, pp. 451–52) offers an alternative explanation:

Paradoxically, verificationists are more interested than anyone else in falsifiability. The reason is that it is all too easy to accuse verificationists of being satisfied with verifications of tautologies since a tautology will always fit the facts. Thus, in order to avoid the embarrassment of claiming one's theory has been verified, only to find out that it was a tautology, verificationists now require every theory to be falsfiable. This is a very useful methodological rule since tautologies are never falsifiable—i.e., as a matter of logic they cannot be false. In other words, the requirement of falsifiability is the solution to a verificationist's methodological problem. I still cannot see any other reason to advocate a "methodology of falsificationism."

This explanation assumes either ignorance or deception—something that cannot be attributed to Hutchison or Blaug.

22. For those interested in further readings on econometrics, a partial bibliography follows: Armstrong (1978); Feige (1975); Fogel (1965); Garb (1964); Glymour (1985); Haavelmo (1944); Hendry (1980); Leamer (1983, 1985); Mayer (1980, 1975); Morgenstern (1976b, 1963); Niehans (1981); O'Brien (1974); Poirier (1988); Rowley and Hamouda (1987); Shupak (1962); Simon (1960); Stone (1980); Swamy et al. (1985); Tintner (1953, 1968); Woodman (1972).

An elementary introduction is provided by *The Economist*'s "The Art of Crunching Numbers" (1987). Lawrence Klein treats methodological questions quite systematically in all of his econometric texts. See especially his *Economic Theory and Econometrics* (1985), part I: "Econometric Methodology." Rowley and Hamouda (1987, p. 59) write, for example, that "statistical reliability implies that some probability concept and an acceptable testing method have been prespecified."

In addition, it might be added that prediction cannot be divorced from history: econometric models are "backward-looking" (Streissler, 1970, p. 18). The "Lucas critique," however, is one attempt to remedy (or at least to call attention to) one component of the "backwardness." Because ideas about policy underlie the equations of econometric models, a change in policy means that the parameters of the model will correspondingly change. Thus, the Lucas critique recognizes that a policy change will invalidate the underlying econometric model and appeals to economists to model expectations more carefully.

23. In his "Writing and Reading in Economics" Salant (1969, p. 545) comments: "What has made it trying is that too much of the writing I have read is clumsy or worse: nearly incomprehensible." Salant's article is well worth reading (and is also valuable for non-native speakers). His discussion of the sloppy usage of "anticipations," wrongly equated with expectations, is excellent. It may be added that the latest malapropism to become popular among economists is "inflationary expectations" used as a synonym for "expectations of inflation."

24. Berkson and Wettersten (1984) prefer to emphasize Popper's *psychology* of learning. They (1984, p. 55) argue quite convincingly that

one thing the positivists did not see was that Popper was addressing a problem different from theirs. His primary question was: how do we learn? His problem therefore, was essentially one of discovery. The positivists' problem, on the contrary, was one of justification or choice of theories—not "How can I increase mankind's knowledge?" but rather, "In what theory should I put my trust?"

9

Lakatos and Kuhn: Science as Consensus

Why Economicus Academicus Chose Lakatos as His Darling

Lakatos is currently the most popular philosopher of science among economists. According to Rosenberg (1986b, p. 127) Milton Friedman's positive methodology is being supplanted by Lakatos's methodology of scientific research programs (MSRP). At any rate, the Kuhnian wave of the seventies is being swallowed up by the Lakatosian program. Such illustrious personages as Mark Blaug, A. W. Coats, Neil de Marchi, Spiro Latsis, and D. P. O'Brien, among others, embrace and advocate the MSRP.

Lakatos was organizing a research colloquium on scientific research programs (SRPs) in physics and economics to be held in Nafplion, Greece, in September 1974 when his death intervened. The colloquium took place without him. The portion dedicated to economics yielded the book *Method and Appraisal in Economics* (Latsis, 1976a), edited by Latsis and containing a collection of papers in which the Lakatosian MSRP is applied to economics. Since that publication the interest in the MSRP has continued to grow among economists.[1]

Why is Lakatos so popular among economists (especially historians of economic thought)? It is certainly not because Lakatos was an expert on economics or even the philosophy of the social sciences. He limited his comments to the natural sciences and was a mathematician by training. The only reference to economics in his works appears to be a footnote in his well-known essay "Falsification and the Methodology of Scientific Research Programmes" (1970, p. 179): "The reluctance of economists and other social scientists to accept Popper's methodology may have been partly due to the destructive effect of naive falsification on budding research programmes." But this comment, as T. W. Hutchison (1977c, pp. 50–51) so aptly notes, is as credible as Sir Karl Popper's discussion of "Newtonian revolutions" in economics:

In fact it seems difficult to escape the suspicion that Lakatos' 'naively' and unfairly threatened, 'budding research programmes' in economics are as insubstantial, or even mythical, as Sir Karl's 'Newtonian revolution'. Taking a critical look at the history of political economy and economics one simply fails to find timorous, cautious economists hastily retreating and abandoning their 'budding research programmes' to the first crude and 'cruel' attacks of 'naive' and undiscriminating falsificationists. Quite the reverse, and dangerously and even reprehensively so. The history of the subject is, in fact, full of exaggerated theoretical claims put forward in order to sell particular professed policies of one political stripe or another; and these claims have often been tenaciously and dogmatically maintained, with the aid of every kind of conventionalist stratagem, for decades, and indeed half-centuries on end, regardless of evidence or the lack of it.

Consonant with Hutchison's theme of methodological opportunism is Hand's (1985b, p. 2) assertion that Lakatos's popularity stems primarily from the fact that the MSRP reduces the role of falsification (refutation) of theories in science. Blaug (1976, p. 155) reaches the conclusion that Lakatos represents a compromise between Popper's ahistorical position and Kuhn's sociological relativism, "but a compromise which stays within the Popperian camp." But Lakatos's MSRP is not typically Popperian because it stresses metaphysical programs. (See part I, chapter 4.) Alexander Rosenberg (1986b, p. 136) offers the simple explanation that economists have the nasty habit of attaching themselves to degenerating philosophy of science research programs. In this sense the fascination with Popper, then Kuhn, and finally Lakatos represents a simple chronological succession that lags the developments in the philosophy of science.

Mirowski (1987a, p. 296) has cut through to the substance of Lakatos's MSRP:

> The Lakatosian method of "rational reconstruction" is in fact a thinly disguised blueprint for the justification of the status quo in any intellectual discipline, because it freely advises the historian to ignore any contradictory evidence which might call into question a presumption of pure and unhindered progress in a science. It encourages a carelessness about the sources. . . .

It is interesting to compare this with what advocates of the MSRP designate as the advantages of using a Lakatosian approach to write the history of economic thought. The following example is taken from Fisher's recent work *The Logic of Economic Discovery* (1986, pp. 55–56), written from a Lakatosian perspective:

> Our understanding of the current state of economic theory is enhanced by the history of economic thought. A Lakatosian history of thought has a peculiar efficiency in this respect because it captures exactly those elements from the history of the science that are necessary for the understanding of current theory. Because of this the Lakatosian approach has sheer pedagogical power with respect to both economic theory and the history of economics. Furthermore, the Lakatosian approach embraces valuable lessons.

The "peculiar efficiency" that Fisher mentions is clearly Lakatos's methodological advice to the scientist. Explains Musgrave (1973, p. 400):

> Moreover, Lakatos's rule, stated abstractly, is clearly not one which scientists *ought* to follow. Lakatos recommends scientists to select certain of their hypotheses, christen them a 'hard core', and decide not to modify or renounce them in the face of empirical difficulties. He tells us little about how such hypotheses are to be selected. As it stands, therefore, his methodology gives *carte blanche* to any group who want to erect their pet notion into a dogma.

It is true that economists seek to find supporting evidence for their theories rather than give reasons that the theory is weak. This has been confirmed by Canterbery and Burkhardt (1983), who surveyed 542 empirical studies in economic journals appearing between 1973 and 1978 and arrived at the conclusion that economists do not try to falsify the hypotheses tested. It is no wonder, then, that Lakatos has become economists' methodological darling. Economists need not change the rules of the game and can declare they have the backing of two famous philosophers of science (Lakatos and Popper), and that they practice falsification: all this creates such an aura of scientific authority.

Semantics Revisited: How "Paradigm" and "Research Program" Have Come to Mean Anything and Everything

Mark Blaug has argued that the term *paradigm* should be "banished from economic literature, unless surrounded by inverted commas" (1976, p. 149). Economists often do use *paradigm* interchangeably with *Weltanschauung* (a sense in which Kuhn does not want it to be used), and they have felt quite free to apply the concept rather indiscriminately, citing Kuhn's (1977l, p. 295 n.4) comment that paradigms are characteristic of prescientific stages of development as well as scientific stages.

Two examples serve to show how this terminology borrowed from the philosophy of science acts not to clarify but serves, rather, to obscure the issues. The first is from a discussion by Arouh (1987, p. 396):

> The neoclassical synthesis, by suppressing Keynes' methodological and substantive criticism, regressed to the "classical" MSP (Weisman, 1984, pp. 418–419). This new classicism takes the methodological form of positivism, instrumentalism, and naive falsificationism, while at the substantive level it takes new forms of rationalism. Keynes's contribution is seen as a macro-view of the economy that implies imperfections and rigidities in market mechanisms and their coordination. Once these are removed, hydraulic Keynesians enter the "classical" world of timelessness.

The second example is Wiles's (1984a, pp. 323–24 n.5):

> This word [paradigm] was introduced by Kuhn (1970). It had various meanings in that book, which the author only sorted out later (Kuhn, 1974). The

definition here is a sentence or a few sentences *describing* the essence of some part of the natural or social order (e.g. Ptolemy, Copernicus, Keynes, Friedman). A paradigm is positive, and neither normative nor axiomatic. It may not be ideologically loaded; if it is, its users will of course be intellectually and perhaps also otherwise unscrupulous in its defence.

These two examples show how clearly *program, paradigm,* and other terminology borrowed from philosophy do not illuminate; rather, they obscure and clutter thought. The usage often does not remain true to its original meaning. Arouh (1987, p. 395), for instance, defines MSP as a "methodological and substantive position" but seems, nonetheless, to be using it as many economists use Lakatos's MSRP, to mean a school of thought. In both cases, the reader could be given a clue about what the author really means if he would only describe in plain language what he means.

But plain talk is just what economists intend to avoid. *MSRP* and *paradigm* sound scientific and hence more persuasive. As Hollinger notes, Kuhn's terminology has been abused in general, and then not only by economists. He (1973, p. 372) reminds us that historians have called the decision to pull Americans out of Vietnam a paradigm switch. (Are we to assume then that World War III will be a major paradigm switch?)

Blaug was wrong: *paradigm* and *program* should be banished from economic literature once and for all.[2]

Paradigms and SRPs Applied

Paradigms and research programs are essentially the same thing: as Lakatos (1970, p. 155) concedes, normal science with its accepted paradigm "is nothing but a research programme that has achieved monopoly." Then it is no wonder that economists encounter the same problems in attempting to adopt either methodology to economics.

In his *Structures of Scientific Revolutions,* Kuhn focuses on several key concepts: "scientific revolution," "paradigm" (later renamed "disciplinary matrix" and "exemplar"), and "normal science." Considering the central place of revolutions in his work, one would expect that the more radical element in economics would have taken up Kuhn's theme eagerly. Indeed, an entire volume of the Union for Radical Political Economy (URPE) journal was devoted to Kuhn's paradigms ("Radical Paradigms in Economics," 1971). Ward's *What's Wrong with Economics?* (1972) also draws on the Kuhnian philosophy. Nonetheless, the mass of literature in economics adopting Kuhn's methodology to economics has come from mainstream economics.[3] Kuhn, it might be recalled, never intended his work to be adopted by social scientists and has not been very enthusiastic about the broader interpretations and extensions of his work (Kuhn, 1977l, p. 459).

It is not surprising that the incorporation of a Kuhnian framework, or the MSRP, into the history of economic thought has been accompanied by much disagreement over what constitutes a paradigm or a revolution in the history of economic thought. Many economists believe that all of the neoclassical tradition since Adam Smith constitutes a program/paradigm. Blaug (1978, p. 722), among others, argues: "There is hardly any doubt, therefore, that Keynesian economics marked the appearance of a new SRP in the history of economics." Fulton (1984), on the other hand, argues that the MSRP should be applied to individual economic theories and not to an entire discipline. Moreover, in the cases where there is agreement on what constitutes a program/paradigm, there is often no consensus on the specification of the hard core of the program. Comments Fulton (1984, p. 191): "If the application of the MSRP (paradigm) to economics can result only in such wide divergences of opinion, then there would appear to be little point in its employment in appraisals of economic theories." Appraisal of theories is, after all, economists' chief reason for their interest in Popper, Kuhn, and Lakatos.

The Drive to Be "Normal Scientists"

Most economists read Kuhn's normative formula for scientific success to be "restrain criticism, form a consensus—a paradigm" Paul Feyerabend (1970a, p. 198) offers revealing insights into this matter:

> More than one social scientist has pointed out to me that now at last he has learned how to turn his field into a 'science'—by which of course he meant that he had learned how to *improve* it. The recipe, according to these people, is to restrict criticism, to reduce the number of comprehensive theories to one, and to create a normal science that has this one theory as its paradigm. Students must be prevented from speculating along different lines and the more restless colleagues must be made to conform and 'to do serious work.' *Is this what Kuhn wanted to achieve?*

Feyerabend's view is reinforced by Kunin and Weaver (1971, p. 391), among others:

> Of the social sciences, economics offers the greatest potential for the fruitful application of Kuhn's schema, because professional economists seem to have achieved the strongest consensus as to what constitutes a meaningful set of conceptual categories and relationships to guide their scientific activity.

It is well worth repeating that Kuhn responds less than enthusiastically to this interpretation of his work. He (1970b, p. 245) answers Feyerabend in the following way:

> If, as Feyerabend suggests, some social scientists take from me the view that they can improve the status of their field by first legislating agreement on fundamentals and then turning to puzzle solving, they are badly misconstruing my point. . . . Fortunately, though no prescription will force it, the transition to maturity does come to many fields, and it is well worth waiting and struggling to attain. Each of the currently established sciences has emerged

from a previously more speculative branch of natural philosophy, medicine, or the crafts at some relatively well-defined period in the past. Other fields will surely experience the same transition in the future. Only after it occurs does progress become an obvious characteristic of a field. And only then do those prescriptions of mine which my critics decry come into play.

Yet, economists have always suffered from stress because they cannot agree. Consider, for instance, James Mill's argument (1966, p. 382) in his essay "Whether Political Economy Is Useful," written in 1836:

> Even with regard to the supposition on which they mainly build, that there is such a diversity of opinion among political economists as raises a presumption against their doctrines, the fact is the reverse. Among those who have so much knowledge on the subject as to entitle their opinions to any weight, there is wonderful agreement, greater than on almost any other moral or political subject. On the great points, with hardly any exception, there is general accord; and even on those points on which controversy is maintained, the dispute is about words, the ideas being in almost all cases the same. Take a summary view of the subject. In the great doctrines concerning production, distribution, exchange, and consumption, you find perfect concurrence; it is only as to some of the minor questions involved in these great doctrines that there is any dispute; and I might undertake to show that in few instances is even that dispute other than verbal. . . . There is no branch of human knowledge more entitled to respect; and the men who affect to hold it in contempt afford indication only against themselves.

In other words, the experts who have the knowledge agree and hence are right—circular argument.

Like booms and busts, there have been cyclical pronouncements of consensus and controversy. Cairnes (1965, pp. 19–20), for instance, wrote in 1888:

> It is now a quarter of a century since Colonel Torrens wrote as follows: "In the progress of the human mind, a period of controversy among the cultivators of any branch of science must necessarily precede the period of unanimity. With respect to Political Economy, the period of controversy is passing away, and that of unanimity rapidly approaching. Twenty years hence there will scarcely exist a doubt respecting any of its fundamental principles." Five-and-thirty years have now passed since this unlucky prophecy was uttered, and yet such questions as those respecting the laws of population, of rent, of foreign trade, the effects of different kinds of expenditure upon distribution, the theory of prices—all fundamental in the science—are still unsettled, and must still be considered as "open questions," if that expression may be applied to propositions which are still vehemently debated, not merely by sciolists and smatterers, who may always be expected to wrangle, but by the professed cultivators and recognized expounders of science. So far from the period of controversy having passed, it seems hardly yet to have begun— controversy, I mean, not merely respecting propositions of secondary importance, or the practical application of scientific doctrines (for such controversy is only an evidence of the vitality of a science, and is a necessary condition of its progress), but controversy respecting fundamental principles which lie at

the root of its reasonings, and which were regarded as settled when Colonel Torrens wrote.

About one hundred years later Sir Robert Hall (1959, p. 647) remarks: "But perhaps the most compelling reason why I think we cannot rest at all content with the present position is the fact of the very wide differences of opinion which appear to exist between economists on quite fundamental aspects of policy."

It might be instructive to ask to what extent economists concur. That has been answered by Kearl et al. (1979, p. 36): "Consensus tends to center on micro-economic issues involving the price mechanism while the major areas of disagreement involve macro-economic and normative issues." In addition, a survey of the United States, Austria, France, Germany, and Switzerland (Frey et al., 1984, pp. 993–94) yielded:

> The answers given to the 27 propositions by the over 900 economists in five countries exhibit the highest degree of consensus in one central aspect; namely, that the price system or market is taken to be an effective and desirable social choice mechanism. The propositions about which there is most disagreement were (a) those in which there is an abnormally high frequency of "no answer" responses, which may be attributed to differences in economic policy traditions and unfamiliarity with the terms used; (b) outspokenly normative propositions about income distribution and government spending; and (c) propositions on at the present hotly debated issues such as monetarism or supply-side economics. . . . The analysis further shows that a major cause for dissension is the differences in views between the economists in the five countries surveyed, attributable to the differences in culture and history as well as to the current economic and political conditions. Economists have had varying experiences with respect to the economic policies practiced in their countries, and therefore have different points of reference. The American, German, and Swiss economists tend to support more strongly the market and competition than their Austrian and French colleagues, who rather tend to view government interventions into the economy more favorably.

Despite basic agreement on institutional and other factors, economists continue to strive toward perfect consensus. But this phenomenon has traditionally been confined to one particular school of thought, and then essentially to the ideas of one's contemporaries within that one particular school. Ernan McMullin (1979, p. 61) describes how different it is in natural science:

> Scientists are often unaware of, and not particularly interested in, the long-term history of their field. *But they are ordinarily very keenly aware of the short-term history of the theoretical model with which they are working and of its rivals, if it has any.* (my emphasis)

Yet, economists ignore rival theories and are often embarrassingly ignorant of the short-run development and history of their models and theories. For example, it is possible to walk away with a Ph.D. today and not know what the difference is between Keynes's and Keynesian theory or to know that there are various definitions and uses of rational expectations.[4]

Arjo Klamer, in his *New Classical Macroeconomics: Conversations with the New Classical Economists and Their Opponents,* records interviews of representatives of the rational expectations school and their opponents, arriving essentially at the conclusion that economists of different schools talk past one another. Part of Klamer's (1984b, p. 76) interview with Thomas J. Sargent is reproduced below.[5]

(K) You talk a lot about uncertainty. Post-Keynesians do, too.

(S) I don't know that well enough to comment. I read some of it, but not enough.

(K) Is there a communication problem between people like Paul Davidson and you?

(S) I think you can overdo these communication gaps. There's extensive communication with the theorists who are working on uncertainty.

(K) But they reason in your terms, that is, in probabilistic terms. Some Keynesians say that it is impossible to capture uncertainty in those terms.

(S) I don't have anything to say about that . . .

(K) What about Marxists?

(S) I read them a long time ago. I don't know enough about them.

It is then no surprise when we learn in George Stigler's Nobel lecture (1983, pp. 541–42): "The direct confrontation of two alternative theories, each seeking to explain the same body of observable phenomena, is not common in economics. (It is perhaps encountered more often in macroeconomics than in microeconomics.)" Kuttner (1985, p. 84) sums up in his "The Poverty of Economics": "At a time when the other social sciences—sociology, psychology, political science, history—are letting many flowers bloom and many schools contend, only economics has such a fear of dissension." Economists, in fact, have a history of eschewing criticism. Keynes (1972a, pp. 199–200) paints Marshall's character as follows:[6]

First, Marshall was too much afraid of being wrong, too thin-skinned towards criticism, too easily upset by controversy even on matters of minor importance. An extreme sensitiveness deprived him of magnanimity towards the critic or the adversary. This fear of being open to correction by speaking too soon aggravated other tendencies. Yet, after all, there is no harm in being sometimes wrong—especially if one is promptly found out. Nevertheless, this quality was but the defect of the high standard he never relaxed—which touched his pupils with awe—of scientific accuracy and truth.

If one is interested in being informed and in furthering the discipline, then it would seem especially important if a scientist—and then a Nobel laureate—admits that his or her earlier work was wrong or dissatisfactory. Yet John Hicks's IS–LM apparatus is still being taught without reference to the fact that he (1980–1981, p. 139) states that "that diagram is now much less popular with me than I think it still is with many other people." Hicks discusses the deficiencies of the IS–LM apparatus in his *Crisis in Keynesian Economics* (1974) as well. (See also Leijonhufvud 1983a.) Many an economist has noted that errors and qualifications of economic theory seem to go ignored by the profession. Thus, even when criticism is exercised, it seems to have little impact.

All of this would suggest that a revival of J. S. Mill is in order. In "On Liberty" Mill (1972, p. 97) explains:

> He who knows only his side of the case, knows little of that. His reasons may be good, and no one may have been able to refute them. But if he is equally unable to refute the reasons on the opposite side; if he does not so much as know what they are, he has no ground for preferring either opinion.

Yet this "clannishness," this fear of dissension in economics, this self-imposed isolation, can be explained. It, of course, has political and social roots. Perhaps Joan Robinson (1962, p. 24) sets forth the most convincing reason for this phenomenon: "[W]hen a writer's personal judgement is involved in an argument, disagreement is insulting." And, moreover, asserts Robinson (1962, pp. 24–25): "The lack of an agreed and accepted method for eliminating errors introduces a personal element into economic controversies which is another hazard on top of all the rest."

No matter the reason, the drive toward consensus has always been present in economics, and has gained steam since the publication of Kuhn's *Structure of Scientific Revolutions*. Before I come back to consensus and its significance for economics, science, and the philosophy of science, I want to treat the applications of "normal science" and "revolutions" to economics and provide a brief history of the professionalization of the discipline.

How "Normal" Are Economists?

How "normal" are economists? To the point: they are not "normal" at all. "Normal" in Kuhn's terms assumes the existence of (1) one reigning theory, (2) a well-defined universe of discourse (that is, a set of fundamental entities, concepts, methods, theories, and so on), (3) a consensus supporting the theory that is *not* imposed or legislated but is natural, (4) periods of non-"normal" science—that is, crises that develop into revolutions—triggered by *internal* and not external events. Kuhn (1970c, pp. vii–viii) in fact developed the term *paradigm* (and with it *normal science* and *revolutions*) to distinguish the social from the natural sciences:

> [S]pending the year in a community composed predominantly of social scientists confronted me with unanticipated problems about the differences between such communities and those of the natural scientists among whom I had been trained. Particularly, I was struck by the number and extent of the overt disagreements between social scientists about the nature of legitimate scientific problems and methods. Both history and acquaintance made me doubt that practitioners of the natural sciences possess firmer or more permanent answers to such questions than their colleagues in social science. Yet, somehow, the practice of astronomy, physics, chemistry, or biology normally

fails to evoke the controversies over fundamentals that today often seem endemic among, say, psychologists or sociologists. Attempting to discover the source of that difference led me to recognize the role in scientific research of what I have since called "paradigms."

Economists have always had a penchant for taking analogies from the natural sciences, but in this case none of the Kuhnian terminology is applicable. Not even *preparadigmatic* should be used to describe economics, if by *preparadigmatic* is meant a science not yet liberated by the correct method so that it can also predict as well as physics and so that the methodological and theoretical squabbles come to an end.[7]

In economics there is no paradigm or program (theory is often what Kuhn means) that is unquestioned by all economists. Not even the *problem* is defined unanimously—inflation being more of a problem for monetarists, unemployment for the various Keynesians, stochastic disturbance for the rational expectations theorists. (And these groups all belong to orthodox economics.) Moreover, theory change in economics is instigated, at least as I see it, by some major external event in the economy that is not explainable by the current theories, for example, by the Great Depression or the simultaneous high inflation and high unemployment of the seventies. Finally, Kuhn (1970c, p. 90) argues: "Almost always the men who achieve these fundamental inventions of a new paradigm have been either very young or very new to the field whose paradigm they change." In economics, where one really needs to have a good knowledge of economic theory, mathematics, economic history, and the workings of economic institutions in order to be a good economist, an argument that the *younger* economists are better than the older ones necessarily stretches the imagination.

Why No Normal Scientist Would Ever Want to Be Either "Normal" or "Revolutionary"

> The widespread academic disdain for any semblance of mere yeoman service in the cause of furthering knowledge is exemplified by Thomas Kuhn's description of "normal science" as "puzzle-solving": a mundane activity, similar to the solution of a crossword puzzle, or the successful negotiation of a maze by a rat. After all, who wants to grow up to be a normal scientist? (Mirowski, 1986d, p. 2)

It is wrong to believe that every scientist strives to be "revolutionary"—at least if history counts, for revolutionary scientists in the past have often been ostracized. Just by glancing through the *Encyclopedia of Philosophy* portraits of revolutionary scientists, one comes upon ample evidence that being revolutionary is not fun. For instance, Galileo (Drake, 1967) and Einstein (Whitrow, 1967) had their work hindered by "external factors." Galileo (1564–1642) had troubles his entire life with the Catholic church, and was in fact finally condemned to lifelong imprisonment because he refused to renounce the Copernican doctrines. Einstein (1879–1955), despite his brilliant reputation, suffered from the growing hostility against Jews and pacifists in Germany in the 1930s. He was abroad in California when Hitler came to power and hence could take U.S. citizenship and make a new life. Had he remained in Germany, he may have become a casualty of the war.[8]

Major revolutions in science inherently embody the seeds of social change. Their creator or source, hence, is automatically controversial. That seems to be inescapable. Thus, it is no surprise that often prestige comes to revolutionary scientists only posthumously.

There are also major conflicts within the system. For instance, it is not always easy to recognize genius. (See White's *Rejection,* 1982.) What is more, fellow colleagues, especially professionals, do not like to be outshined. The nature of a purely meritocratic system dictates that the less-intelligent scientists give up their positions to the brighter, if science is to be furthered. These cases are no doubt extremely rare; the only case to come to mind is Isaac Barrow (1630–1677), who in 1669 resigned his chair in mathematics at Cambridge (held 1663–1669) to his pupil, Isaac Newton (1642–1727). On the other hand, being just a "normal" scientist brings with it far less prestige (in the long run). Yet, being "normal" is not controversial; chances are high that one will be well-received by colleagues.

In short, what does *homo academicus* want? He wants to be able to research and teach freely, to be well paid for his services, to secure a safe position, to be respected by his colleagues, and to have his work recognized by the scientific community. *Homo academicus* wants to be thought of as a "revolutionary" thinker but, at the same time, to reap the benefits of the uncomplicated and uncontroversial life of the "normal" scientist. There are contradictions here, and tensions, which are resolved by the incentives rewarding or discouraging excellence offered by the state and the scientific community (see John Gardner, 1961).

Whether *normal* or *revolutionary* in Kuhn's sense can or should be used to describe economists is dubious. Change in economic ideas is different from that in physics, inevitably linked to political considerations. Whereas Kuhn's "revolutionary science" refers to the replacement of one theory with another one, what is revolutionary to an economist is generally not replacement of the model (economists' word for theory) but simply a slight modification of the structure of neoclassical economics. Hence, the usage is misplaced.[9]

A Short Digression on the Development of Economics as a Profession

[T]he professionals that devote themselves to scientific work in a particular field and even all the professionals who devote themselves to scientific work in any field tend to become a sociological group. This means that they have other things in common besides the interest in scientific work or in a particular science per se. In most cases they teach the science which they are trying to bring up and to make their living by teaching. Naturally, this will tend to evolve a social and economic type. *The group accepts or refuses to accept co-workers also for reasons other than their professional competence or incompe-*

tence. In economics this grouping took long to mature but when it did mature it acquired much greater importance than it did in physics. (my emphasis)

These are the words of Joseph Schumpeter (1954, p. 47). Because Kuhn rightly stresses the role of the scientific community in science, we want to take a closer look at the "social and economic type" of economist that has evolved.

Because American economists currently enjoy a reputation as leaders in their discipline, this discussion will focus chiefly on the development of American economics. Harry G. Johnson (1977, pp. 22–23) puts American economics into historical perspective in this way:

> As a result of victory in the second World War, and its postwar ramifications—notably the submergence of Europe and the rise of Russia—American society regained its self-confidence. For closely-related reasons, American economics, which had been something of a condescendingly patronized country-bumpkin cousin of European city-slicker economic sophistication, rapidly became the leader in the world profession of economics. In the process, the distinguishing characteristics of American economics became, naturally enough, the distinguishing characteristics of good professional economics.

Leaving aside the question of whether American economists really lead or not,[10] American economics was (and remains) an international product, shaped very much by English and German-Austrian thought.

American economics grew out of two opposing schools: the economics of the German historical school and that of the English marginal utility theorists.[11] These two traditions, if perceived as economics as historical process, and alternatively, as economics as mathematical abstraction, have never fully reconciled themselves in the one hundred years in which economics in the United States has been recognized as a profession.

Prior to the 1870s, American economists were self-trained. Almost no university courses existed on the subject, and the few that did were usually subsumed under moral philosophy. When political economy—the normal terminology for economics prior to 1900—began to emerge as a discipline in its own right, it was placed as a subfield of history, literature(!), mathematics, or jurisprudence (Parrish, 1967).

Because of a dearth of course offerings in economics on U.S. soil, students of the latter nineteenth century turned to Europe—in most cases to Germany. The single most important reason that Germany was so popular was that political economy was a respected discipline to study there—not so in the United States until after 1880. German education was highly respected in the United States; German scholars were recruited by the U.S. universities. In short, the German university system influenced the development of the American university system to no small extent. As Parrish (1967, pp. 3–4) indicates,

> There were many reasons why American students in almost all disciplines at this time went to Germany for advanced training. The German universities placed much more stress on advanced work than institutions in this country.

They utilized the seminar system which permitted professors more specialization in subject matter and more time for research and writing. The seminars were kept small, permitting an intimate teacher-student relationship with stress on independent reading and research by the student. In contrast many prominent American college teachers were bogged down with large classes of undergraduates.

Most German economists at that time were members of the German historical school: during the winter semester 1889–1890 Schmoller could be found at Berlin, Knies at Heidelberg, Roscher and Bretano at Leipzig. Carl Menger, the opposition, was at Vienna (Rowe, 1890).[12]

The "German era" of American economics peaked in the early 1890s but was in decline by 1900.

The growing criticism of the United States and the growing insistence of German economists on Germany's need for more "Lebensraum" cooled enthusiasm. Interest of American scholars shifted to the leading British economists whose theoretical contributions became widely adopted in this country. By 1910 the migration to Germany had almost ceased. (Parrish, 1967, pp. 5–6)

Meanwhile, economics was becoming established in the United States. The first professorship in political economy went to Charles Dunbar in 1871 at Harvard. In 1886 the American Economic Association (AEA) was founded by Richard T. Ely of Johns Hopkins University. The usage "political economy" slowly grew pejorative: it was not scientific enough and economists did not want to be designated as "political." They sought autonomy from the other social sciences. After the 1890s economics was recognized as a profession.

It has certainly been forgotten that the founder of the American Economic Association was educated in Germany and represented the German historical school of economics. At Johns Hopkins University he and Simon Newcomb, Ely's adversary and a mathematician who viewed economics from a perspective close to Jevons's, quarreled publicly over the nature of economics. Barber's (1987, p. 180) summary captures the situation:

Newcomb occupied a professional chair in mathematics; Ely held the title of "associate" in political economy, an untenured position. In their approaches to political economy, the two men were at opposite poles. Ely, writing from the German historical perspective, challenged laissez-faire and called for the scholar to be engaged actively in setting the world to rights. In his vision, economics should be purged of doctrines linked to natural laws that were beyond human manipulation and of conclusions arrived at by deductive procedures. Newcomb, on the other hand, drew inspiration from Britain's William Stanley Jevons and championed the value of mathematical procedures in economic inquiry. Though a staunch advocate of the benevolence of the invisible hand, Newcomb maintained that economics should aspire to become a positive science, free of contamination by normative judgements.

It is safe to say that Newcomb won that round because Ely was not awarded tenure and consequently left Johns Hopkins University. Moreover, American economics has surely developed more as Newcomb envisaged it should. Later the AEA publications became the mouthpiece for orthodox economics. Despite a much more quantitative turn, the Ely perspective and its variations live on, and their journals and literature have proliferated with time.

The 1880s and 1890s brought especially hard times for American economists. A rash of investigations and legal suits against economists broke out as the policy measures they advocated clashed with the interests of big business (a major financial source for the universities). Ely and his students suffered the greatest consequences because they were civil activists and, moreover, champions of *Lehrfreiheit*. Furner (1975) dedicates most of her *Advocacy and Objectivity* to an analysis of the tension between reform interests of economists and the establishment of scientific authority during this era. The scenario is aptly painted by a

> cartoon on the cover of *LIFE* [that] showed the dignified figure of a university president and former economist leaving the college gate with his suitcase, while a plump trustee nailed a want ad to the ivy-covered pillar. The message read: "WANTED: By the Corporation of Brown University, a young man of submissive disposition as president. A reasonable amount of scholarship will not be a disqualification, but the chief requisite will be an obsequious and ingratiating behavior toward millionaires and an ability to RAKE IN THE DOLLARS. Opinions and first-class board furnished by the Corporation. No gentleman encumbered with a back-bone need apply." (Furner, 1975, p. 205)

It is no wonder, then, that economists—like other social scientists—have always felt inadequate: their attempts to cope with an inferiority complex certainly deserve a place in their history. They do in fact enjoy a rather miserable reputation. (See Viner 1963 and Grammp 1973.) Thomas Carlyle (in Viner, 1963, p. 8) summed up his feelings about economists in the following way: "Of all the quacks that ever quacked, political economists are the loudest." J. M. Keynes (1972b, p. 332) concluded in his essay "Economic Possibilities for Our Grandchildren": "If economists could manage to get themselves thought of as humble, competent people, on a level with dentists, that would be splendid!" And the striving to be recognized as true professionals and scientists seems never ending. Myrdal (1953, p. xiii) bemoans the fact that

> every economist is painfully aware that there exists widespread doubt about the supposed 'scientific' character of economics. The distrust is, indeed, well founded. A branch of knowledge which works with a whole set of premises missing is hardly reliable.

Kuttner (1985, p. 76) records the following concession:

> Charles Schulze, a senior fellow at the Brookings Institution and a recent president of the AEA, says, "When you dig deep down, economists are scared to death of being sociologists. The one great thing we have going for us is the premise that individuals act rationally in trying to satisfy their preferences. That is an incredibly powerful tool, because you can model it."

Economics' searching for an identity, its feelings of inadequacy, its paranoia, its infrangible link with policy and the subsequent influence of (control by) outsiders, its tendency toward quantification and penchant for modeling are all intertwined threads of the fabric of economics' history. Nothing better demonstrates this than the story behind the development of quantification, the most salient trend in American economics.

Much of the history of twentieth-century American economics has been marked by an increasing quantification, growing noticeably after World War I and especially after the development of macroeconomics in the 1930s (Spengler, 1961, p. 175). Galbraith (1973, pp. 1–2) remarks that this mathematical technicality has freed economics from outsider control:

In its first half century or so as a subject of instruction and research, economics was subject to censorship by outsiders. Businessmen and their political and ideological acolytes kept watch on departments of economics and reacted promptly to heresy, the latter being anything that seemed to threaten the sanctity of property, profits, a proper tariff policy, a balanced budget, or which involved sympathy for unions, public ownership, public regulation or, in any organized way, for the poor. The growing power and self-confidence of the educational estate, the formidable and growing complexity of our subject and, no doubt, the increasing acceptability of our ideas has largely relieved us of this intervention. In leading centers of instruction faculty responsibility is either secure or increasingly so. But in place of the old censorship has come a new despotism. That consists in defining scientific excellence as whatever is closest in belief and method to the scholarly tendency of the people who are already there. This is a pervasive and oppressive thing not the less dangerous for being, in the frequent case, both self-righteous and unconscious.

Mathematical development in theory allowed economics to live in the splendid isolation economists so longed for. And it *was* longed for: social scientists once again became victims of censorship during the McCarthy era. Harry G. Johnson (1977, p. 25) stresses the importance of an especially dark era in U.S. history.

There has, however, been another important influence at work, virtually ignored because of an academic conspiracy of silence. This was the traumatic impact on the American university system of the McCarthy communist witch-hunting period, which served notice on prudent American economists to confine their attention to "scientific" problems and steer clear of issues that might raise suspicions about loyalty to the American nation. The consequence of prudence of this kind was extreme vulnerability to "radical" student demands for "relevance" and for "political economy" rather than "(mathematical) economics," demands which both appealed to an inner consciousness of guilty professional neglect, and could not generally be resisted by reference to any serious professional thought or research effort on the issues in question. The result was partial institutional capitulation to the demand to replace tests of professional competence by liberty to qualify by the undisciplined display of the proper dissenting emotions, and thus to risk jeopardizing the reputation for professional competence on which American post-World War II leadership of world economics has been based.

In addition, quantification gave economists international notoriety—the reputation as number one—as H. G. Johnson (1977, p. 24) notes, but not without cost:

> America's rise to world leadership in economics has been closely associated with—and indeed well advertised as being attributable to—the rapid conquest of the American graduate program and teaching by mathematical economics, largely inspired by the powerful example of America's first Nobel Laureate in Economics (i.e., Paul Samuelson), and (to a lesser extent and with less of a break from the previous American tradition) by econometrics, largely the work of the Cowles Commission. The conquest did not occur without prolonged and bitter resistance from the inheritors of the older tradition, who have remained both extremely skeptical about the usefulness of mathematical economics for its own sake, and critical of the tendency of econometrics to avoid the dirty work of grubbing for facts.

Today it is difficult for students to imagine that economists ever bore ill feelings toward mathematical economists. Morgenstern's biography (Coser, 1984, p. 143), however, reminds us how false this impression is.

> In view of the subsequent success of game theory on the academic scene, it is interesting to note that, when Morgenstern first presented its guiding ideas to graduate students in economics at Princeton University in the early 1950's, he met with indifference, even hostility. He was perceived as an eccentric and, as Martin Shubik puts it, "as an essentially 'high risk' member of the faculty to be involved with." "He was an outsider," Shubik continues; "no one in economics really understood what he did." In addition, graduate students were intimidated by the formality that Morgenstern had brought from Vienna to Princeton. By and large colleagues and students seem to have felt that the best way to cope with their unease in the presence of Morgenstern was to treat him patronizingly and with some derision. They may have felt that he had something important to say, but for the time being he had better be kept at a distance. Many graduate students in economics seem also to have believed that to study with more traditional professors would offer a safer route to future appointments. For all these reasons, most students interested in game theory were initially to be found in mathematics, not in the economic department.

Now perhaps we are experiencing the other side of the coin: since the sixties, quantification seems to have become a goal in itself.

Indeed, the blessings have been mixed; this trend toward quantification must be understood as a partial therapy for some of economics' ailments. Quantification may well have eliminated the "censorship crisis," but the language barrier that came with it makes communication with the outside world (for example, desired dialogue with politicians) all but impossible. (In some cases today it renders communication among economists impossible, espe-

cially between the older and younger generations.) Quantification and specialization let American economics rise to international fame and gave it the
likeness of physics, solving the "crisis of respect" and hence the "identity
crisis" (at least superficially). But mathematics is a tool and not an end in
itself. It is a valuable tool: perhaps once quantification has been put into
historical perspective, its practical use will become as valued in the future as
its limitations will be respected.[13]

There is a dearth of information about economics as a profession since
1960. One can, however, gain a picture of the typical economist by piecing
together economists' discussions and studies on the discipline's power structure, the position of women, foreigners, and minorities, its publication explo-
sion, and its educational program and philosophy.[14]

The complaints about the structure of power in economics have grown
almost as loud as those against "quantification for quantification's sake."
Ward's *What's Wrong with Economics?* is in many ways an analysis of the
power structure in economics. Ward (1972, pp. 29–30) describes power in
economics in the following way:

> The power inherent in this system of quality control within the economics
> profession is obviously very great. The discipline's censors occupy leading
> posts in economics departments at the major institutions, and their students
> and lesser confrères occupy similar posts at nearly all the universities that
> train new Ph.D.'s. The lion's share of appointment and dismissal power has
> been vested in the departments themselves at these institutions. Any econo-
> mist with serious hopes of obtaining a tenured position in one of these depart-
> ments will soon be made aware of the criteria by which he is to be judged. In
> a word, he is expected to become a normal economic scientist.
>
> Of course it is not true, as the last paragraph may seem to imply, that this
> decision as to whether to become a normal science economist is made at the
> stage in his career at which the economist has obtained his last degree. For, as
> in all normal sciences, the entire academic program, beginning usually at the
> undergraduate level but certainly at the graduate, consists of indoctrination
> in the ideas and techniques of the science. As much as anything, this is a self-
> selection process. Those who do not accept the basic ideas of the science will
> not proceed very far with its study. Standards for admission to economic
> science are not terribly demanding, comparatively speaking; consequently
> those who drop out of the system, at least in the author's experience, are not
> typically intellectual failures. Rather, they are those who have become
> "turned off," and their most common complaint is lack of relevance, not
> difficulty.

Leontief's response (1982, p. 107) is deeply troubling because it comes from a
mathematical economist and a Nobel laureate:

> That state (the abstraction crisis) is likely to be maintained as long as tenured
> members of leading economics departments continue to exercise tight control
> over the training, promotion, and research activities of their younger faculty
> members and, by means of peer review, of the senior members as well. *The
> methods used to maintain intellectual discipline in the country's most influen-*

tial economics departments can occasionally remind me of those employed by
the marines to maintain discipline on Parris Island. (my emphasis)

Stanfield (1979, p. 141) tells us nothing new about the attitude toward radical
elements in economics but toward the critical: "Taking the social sciences
generally, research that takes for granted the extant class structure and the
habitual way and means is surely favored over research reflecting radical,
critical, or utopian viewpoints." In short, the power structure of economics
today bears an uncanny resemblance to that of the nineteenth-century German university: a nonmeritocratic, undemocratic system marked by Prussian
authoritarianism. Full power in that system, as Ash (1980, pp. 258–59) describes, was held by the tenured lot only, and it was won not through scholarship but through "academic nepotism":

> To exercise such privileges to the full, however—indeed, to reach the level of
> prestige and security necessary to carry out innovation of any kind within the
> system—one had first to become a full professor. . . . The conservative effects
> of this were mitigated, however, by the tendency of younger scholars to present
> themselves as the "students" of their professorial "masters," who responded in
> the same spirit by pressing for the advancement of the more promising among
> them. A well-known example of the role of such personal allegiances was the
> appointment of Georg Elias Müller to a chair of philosophy at Göttingen in
> 1881 on the recommendation of his teacher and predecessor Lotze, even
> though Müller was relatively young at the time (31), lacked teaching experi-
> ence, and had not been the faculty's first choice for the position.

Fallon (1980, pp. 5–6) describes Leibniz's rejection of the German university
because of its anti-intellectualism:

> At the end of the eighteenth century most universities in German-speaking
> Europe could be characterized as sites of rote disputation inhabited largely
> by pedants. Many genuine intellectuals regarded them with disdain. This
> situation had existed for at least a century, if one judges from instances such
> as the behavior of Leibniz, who essentially wrote the universities off in the
> seventeenth century and prevailed upon the elector of Brandenburg to found,
> instead of a university, an academy of science. Leibniz referred to universities
> as monk-like institutions concerning themselves with sterile fancies. Leibniz's
> academy, the *Societät der Wissenschaften,* was founded in 1700 in Berlin, but
> fared little better than the universities after Leibniz died in 1716. King Freder-
> ick William I (1713–40) treated its members contemptuously, even appoint-
> ing three of his court fools presidents of the academy. The academy was later
> rehabilitated under Frederick the Great (1740–86) as the Prussian Academy
> of Arts and Sciences. (*Preussische Akademie der Wissenschaften*).

At the end of the nineteenth century the German university was in such ill
repute that the word *university* (in German, *Universität*) almost fell out of use:

> Universities in general were in such disrepute among intellectuals at the
> beginning of the nineteenth century that the Prussian scholars and reformers
> who sought a new institution in Berlin actively avoided using the word "uni-
> versity" in their essays and arguments. Humboldt's own memorandum from

this era refers to "the higher scientific establishments of Berlin" by which he clearly means that one institution which he later called the University of Berlin. In the end, it was probably the transient successes of Göttingen, Halle, and Jena that made adoption of the name university possible. The title of university put this brilliant new institution in a historical continuum, placing it in a strategic position to infuse its sister institutions of similar name in Germany and around the world with vital strength. A strong argument can be made, for example, that the development of universities in the United States would have been dramatically, perhaps drastically, different if Humboldt had named the institution in Berlin an academy, seminary, or institute. The seemingly simple act of calling the new institution the University of Berlin may have been one of history's close calls. (Fallon, 1980, pp. 30–31)

Philosophers of science often mention the detrimental effects of external control of science: the Lysenko affair is the paradigm case. T. D. Lysenko was a Soviet agronomist who, by wooing high Soviet party officials (including Stalin), replaced Soviet theories of modern genetics with his own false theory of the "shattering" of plant heredity. From 1948 to 1953 research on and teaching of genetics was banned. The university and scientific journals were controlled by Lysenkoites into the mid-1960s. During Lysenko's control, geneticists lost their jobs; those who opposed the measures were arrested. (See Medvedev, 1969.) External control of science also occurs in the West: during World War II Lord Cherwell, science adviser to Churchill, obtained almost full control over science in Britain (Snow, 1962). The Nazi control of science and university is well documented (Ritter, 1946). The McCarthy era, already mentioned for its negative effects on economics, is a third example. Given Leontief's and others' descriptions of the power structure in economics, one is forced to conclude that external control of economics has been replaced by an equally pernicious internal control.

The status of women, minorities, and foreigners is closely related to the power structure of the discipline (as well as the state of the law). Starting with a profile of women in economics, one is immediately confronted with the extremes: among economists have been some of the most outspoken advocates of women's rights ever (for example, J. S. Mill), as well as outright misogynists.

Letwin (1963, p. 85) describes the first English lectures on economics, which were based on Aristotle's *Economica*. This work is *not* the basis for today's economic theory; studying trade then was not respectable. Nonetheless, this is the structure upon which modern American economics has been built:

.At least some of the seventeenth century Englishmen who heard lectures on the *Economica* did not consider it a dead letter. As late as 1669 a Cavalier Baron published a long tract *Observations and Advices Oeconomical,* which assumes that the family, though small, is 'an aggregation capable of Government', defines 'Oeconomy' as 'the art of well governing a man's private house and fortunes', and offers such wisdom as that a man ought to pick for his spouse a lady who 'may be no less useful in the day than agreeable at night'.

Alfred Marshall represents, interestingly enough, both poles. A strong supporter of women as a young man, he nonetheless displayed truly chauvinistic tendencies in his later years. Keynes notes that Marshall collaborated with his wife on works to such an extent that the author plate should also bear her name. The following passage appears in Keynes's (1972a, pp. 176–77) biography of Marshall:

> In 1876 Alfred Marshall became engaged to Miss Mary Paley, a great-granddaughter of the famous Archdeacon. Miss Paley was a former pupil of his and was a lecturer in Economics at Newnham. This first book, *Economics of Industry,* published in 1879, was written in collaboration with her; indeed it had been, at the start, her book and not his, having been undertaken by her at the request of a group of Cambridge University Extensions lecturers. They were married in 1877. During forty-seven years of married life his dependence upon her devotion was complete. Her life was given to him and to his work with a degree of unselfishness and understanding that makes it difficult for friends and old pupils to think of them separately or to withhold from her shining gifts of character a big share in what his intellect accomplished.

Midway in his life, reminisces Keynes (1972a, p. 220), Marshall underwent a change of heart:

> The controversy about admitting women to degrees, which tore Cambridge in two in 1896, found Marshall in the camp which was opposed to the women's claims. He had been in closest touch with Newnham since its foundation, through his wife and through the Sidgwicks. When he went to Bristol he had been, in his own words, 'attracted thither chiefly by the fact that it was the first College in England to open its doors freely to women'. A considerable proportion of his pupils had been women. In his first printed essay (on 'The Future of the Working Classes', in 1873) the opening passage is an eloquent claim, in sympathy with Mill, for the emancipation of women. All Mill's instances 'tend to show', he says in that paper, 'how our progress could be accelerated if we would unwrap the swaddling-clothes in which artificial customs have enfolded woman's mind and would give her free scope womanfully to discharge her duties to the world.' Marshall's attitude, therefore, was a sad blow to his own little circle, and, being exploited by the other side, it played some part in the overwhelming defeat which the reformers eventually suffered. In his taking this course Marshall's intellect could find excellent reasons. Indeed, the lengthy fly-sheet, which he circulated to members of the Senate, presents, in temperate and courteous terms, a brilliant and perhaps convincing case against the complete assimilation of women's education to that of men. Nevertheless, a congenital bias, which by a man's fifty-fourth year of life has gathered secret strength, may have played a bigger part in the conclusion than the obedient intellect.

This "congenital bias" of which Keynes speaks—no doubt in today's jargon "male menopause syndrome"—still did not prevent Keynes from concluding that Marshall was not sexist.

There have been no American J. S. Mills or young Marshalls. Samuelson (1972, p. 156) refers to J. S. Mill's "feminism" as "radical philosophy." With

162   ECONOMICS AND THE PHILOSOPHY OF SCIENCE

time, things have in fact only gotten worse for American women who want to become economists. The number of Ph.D.'s awarded to women has fallen appreciably since the 1930s. The latest study on this (Spellman and Gabriel, 1978, p. 186) shows that

> Froman first collected data on the sex of 1929–40 degree participants. During those years, an average of 8.0 percent of the Ph.D.s were women, and the percentage ranged annually from 4.7 to 12.9 percent. The proportion of doctorates in economics who have received their degrees since 1940 and are female has declined substantially. An average of only 5.2 percent of the 1940–74 Ph.D. recipients were women. The percentage ranged from an annual high of 11.2 percent in 1947 to a low of 2.5 percent in 1940. . . . The undergraduate training of women economists has been very concentrated as one-fourth of the degrees have been from the seven sisters institutions. . . .

They (1978, p. 186) conclude that "over 90 percent of these economists who received their Ph.D.s in the last twenty-five years are male." Fifty percent of American (and Canadian, Finnish, and French) students are women (Scully, 1986). In some cases, such as the University of North Carolina at Chapel Hill, where women make up two-thirds of the student population, the ratio is considerably higher ("Women Pay for Success," 1978).

The case of Joan Robinson is well known and can be stated concisely. In a biography written shortly before her death, Skouras (1981, pp. 216–17) concludes that

> Joan Robinson has two unique distinctions in the economics profession. She is the only great economist that has ever lived who is not a man. She is also the only great living economist who has not been awarded the Nobel prize. These two are the great scandals of the economics profession.

In 1971 a study of hiring practices at American universities (Lewin and Duchen, 1971, p. 894) revealed that "the underutilization of potentially qualified women in science represents lost opportunities to society and to science in terms of the overall goal of advancing science." Although (to my knowledge) no study focuses solely on economics, there seems no reason to doubt that the above generalization for the natural sciences is just as valid for economics.[15]

Whereas the economics profession is disturbingly male, it embraces (male) foreigners with enthusiasm. Grubel and Scott's (1967, p. 144) study concludes as follows: "Our analysis of the American economics profession in 1964 has revealed that the proportion of foreign-born is 12 per cent, which is about twice as high as that for the U.S. population in general." Craver and Leijonhufvud (1987, p. 173) characterize the Continental influence on American economics:

> Since 1948, when Joseph Schumpeter was elected President of the American Economic Association, ten presidents of the Association have been of European birth, eight were established scholars coming to the United States, seven of these from continental Europe (Schumpeter, Gottfried Haberler, Fritz Machlup, William Fellner, Wassily Leontief, Jacob Marschak, and Tjalling

Koopmans), and two of the latter (Leontief and Koopmans) have been recipients of the Nobel Memorial Prize in Economics.

According to Grubel and Scott's findings (1967, p. 145), many of those foreigners found a good life in the States: "The data reveal that economists with Western European and Canadian high school training who obtain U.S. Ph.D.'s and stay on at U.S. universities earn a higher income on average than economists who were born and trained in the United States."

There is little to say about minorities, but that in itself says a great deal. There are (to my knowledge) no studies on minorities in economics: minorities simply are not visible.

Already hinted at is another trend in economics: the tremendous proliferation of its literature. Spengler (1968b, pp. 171–72) characterizes the situation in 1968 in the following way:

A crude index of the growth of infrastructure is supplied by a comparison of the number of journals consisting entirely or partly of economical articles in English in 1886–1900 with the number of corresponding journals in existence around 1960; it was 9 in the former period and 89 around 1960. The average annual number of articles published in 1960–63 was about 16 times that published in 1886–1924. It is probable that the number of articles submitted for publication grew in even greater measure.

Enlargement of the infrastructure of communication affects the life of that which is transmitted and intensifies competition among transmitters as well as weakens resistance on the part of potential acceptors. Today economic information is diffused much more rapidly than it was 50–75 years ago. Vulgarized economic information as well as misinformation probably is distributed even more rapidly through so-called highbrow journals. The sheer extent of the flow of information can produce swamping effects. In the absence of an efficient information-retrieval system, it becomes costly to draw on the current flow, perhaps so costly that it is cheaper to generate the information anew. Contributions, significant and insignificant, tend to be forgotten, neglected or hidden from the minds of all but the authors' contemporaries. Preoccupation "with the latest fashion causes much that is valuable in former work to be discarded and forgotten, only to be 'rehabilitated' when fashion changes again." Progress itself is slowed down by the resulting high ratio of waste to gross output of "information."

The Economist ("Why Scientific Fact Is Sometimes Fiction," 1987) recently reported: "The number of science journals has doubled every 15 years since 1750, which makes it ridiculously hard for scientists to keep up in their field." One result of this flood of information, cited by *The Economist,* is that "in the past 15 years, 16 big frauds in scientific research have come to light, most of them in biology and medicine." Since Spengler wrote about the communication crisis over twenty years ago, information has more than doubled. Sir Robert Hall's prescient lamentation of the situation (1959, p. 651) was made thirty years ago: "It is becoming more and more difficult for the practical economist to learn from his academic colleagues or, I suspect, for any but the most industrious in the academic field to keep up with developments in other branches than his own." Thus, a vicious circle is established. The more that is published, the more control is needed (that is, in the form of good referees). But because the referees then become increasingly more burdened by the struggle of refereeing more and more articles and keeping up with the ever-growing mass of literature in their own field, they become less efficient in general.

This problem is related to the final topic, economic education, because "publish or perish" has become the iron law of promotion of both graduate students and young professors. Winch (1962, p. 198) makes a startling comment about an academic discipline: "Economics may have produced some great scholars but it is not a scholarly discipline." Viner (1958, p. 369) offers this remedy:

> What I propose, stated briefly and simply, is that our graduate schools shall assume more responsibility than they ordinarily do, so that the philosophers, economists, mathematicians, physicists, and theologians they turn out as finished teachers, technicians, and practitioners shall have been put under some pressure or seduction to be also scholars.

One reason for this proposal (Viner, 1958, p. 380):

> Men are not narrow in their intellectual interests by nature; it takes special and rigorous training to accomplish that end. And men who have been trained to think only within the limits of one subject, or only from the point of view of one subject, will never make good teachers at the college level even in that subject. They may know exceedingly well the possibilities of that subject, but they will never be conscious of its limitations, or if conscious of them will never have an adequate motive or a good basis for judging as to their consequence or extent.

This, of course, is a plea for a good liberal arts education to counteract *Fachidiotie*. Moreover, it indicates how valuable methodology is for economics because methodology emphasizes limitations, valid argument, reflection upon purpose.

Lekachman (1976, p. 108) describes the goal of graduate education in economics today as one that "enables its survivors to rise above the common sense with which no doubt they were endowed at birth." In a very recent study, Colander and Klamer (1987, pp. 109–10) obtained the striking result that "what students believe leads to success in graduate school is definitely techniques; success had little to do with understanding the economy, nor does it have much to do with economic literature." Attacks on the way future economists are trained, it may be added, are cyclical in nature and not at all new to the discipline. (Veblen's (1948a) vituperation against the businesslike nature of the university has never been equaled by anyone else in the profession.) Consider how topical Richard T. Ely's (in Furner, 1975, pp. 39–40) enumeration of the duties of an economics professor in 1876 is (at Columbia, then the leading school in economics). He explains rather cynically that all you need to do as professor is "buy Mrs. Fawcett's *Political Economy for*

Beginners; see that your pupils do the same; then assign them once a week a chapter to be learned; finally, question them each week on the chapter assigned the week before, using the questions found at the end of the chapter."

As if this were not enough, Clower (1974, p. 1) reflects on the quality of the economists turned out of the system:

> No doubt the economics profession includes some arrogant incompetents. As the Nobel Laureate James Watson has remarked: "One could not be a successful scientist without realizing that, in contrast to the popular conception supported by newspapers and mothers of scientists, a goodly number of scientists are not just only narrow-minded and dull, but just also stupid."

There is at least a grain of truth to this: economists are known for their shamelessness. Economics' first American Nobel laureate, Paul Samuelson (1972, p. 166), for instance, is simply shameless when discussing his accomplishments.[16] In the passage below he discusses Keynes's *General Theory:*

> And I think I am giving away no secrets when I solemnly aver—upon the basis of vivid personal recollection—that no one else in Cambridge, Massachusetts, really knew what it was about for some 12 to 18 months after its publication. Indeed, until the appearance of the mathematical models of (it) there is reason to believe that Keynes himself did not truly understand his own analysis.

And consider *Business Week*'s ("The Furniture Movers," 1978) description of the annual American Economic Association's meeting in 1978:[17]

> There was no evidence of either humility or competence at the AEA meeting. Nor did any economist or group of economists offer anything resembling a new idea for addressing the major policy dilemma of the industrial West— the apparent incompatibility of full employment and price stability. Instead, the sessions were dominated by papers seeking to refine methodologies [that is, techniques] that already have been proven ineffective.

One wonders if economics wasn't named "the dismal science" because all economists are destined to become disenchanted with its subject matter. Keynes (1972a, p. 200) writes of the "Father of English Economics," Alfred Marshall:

> Near the end of his life, when the intellect grew dimmer and the preaching imp could rise nearer to the surface to protest against its lifelong servitude, he [Marshall] once said: 'If I had to live my life over again I should have devoted it to psychology. Economics has too little to do with ideals. If I said much about them I should not be read by business men.'

End of digression. What can we conclude from this? The average American economist is white, middle-class, male, swamped with literature to absorb or referee, intimidated by his superiors and colleagues, goaded along by two creeds—"publish or perish" and "conform or perish"—and unhappy with the state of current economics. In addition, the typical graduate student—also male, middle class, white—believes success means techniques and not an

understanding of the economy *or even of the literature,* even though, as contrary as it may be to common sense, he knows that publications are what will
lead to recognition within the discipline. Here we have the "economic and
social type," to return to Schumpeter.

Schumpeter's view that economists accept or reject their potential colleagues for reasons other than their professional competence or lack thereof is
quite a controversial thesis, as is his view that sociological factors play a
greater role in economics than in physics. This last thesis is, of course, exactly
what Popper was attempting to refute. That they were put forth by a fellow
Austrian, who unlike Popper possessed expert knowledge of economics,
makes the view all the more interesting.

Which Austrian is right? Perhaps both are right in a limited sense:
Schumpeter because external aspects are and have been overpowering internal science in economics, Popper because there is equal potential for the
social and natural sciences to be objective.

But both are wrong in a much more substantial way. This long digression
on the development of the profession indicates there is indeed a consensus
among economists: both the profile of the average American economist and
Schumpeter's description of the sociological factors in economics point to the
fact that economics is more club than profession. If Schumpeter is right that
economists choose colleagues for reasons other than competence (and the
vignettes on women, minorities, and the power structure in economics seem
to indicate this), then this choice is not bound by reason (that is, rationalism,
or intellectualism) but by predilection or arbitrariness. Economists as academicians and scientists are bound by the meritocratic principles of fairness and
rational decision making: this is in fact law.[18] (When Schumpeter was writing
in the forties and fifties, equal opportunity laws that declare arbitrary employment decisions illegal were in their infancy.) Schumpeter actually misses the
point that those professionals' very professionalism rests on their adherence
to rational scientific discussion, judgment, and decision making—a point not
missed by Popper, but one that he does not develop very well. Popper does in
fact assert that the attitude of critical rationalism is the scientific attitude, but
he often takes this attitude for granted. Schumpeter's description of the profession, however, clearly shows that this principle of scientific rationalism that
Popper assumes binds all scientists is contravened.

We have seen that a consensus exists and that economists strive to reach a
consensus. Certainly, consensus can be born of clubs and maintained by pres-
sure of all sorts. But such consensus is not *scientific.* Although philosophers of
science normally shun discussions of the external history of science (unless it
happens to be Soviet history), Kuhn finds the inclusion of the external history
of science into the history of science important for the very reason that it
emphasizes the social problems of science. Kuhn's focus on the *community* of
scientists and on consensus and change is crucial because it brings us to the
rather obvious question: When no mechanical rules or absolute criteria determine what scientific knowledge is, is the consensus of the scientific community
scientific? In Popper's words, because of the fallibility of observation state-

ments, their "corroboration" by scientists opens up the *decision* to accept or reject empirical data. Is this decision rational? Kuhn argues that the scientist *decides* whether a theory is good or not by means of five standard criteria. Is this *decision* rational? Yet, Schumpeter, despite the fact that he pushed for high academic standards, would have a less competent scientist making these decisions, given that a lesser qualified scientist is accepted into the group. The philosophy of science leads us to an admission that science rests ultimately on human decision—something that causes some philosophers to raise cries of "irrationality." It is not: it only shows that rationality in the last analysis lies with the scientist him- or herself.

Communis Opinio Doctorum

Imposing consensus is not Kuhn's message. That the pathological drawbacks of "normal science" appear to apply to economics is due to—and in fact is exaggerated and exacerbated by—economists' unnatural consensus, the result of a rigid disciplinary structure and a clublike mentality. The drawbacks of this—lack of a critical attitude, change resembling a bandwagon effect, acceptance and spreading of dogma, technicianlike scientists—may seem to be shortcomings of "normal science," but these symptoms are less than "normal." "Normal" activity in Kuhn's sense is not stagnation; nor is it "mumpsimus," the stubborn persistence in an error after it has been exposed (Robinson, quoted in Arouh, 1987, p. 395). Under no circumstances does it imply that scientists suspend the standards of reasoned discussion, judgment, evaluation of students or colleagues, and so on. (This is scientific rationalism's first tie to human freedom: adherence to a principle of fairness.) Science—economics or any other—must be open to new ideas and criticism or there can be no revision of argument. Whereas deflection or muzzling of criticism signals that one's position is not tenable, rational argument embodies a dynamic potential for intellectual growth: it invites attack, welcomes revision, opens itself to learning. And it assumes pluralism. After all, being open to criticism means exposing oneself to all other positions, and hence other positions must be allowed to exist. (This is scientific rationalism's second tie to human freedom: adherence to the principle of democracy.)

The control over ideas and stifling of competition among theories and schools of thought (and that by people who champion competition in the marketplace and insist that "variety is the spice of life") is perhaps the most anti-intellectual characteristic of the discipline. No argument exists to support a striving toward and imposition of consensus per se in economics. Forced consensus may be psychologically more palatable than differences of opinion, but collective security does not rate as science. And scientific criticism is not dissension; a natural but limited consensus will form as faulty argument is discarded and weak argument strengthened.

Once we recognize that there is no final arbiter in the appraisal of theories, the concepts of "rationality," "objectivity," "science," and so on gain new meaning. They are no longer grounded in demarcation or the appeal to certain empirical data. Indeed, it was hoped that the new logic and empiricism would provide a certain foundation for science. Popper's use of the *modus tollens,* his development of a "situational logic," and his emphasis on testing clearly show that in this sense he adhered to the tradition of the logical positivists: science becomes objective and rational because logic provides us with a *proof.* But it is the very fact that rationality and objectivity cannot be captured by a logical or deductive schema, and that they ultimately rest on human decision, that makes it so problematic.

This points to the fact that the special successes of science require a balancing act. Problems and controversies are not settled by appeal to a rule or to empirical data. They demand instead a balance between the use of tools such as mathematics, logic, statistics, and history and a respect for their limits, a balance between pluralism and tradition, between the proliferation of revolutionary ideas and normal science (science marked by incremental change), between what is criticized and what is given. Science, rationality, objectivity can no more be definitely captured, legislated, or demarcated than can the concepts of freedom or democracy because their existence hangs on both professional training and on an attitude that itself is bound up with freedom and democracy.[19]

The major problem with calling scientific rationalism critical is that everyone today calls himself or herself "critical." And this brings special snares for the economist because one meaning of "critical" is to be against something. Yet, scientific rationalism rests on a spirit of tolerance, a respect for well-argued positions even if they do not correspond to one's own view or vision, a belief that we can learn from others. What being rational boils down to is in part a tolerant, courteous attitude. John Henry Cardinal Newman (in Fearnside and Holter, 1959, p. 172) described this gentle person as follows:

> He is never mean or little in his disputes, never takes unfair advantage, never mistakes personalities or sharp sayings for arguments or insinuates evil which he dare not say out. . . . He has too much good sense to be affronted at insults, he is too indolent to bear malice. . . . If he engages in controversy of any kind, his disciplined intellect preserves him from the blundering discourtesy of better, perhaps, but less educated minds; who like blunt weapons, tear and hack instead of cutting clean, who mistake the point in argument, waste their strength on trifles, misconceive their adversary, and leave the question more involved than they find it.

The art of criticism involves, then, not only a willingness to be exposed to criticism but also respect for the other parties.[20] (Mutual respect is the final tie to human freedom.) It was this attitude that attracted Popper to the Vienna Circle, this attitude that allowed logical positivism to evolve into logical empiricism.

Yet, American economists are not trained to be critical or to learn all sides of the issue. Stigler (1969, p. 221) asserts that students are in fact taught to *disrespect* perspectives different from that of the particular institution where they are being educated:

> Every major center of graduate instruction in economics has a degree of engagement in current economics. It has a faculty which is active in research and publication, and inevitably a certain amount of taking sides occurs in the graduate courses. Young Ph.D.'s come out prepared to read "good" economics uncritically and "bad" economics hypercritically.

Not only that, scholarliness does not seem to be valued in economics. One gets the feeling from economists that good economists are "naturals" rather than the product of diligence and training. J. M. Keynes (1972a, pp. 173–74) wrote in 1924:

> The study of economics does not seem to require any specialised gifts of an unusually high order. Is it not, intellectually regarded, a very easy subject compared with the higher branches of philosophy and pure science? Yet good, or even competent, economists are the rarest of birds. An easy subject, at which very few excel! The paradox finds its explanation, perhaps, in that the master-economist must possess a rare *combination* of gifts. He must reach a high standard in several different directions and must combine talents not often found together. He must be a mathematician, historian, statesman, philosopher—in some degree. He must understand symbols and speak in words. He must contemplate the particular in terms of the general, and touch abstract and concrete in the same flight of thought. He must study the present in the light of the past for the purposes of the future. No part of man's nature or his institutions must lie entirely outside his regard. He must be purposeful and disinterested in a simultaneous mood; as aloof and incorruptible as an artist, yet sometimes as near the earth as a politician.

Keynes was always quite "myopic" (E. and H. Johnson, 1978, p. 16) about the advantages of his education. His remarks—that the subject of economics is easy, that there are only a few who master it, that the outstanding economists must be wise in the broadest sense—suggest that the study of economics *should* require that "specialised gifts of an unusually high order" be developed and thus the discipline should be very demanding of its students.[21]

Scientific rationalism presupposes honesty (tempered by tactfulness) and a commitment to a higher goal: the advancement of science. This calls for modesty and a certain selflessness. Yet, there has been much ado about intellectual dishonesty in economics. Wolozin (1974, p. 196) asserts, for instance, that Paul Samuelson has implied he favors lying about economic matters so long as the lies serve a good cause and are plausible (*Newsweek,* 5 February 1973, p. 46; in Wolozin, 1974, p. 196). But Wolozin is not talking about lying at all but about the use of analytical abstractions that simplify so that the theory no longer fully corresponds to reality. (See Cartwright's *How the Laws of Physics Lie,* 1983, where a similar argument is developed for physics.) What Wolozin is talking about is the use of *homo economicus* or *ceteris paribus,* which amounts not at all to dishonesty but to the use of such simplifications (whose limitations need to be investigated and respected as with all such devices). It is, rather, the type of dishonesty discussed by Schumpeter (1954, p. 43) that is pathological for science:

> It is true that in economics, and still more in other social sciences, this sphere of the strictly provable is limited in that there are always fringe ends of things that are matters of personal experience and impression from which it is practically impossible to drive ideology, or for that matter conscious dishonesty, completely.

In the note discussing "conscious dishonesty" he tells us:

> The role of what above is meant by conscious dishonesty is greatly enhanced by the fact that many things that do amount to tampering with the effects of logic do not in our field necessarily present themselves as dishonesty to the man who practices such tampering. *He may be so fundamentally convinced of the truths of what he is standing for that he would rather die than give new weight to contradicting facts or pieces of analysis. The first thing a man will do for his ideals is lie.* Now we do not interpret this element in the case as we do when speaking of ideological bias, but of course it reinforces the baleful influence of the latter. (my emphasis)

Schumpeter is again describing the practice of science. His viewpoint would suggest that although Popper assumes that all scientists are bound by this attitude of scientific rationalism, practice confirms the opposite. And Schumpeter's description of science contains at least a kernel of truth. In a recent article in *The Chronicle of Higher Education* (O'Toole, 1989) the author, a biologist, reported that she found inaccuracies and misstatements in a published paper. The authors of the paper acknowledged but refused to correct their errors because they insisted that only false assertions made *intentionally* require correction. Even an investigation did not change this. The author (p. A44) reaches the conclusion:

> If this case is a manifestation of accepted practices of biologists, neither researchers themselves, nor the public paying for the research, can rely on published claims. . . . If scientists really regard misstatement of results as equivalent to unidentified scientific errors, the literature will soon be meaningless.

The author lost her job because she insisted on accurate presentation of results. The United States Congress has now decided to step in to investigate fraud and other misconduct in science (Jaschik, 1989), including the case mentioned above. When publishing misinformation becomes established practice, the purpose of science is completely defeated. It appears that Popper's assertion that "errors are systematically criticized" and corrected (part I, chapter 1) in science cannot be taken for granted. Bartley does not, but this is an unpopular stance because it stresses professionals' personal responsibility to science.

Scientific rationalism presupposes not only honesty, courtesy, and tolerance but also a good education. Toulmin (1972b, p. 479) argues that "the rationality of natural science and other collective disciplines has nothing intrinsically to do with formal entailments and contradictions, inductive logic, or the probability calculus." But it does. Scientific rationalism does presuppose education or learnedness. One must be able to know whether the competing positions are being competently argued in order to criticize them. Just as sound personal judgment is based on knowledge and experience, the rational consensus of a scientific community is based on all available information and evidence. Because science has to do with problems and problem solving, rationality also presupposes that we know that whether a problem arises or not and how it is solved is a function of what tools we use to solve the problem.

Both academia and science have always had explicit and implicit values: simplicity, consistency, accuracy, coherency, clarity, scholariness and thoroughness of research, honesty about findings, and so on. Then it is not a question of escaping values, as economists (and some philosophers) have assumed in the past—for that is impossible—but of acknowledging and embracing the right ones.[22] Hayek (1978, p. 22) echoes this theme:

> At present the postulate that we should avoid all value judgements seems to me often to have become a mere excuse of the timid, who do not wish to offend anyone and thus conceal their preference. Even more frequently, it is an attempt to conceal from themselves rational comprehension of the choices we have to make between possibilities open to us, which force us to sacrifice some aims we wish to realise.

This principle extends beyond economics. Those who wish to ignore the external history of science, to close their eyes to the conditions that allow science to flourish and to their respective responsibility within this environment, are trying to conceal how important scientists' responsibility to science is. Scientific rationalism is an attitude that must be taught and must become a tradition in order to guarantee the existence of the most favorable environment for science. A study of the internal history of science provides us with no guide to success or key to continued growth of science; the external history of science, when worth discussing, records mistakes that we should avoid making in the future and makes us aware of our responsibility to science. Much of the current literature on economic methodology indicates that *external* considerations, or problems with the scientists' environment, are the focus rather than internal aspects that focus on science as the rational evolution of ideas.

I want to conclude on two optimistic notes.[23] The first is that criticism is not dead in economics, as the "crisis" itself indicates, although the literature indicates that it needs to be resuscitated democratically. Kuttner's "Poverty of Economics" (1985, p. 78) provides us with one example of constructive criticism in economics:

> One celebrated methodological and ideological controversy raged over work by Martin Feldstein, a professor of economics at Harvard and a former chairman of President Reagan's Council of Economic Advisors, on the influence that Social Security has on savings rates. In a series of journal articles Feldstein reported econometric findings that Social Security had depressed savings, by about 50 percent, which would have depressed GNP, by many

hundreds of billions of dollars. But at the 1980 meeting of the AEA two relatively unknown researchers, Dean Leimer and Selig Lesnoy, of the Social Security Administration research staff, presented a paper showing a serious technical error in Feldstein's equations. When the error was corrected, the model Feldstein had used showed absurd results. *Without* Social Security, savings would have been negative. Feldstein ultimately revised his own equations to reconcile them with his hypothesis. But the affair was considered damaging to Feldstein's prestige and a good object lesson in the hubris of econometrics.

The second point is that the unhealthy nature of restrictive consensus has been recognized by economists. So has the fact that science does rest on an "attitude," as Popper calls it, an attitude of scientific rationalism. Phyllis Deane (1983, p. 11) puts this in the following way:

> The lesson, it seems to me, that we should draw from the history of economic thought is that economists should resist the pressure to embrace a one-sided or restrictive consensus. There *is* no one kind of economic truth which holds the key to fruitful analysis of *all* economic problems, no pure economic theory that is immune to changes in social values of current policy problems. The scope and method of our discipline needs at all times to be defined in relation to the social problems which give purpose to it and there is room for more than one progressive research programme [theory] in operation at the same time. We ought indeed to profit from the divergent approaches of leading theorists, provided that they and their disciplines avoid the kind of arrogance of which Wagner complained a century ago and are scrupulously open-minded in the conduct of the necessary debates.

She (1983, pp. 11–12) concludes:

> If economics is a science—and it is not as clear as it used to be what *that* is—it is evidently a science whose powers of prediction and control are limited, largely because the phenomena it seeks to explain are subject to persistent change and often for reasons that may lie outside the traditional boundaries of the discipline. In these circumstances successful explanation, or prediction, must depend on the economist's willingness to renew, discard and adapt theories, concepts and analytical techniques and even to adjust the boundaries of the discipline. What we need is not a new orthodoxy on scope and method but a readiness to listen seriously to our colleagues who take opposing views on fundamental principles and to admit to the natural scientists, or the politicians, or the students, who are crying out for economic truth, that the right answers are unlikely to come from any pure economic dogma.

Economists have been looking for the key to scientific success in the philosophy of science. They will find much valuable in the philosophy of science, but a formula for success will not be among the fruits won. If we really want to know what we can do to advance science, we can guarantee that this attitude, scientific rationalism—tolerance, honesty, commitment to the advance of science above personal advance and to the freedom to exercise criticism, a willingness to listen and learn from others, and so on—is not violated and becomes entrenched as a tradition.

Notes

1. See note 1, chapter 7.
2. Such uses abound in the literature. (See note 1, chapter 7.) I did not use the passages from Wiles and Arouh to single them out as major offenders; their work, being recent, was simply at hand. (See the preface.) Furthermore, abuse of this terminology pervades not only economics but the rest of the social sciences and humanities as well.
3. See note 1, chapter 7. The terminology of "revolution" is not new to economics, where discussions of Keynesian and marginalist revolutions were fashionable well before Kuhn was. An excellent discussion of the origins of the concept of scientific revolution is given by I. B. Cohen (1976, 1985).

There has been considerable interest in revolutions and economics, not just as a result of Kuhn's work, and not all of it particularly enlightening. One piece well worth reading is H. G. Johnson (1971); Kuhn goes unmentioned. A second is Hutchison's *On Revolutions and Progress in Economic Knowledge* (1978), which does include references to the philosophy of science. Finally, on "The Concept of the Industrial Revolution," see Pollard (1987).
4. On Keynes and the Keynesians, see Hutchison (1977a). The appellation *Keynesian* is troublesome. Sir Austin Robinson (in Hutchison, 1977a, p. 58) recalls Keynes's remark about the "Keynesians":

> I am reminded of Keynes' comment to Lydia and me at breakfast in Washington in 1944 after he had dined the night before with the Washington Keynesian economists: 'I was the only non-Keynesian there.'

See also L. Klein (1968); Leijonhufvud (1966, 1968); Shackle (1967). On rational expectations (RE), see D. Redman (forthcoming) for a concise introduction to RE including a discussion of problems with the terminology, a short history of the evolution of RE, and a comprehensive, annotated bibliography; Sargent (1986) for an up-to-date, nontechnical discussion of the new classical version of RE; Klamer (1984b) for conversations with the new classical economists and their opponents; and Leijonhufvud's "What Would Keynes Have Thought of Rational Expectations?" (1983, p. 200), where he reaches the interesting conclusion that "obviously, Keynes had an adequate working knowledge of that 'unpleasant monetarist arithmetic'! A more detailed reading of Keynes and Sargent only makes the agreement between the two even more remarkable." The reference is to Sargent and Wallace's "Some Unpleasant Monetarist Arithmetic," which appears in Sargent (1986, pp. 159–90) as chapter 5.
5. Donald McCloskey and Arjo Klamer are the leading members of a growing group of economists interested in the role of rhetoric in economics. Rhetoric for these economists is "the study of how people persuade" (McCloskey, 1985, p. 29).

The view that economics and rhetoric are intertwined actually has roots in the origins of the discipline. Adam Smith dwelled on rhetoric. See his *Lectures on Rhetoric and Belles Lettres* (1983 [1748–1749]). Letwin (1963) fully incorporates the importance of the development of rhetorical devices into his history, *The Origins of Scientific Economics* (see especially pp. 47, 49, 137, 182–83, 203). Rhetoric, he argues (1963, p. 97), was needed to gain the detached mien befitting the scientist at work and to combat the image of the economist as self-interested partisan:

> Roger North's prescription for objectivity, in short, is a taut deductive system that infers its conclusions from a set of simple principles. This is the method

of 'demonstration', as that term was understood in classical logic; it is precisely the method of Euclid's proofs, and as Roger North suggested, nobody would accuse Euclid of maintaining any of his theorems because they suited his self-interest. The only relevant critique is that the premises are false or inadequate or the chain of reasoning imperfect; failing this, the conclusions are binding, no matter what may be the character of the author.

But this method for guaranteeing objectivity is also the method *par excellence* of scientific theory. The theory consists of a deductive system linking a few fundamental principles with a set of conclusions, as modern price theory, for instance, links fundamental propositions about utility and costs of production to conclusions about equilibrium conditions of an economy; and here too the only pertinent questions are whether the premises are correct and adequate and the chain of reasoning flawless.

In the search for a way of dispelling the problem of special pleading, a scientific method was hit on. *The needs of rhetoric brought forth the method of economic theory.* (my emphasis)

6. Marshall was not alone. Isaac Newton was even thinner-skinned than Marshall. Shapere (1967, p. 490) provides a character portrait of Newton.

In personality Newton was quiet and only rarely given to laughter. He did not communicate his ideas readily and was far from being a quick or facile speaker. Morbidly sensitive to criticism, he was so possessive about his own work that not only was he reluctant to publish it, but when he did, he only rarely acknowledged the specific contributions of others. This was the case despite his statement—made as much in an attempt to pacify Robert Hooke as in modesty—that what he accomplished was done because he had "stood on the shoulders of giants." When his priority was questioned or when he was crossed in other ways, he grew resentful and even vindictive, threatening to withdraw entirely from science. On more than one occasion, he used the power of his reputation of office to crush others. Even some of his closest friends, John Locke and Charles Montague (earl of Halifax), suffered from his suspicions.

Consider notes 18 and 20, below.
7. I believe this is what Kuhn means by "preparadigmatic." He assumes that all sciences will in time (and thus have the potential to) develop like physics—something that I believe is indefensible. (See my discussion, part I, chapter 1.) That is, incidentally, why I also reject the perennial discussion in the social sciences of "mature" and "immature" sciences. The very distinction implies that the social sciences can develop exactly as the natural. (Not even *Popper* goes that far, if one reads him carefully enough to note his inconsistencies and reservations.)

One gets the impression from Kuhn and the economic literature that natural scientists never engage in methodological disputes. This is simply not so. In 1980, for instance, a methodological battle raged across the editorial pages of *Chemical and Engineering News*. One excerpt (Northrop, 1980, p. 4) captures the flavor of the discussion:

Contrary to their media image, scientists are human beings, who, during the practice of their craft, make many unconscious choices and subjective decisions which are neither governed by reason nor easily identified. For most of us then, the only way we can begin to approach objective thinking and thus pursue "scientific truth" is to value such truth in and of itself. As soon as something else becomes the goal, then we mere humans find it terribly difficult to resist completely the force of the vested interests expressed by the usurping goal. We use the pH meter rather the way we play the pinball machine and produce instead a skewed or even prejudiced version of the scientific truth.

The philosophizing efforts of chemists do not seem to be much different from those of economists. (For the tip on methodological controversies among chemists I am indebted to a professional chemist, Dr. N. L. Redman-Furey.)

And certainly economists' methodological disputes do mirror those of philosophers of natural science. Consider this passage from Nancy Cartwright's *How the Laws of Physics Lie* (1983, 52–53):

> Most scientific explanations use *ceteris paribus* laws. These laws, read literally as descriptive statements, are false, not only false but deemed false even in the context of use. This is no surprise: we want laws that unify; but what happens may well be varied and diverse. We are lucky that we can organize phenomena at all. There is no reason to think that the principles that best organize will be true, nor that the principles that are true will organize much.

This has a familiar ring. The point is that methodological disputes are not per se a symptom of malaise. Epistemological queries belong to all sciences.

8. Under no circumstances should this be interpreted to mean that the United States was a haven for émigrés. Anti-Semitic feeling existed in the United States, taking on "more subtle forms of rejections" (Krohn, 1987, pp. 30–31).

9. I. B. Cohen (1985) treats the history of revolution in science extensively.

Without intending to belabor the point, using analogies from the philosophy of science (and hence often analogies that refer to natural phenomena) runs its risks. Without an adequate knowledge of the subject matter, major gaffes (including lying with methodology) are committed. Consider two examples. First, Goodwin (1980, p. 615) extends the idea of a crucial experiment to economics: "The second example of an external crucial experiment was the Great Depression of the 1930s, which forced economists to take account of what amounted to a really giant anomaly for prevailing theory." But philosophers of science give cogent reasons for believing that crucial experiments do not exist even in the natural sciences. Moreover, the Great Depression is no crucial experiment: an experiment is a procedure used to find facts, test ideas, and so on. The Great Depression is a historical event.

The second example has to do with Ockham's razor, the principle of parsimony that states that entities should not be multiplied beyond necessity. In a recent work, Wiles (1984a, p. 306) offers this advice for improvement of economics: "The revolution that *we* need *now* is 'anti-Occamical': *entia sunt multiplicanda, sed non praeter necessitatem.*" And a note explains what Ockham's razor is (1984a, p. 324):

> *Entia non sunt multiplicanda praeter necessitatem:* concepts must not be unnecessarily multiplied. It is a common fault of economic theorists to quote the first four words only, and so leave out the human judgement of necessity (Wiles, 1970). There is of course no virtue at all in reducing the number of concepts at the expense of understanding the phenomena. Taken strictly Occam's Razor is not a very important rule. It reminds us too strongly of the

German joke: *Warum es einfach machen wenn es auch komplizierter geht?*—
why do it simply when more complicated methods suffice?

There are several problems here. Ockham's razor is a very reasonable, useful rule of
thumb that is taken for granted today because it is common sense. (William of
Ockham lived from ca. 1285 to 1349, and was fighting against church doctrine.) The
thrust of the principle is economy in explanation. (See Moody, 1967.) But Ockham's
razor does not tolerate compactness or conciseness of expression at the expense of
explanation. Hence, both Latin explanations embody Ockham's intention. Moreover,
the German quip he cites in his favor—"Why do it simply when more complicated
methods suffice?"—would be better translated "Why do it simply when more complicated methods could be found?" The quip is meant to be ironic and thus is a plea for
simplicity. In short, Wiles's thesis is not supported by his argument. Nonetheless, I
think the basic idea, that oversimplification can also damage science, is sound. This
has been developed in an interesting article by Karl Menger (1979), and the principle
has been dubbed "Menger's comb."

   10. On the issue of whether American economists are the best, see H. G. Johnson
(1977) and J. R. Sargeant's "Are American Economists Better?" (1963), which is
answered in the affirmative.

   11. It has been forgotten that economics in both England and the United States
grew out of religion. Checkland (1951, p. 70) described the nascent British profession:

> All these men, among the greatest university names of their generation—
> Whately, Whewell, Sedgwick, Newman, Arnold—were clergymen. Each be-
> lieved economic truth to be a part of the divine order; each had integrated his
> attitude to economics with his theology. And though they sometimes seem
> like men with a method seeking for problems, rather than some later
> specialists—men with problems looking for a method, yet they are worthy of
> remembrance because they saw that the proper focus of the interest of the
> universities in the new discipline ought to lie in its organon.

Furner (1975, p. 75) describes as follows the founding of the American Economic
Association (AEA): "Of the fifty people who joined the AEA at the Saratoga meet-
ing, more than twenty were either former or practicing ministers, including leading
Social Gospel spokesmen such as Gladden, Lyman Abbott, and R. Heber Newton."
The following is a partial bibliography on the history of the development of British and
American economics: *British Economics:* Checkland (1951); A. W. Coats (1961,
1967a, b, 1968, 1975, 1980a; A. W. Coats and S. E. Coats (1973); Grammp (1973); R.
Harrison (1963); E. Johnson and H. G. Johnson (1978); Koot (1980); Letwin (1963);
Viner (1963). *American Economics:* Barber (1987); Colander and Klamer (1987);
Craver and Leijonhufvud (1987); Dorfman (1969a; 1969b); Eagley (1974); Furner
(1975); Grubel and Scott (1967); Horowitz (1974); H. G. Johnson (1977); Mason
(1982); Parrish (1967); Spellman and Gabriel (1978); Stigler and Friedland (1975).

   12. Parrish (1967, p. 5) adds: "Germany's great economists welcomed American
students, took them into their homes, urged them to pursue the subject, encouraged
them in academic careers. American students admired the fact that German economists were free of sectarian and political pressures." Today it could not be more
different: there is no one less desired at a German university than an American. See
Rowe (1890), Seager (1893), and Parrish (1967) for more on the German university
and its influence on American economics. On the German university in general, see
Ash (1980) and Fallon (1980).

13. Certainly, the use of mathematics among American economists cannot be attributed solely to the attempt to gain isolation from external meddling and criticism. H. G. Johnson (1977, pp. 19–20) explains the American empirical bent in the following way:

> The reasons for this distinctive empirical emphasis, to which we shall return, can in my view be found in two characteristics of the American economics profession, which have, at least until the 1960's, distinguished it from the national professions elsewhere. The first is the sheer numerical size of the American profession, and its diffusion over a large geographical area, which have meant that, on the one hand, the vast majority of economists have not had access to a position in an elite university to be won by distinction in the equivalent of "pure" scientific research, but instead have had to justify themselves by "applied' research; on the other hand, they have had to live within a local civilian community and to justify themselves primarily by their teaching of students and their knowledge of the practical world which they profess to teach about.

14. There is a danger in writing about living scientists—hence current source material is scarce. H. G. Johnson (1977, pp. 17–18) prefaces his discussion in "The American Tradition in Economics" in a rather straightforward manner:

> Naturally, a brief paper on this subject will necessarily be impressionistic, and it would be well at the outset to stress that the emphasis will be on the characteristics of economics as practiced in America, rather than on the great names in the history of economics in America. Partly this is a necessity of scholarly and personal caution. To paraphrase an observation from the natural sciences, some ninety percent of all the economists there have ever been in the United States, or on the North American continent, are still alive and professionally active or, if deceased, so recently deceased as to be still remembered with love or hatred by former students now active practitioners of the profession. To guess at the verdict history will eventually pass on their scientific contributions would combine too high a variance of the outcome in the future with too low a probability of popularity in the present to be worth risking.

Boulding (1971b, p. 51) does not mince words about academe's lack of introspectiveness and self-control:

> Scientists, even social scientists, are somewhat averse to self-study, just as the universities are, perhaps because the results might be embarrassing, and these studies, at any rate up to now, have not become fashionable among the dispersers of research funds. An institute for the scientific study of science seems a long way off, yet perhaps there is hardly any other investment in which the granting agencies might indulge that would provide higher returns.

15. Discrimination against women at U. S. universities has been well documented, especially in graduate programs. See, for example, Dziech and Weiner (1984), Lewin and Duchan (1971), E. and R. Holmstrom (1974), Schneider (1987), and Somers (1982). Dziech and Weiner's *The Lecherous Professor* is recommended reading for anyone having to do with institutions of higher education, and especially for those who doubt that discrimination is a mammoth problem or that it is having an impact on the state of

science. On science and gender, see Keller (1985). Weisskopf's *Psychology of Economics* (1955) treats issues of gender and economics from a Freudian viewpoint.

There is a tendency among some economists to blame the "crisis" in economics—or all problems in economics—on Joan Robinson. For instance, consider O'Brien's essay "Whither Economics" (1974, pp. 16–17), in which he cites Shackle (1967, p. 47) and Shubik (1970, p. 416) respectively:

> But how did economics get on this wrong track? My own personal belief is that it starts in the turn which the subject took in Cambridge in the 1930's: specifically with the intervention of Mrs. Joan Robinson into the debate over imperfect competition. Professor G.L.S. Shackle has explained what happened very lucidly.

>> there is . . . a sharp and striking difference in style of attack. The authors we have been considering (previously) were Marshallians. They examined the existing world in a spirit of respect, they brought as much of it intact into their discourse as they could, they valued the contours and features of the landscape they beheld and tried to mould their argument upon them rather than cut a path direct to rigorous conclusions. Mrs. Robinson did the opposite. Clear and definite questions cannot be asked about a vague, richly detailed, fluid and living world. This world must therefore be exchanged for a *model,* a set of precise assumptions collectively simple enough to allow the play of logic and mathematics . . . the model is a work of art, freely composed within the constraints of a particular art-form, namely the logical binding together of propositions.

Another commentator writes:

> The mathematical apparatus she assembled to do this job was such that von Neumann once remarked that if the archeologists of some future civilisation were to dig up the remains of ours and find a cache of books, the *Economics of Imperfect Competition* would probably be dated as an early precursor of Newton.

It is nonsense to blame economics' problems on any one economist in particular.

16. Consider, for instance, what Samuelson (1972, p. 155; 1983, pp. 1–2) has to say about being America's first economics Nobel prize winner:

> In the dark of the late October night, when a reporter phoned my home to ask my reaction to having received the Alfred Nobel Memorial Prize in Economic Science for 1970—and I spell out the title to emphasize that economics is a latecomer at the festive table with an award that is not quite a proper Nobel Prize—my first response was, "It's nice to have hard work rewarded." My children told me later it was a conceited remark. Nonetheless one must tell the truth and shame the devil. It has been one of the sad empirical findings of my life: other things equal—with initial endowments and abilities held constant—the man who works the hardest tends to get the most done, and the one who saves the most does, alas, end up the richest.

17. The analogy to furniture in the title of this article ("The Furniture Movers," 1978) refers back to J. H. Clapham's article "Of Empty Economic Boxes" (1922), in which he disapproves of some of economics' "mental furniture," namely, the idea of constant, increasing, and decreasing rates of return in industry for the reason that industries cannot, in reality, be classified into the respective categories.

18. An infringement of (a student's or a professor's) rights according to American law involves a violation of the principle of rationalism, that is, it involves treatment that is "arbitrary," "capricious," or "malicious." The only restraint upon professional freedom to teach and research is that he or she exercise *reasoned* judgment during the course of dealing with people. Students have the burden of proof of breach of reasoned judgment. The classic case for students is the *Board of Curators of the University of Missouri v. Horowitz*, 435 U.S. 78 (1978). See Machlup's (1971) classic discussion on academic freedom.

The consequence of the rationalism principle for Schumpeter's sociology of the economics profession is to render his view illegal as a violation of constitutional law. Lipsey (1979, p. 17) also seems to advance a thesis similar to Schumpeter's when he writes:

> Science has been successful in spite of the fact that individual scientists have not always been totally objective. Individuals may passionately resist the apparent implications of evidence. But the rule of the game, that facts cannot be ignored and must somehow be fitted into the accepted theoretical structure, tends to produce scientific advance *in spite of what might be thought of as unscientific, emotional attitudes on the part of many scientists, particularly at times of scientific crisis.* (my emphasis)

Yet, an "emotional attitude" is the converse of reasoned judgment. It is a trivial fact that science advances faster when the best are promoted; an emotional attitude can easily block promotion of the most talented, and thus thwart the advance of science.

19. See Popper's comments in part I, and in note 6, chapter 4, in part I.

20. See note 19 and part I, chapter 4. One would not expect tolerance, charity, and urbanity to be foreign to academicians, but history does confirm this. Consider, for example, Humboldt's description of life with academicians as he was establishing the University of Berlin in the early 1880s: "To direct a group of scholars is not much better than to have a troop of comedians under you" (in Fallon, 1980, pp. 25–26). And later he wrote to his wife (in Fallon, 1980, p. 26):

> You have no idea with how much difficulty I have to struggle above all with the scholars—the unruliest and most difficult to pacify of all peoples. They besiege me with their eternally self-thwarting interests, their jealousy, their envy, their passion to govern, their one-sided opinions, in which each believes that his discipline alone has earned support and encouragement.

I can say, based upon the amplitude of reading that brought about this work, that my general impression is that *no* science or discipline—whether social or natural science—has a truly well-developed critical tradition. Furthermore, I suspect that the complaints raised by economists about problems with the discipline are hardly unique. There are so many famous names now voicing disappointment with the profession that it is tempting to *commend* economists for bringing the problems out in the open.

21. In a recent article, "The Status of Master's Programs in Economics" (Thornton and Innes, 1988), American economics master's programs were surveyed and evaluated. Thornton and Innes's conclusion fits well the pattern that has emerged in this work:

> The net effect of the above findings is somewhat disturbing. . . . Prerequisites for admission appear to have been lowered, the common body of knowl-

edge required remains minimal, and the writing of a thesis to demonstrate professional competence is becoming rare. We can only conclude that the master's degree has been devalued, relative to the bachelor's and doctor's degree. Our finding that the degree is often used as a "consolation degree" may simply be symptomatic of this devaluation, though it is likely that it is also a contributing factor. However, we suspect that Bowen's view of the master's degree, as one which should signify "a substantial accomplishment in economics and a level of professional competence as an economist," is a view which is still widely shared today. If this is indeed the case, then the erosion of standards reflected in our survey results must be halted and reversed.

Perhaps it is time for the economics profession to undertake another comprehensive study of its graduate programs—both master's and Ph.D. Such a study is warranted by its potential revelations in many other areas than those discussed here. In any case, the fact remains that it has been three decades since the last such comprehensive study was undertaken, and this reason alone is sufficiently compelling. (p. 178)

22. Writes Frank Knight in his "Free Society: Its Basic Nature and Problem" (1956, p. 296): "Yet, comparatively, science is free, and its freedom rests on the general acceptance by its votaries of a rather high and austere ideal."

23. I have not been able to treat the following applications of other philosophies of science to economics: (1) German structuralism to economics; see Hamminga and Balzer (1986); Händler (1989, 1982); Hands (1985c); Sneed (1971); Stegmüller (1986); Stegmüller, Balzer, and Spohn (1982); (2) Feyerabend's philosophy applied to economics; see McCloskey (1983) and Wiles (1984a, p. 310), who is probably unaware that he is echoing one of Feyerabend's messages: "I am saying *as a taxpayer* that the public funding of science should be more sceptical and more Philistine."

Unusual sources of philosophy of economics appear in Wiener (1964, chap. 7, pp. 87–93), Cheung (1973), and Leijonhufvud's (1973) "Life Among the Econ." I recommend that all social scientists read Scharnberg's (1984) critique of applications of Kuhn's philosophy to the behavioral sciences. For further reading beyond the works in the bibliography here, see D. Redman (1989).

APPENDIX I

A Short History of the Is–Ought Problem

The attempt to derive "ought" from "is" has been labeled the naturalistic fallacy. The assertion that ethical propositions cannot be derived from nonethical ones is attributed to David Hume (1711–1776) and G. E. Moore (1873–1958). The "offspring" of the is–ought distinction are the factual (positive) versus the normative, the descriptive versus the evaluative, and fact versus value gaps.

What is at the heart of the is–ought controversy? The presentation that one normally finds in textbooks depicts the is–ought fallacy in the following way (Bartley, 1984a, pp. 75–77):

> Premise: *I like x.*
> Conclusion: x is good.

This is a invalid argument not because the conclusion that x is good—a value statement—has been derived from an is statement but because there is no relation between premise and conclusion. That is, the logical form A is B → A is C is false.

In valid form the argument looks like this:

> Premise: I like x.
> Whatever I like is good.
> Conclusion: x is good.

But this is no derivation of an ought from an is statement because the second premise is an ought premise, that is, a value statement.

Modern discussions of the is–ought controversy generally avoid the usage of *fallacious,* but not for reasons that deny that the above logical mistake does occur. Ethical naturalism denies that there is a logical gulf between "what is" (statements of fact) and "what ought to be" (statements of value). Without disputing that Hume and Moore were making an important contribution, many philosophers agree with Harrison's (1967) conclusion in his article on ethical naturalism appearing in *The Encyclopedia of Philosophy* that "it is dangerous to use 'you cannot deduce an "ought" from an "is" ' as a slogan."

Searle has probably mounted the most formidable attack against the is–ought distinction. In "How to Derive 'Ought' from 'Is'," Searle (in Rohatyn, 1975, p. 81) offers the following argumentation:

1. Jones uttered the words, "I do hereby promise to pay you, Smith, five dollars."
   1a. Under certain conditions C anyone who utters the words (sentence), "I hereby promise to pay you, Smith, five dollars" promises to pay Smith five dollars.
   1b. Conditions C obtain.
2. Jones promised Smith five dollars.
   2a. All promises are acts of placing oneself under (undertaking) an obligation to do the thing promised.
3. Jones placed himself under (undertook) an obligation to pay Smith five dollars.
   3a. All those who place themselves under an obligation are (at the same time when they so place themselves) under an obligation.
4. Jones is under an obligation to pay Smith five dollars.
   4a. If one is under an obligation to do something, then as regards that obligation one ought to do what one is under an obligation to do.
5. Jones ought to pay Smith five dollars.

Searle purposefully sought a case where the normative and descriptive elements are so intertwined that no is–ought distinction can hold. His discussion revolves around the "performative aspects of language," however, and not around logical error.

But Searle's argument has also come under fire. See Hudson (1969), Harrison (1967), and Rohatyn (1975). The last contains an informative bibliography (pp. 116–25).

From Searle's discussion it is *incorrect* to infer that all facts are value-laden. Searle's contribution is perhaps better summed up by Rohatyn (1975, p. 113) in the following way:

> If we are correct in supporting and sifting out the pros and cons of Searle's "derivation", then it *is* correct to conclude to (or infer) an 'ought' from an 'is', in some cases. The universal ban on moving from the level of description to that of evaluation has therefore been negated, even though the is–ought distinction (like the analytic/synthetic split) remains pedagogically useful, as one introduction to key ethical problems. Undoubtedly there do remain many fallacious deductions of 'ought' from 'is', scattered throughout ethics and its history, for which the is–ought "gap" still serves usefully as an effective barrier or blockade: but the underlying *theory* has severe limitations and to that extent is erroneous.

Perhaps Blaug's example (1980b, p. 130) "murder is a sin" is a clearer example of "is" functioning in a prescriptive sense (that is, communicating "one shouldn't murder"). Another example working on the same principle: "the university is a meritocracy" does entail "the brightest *should* be university professors" so long as a meritocracy is defined as a system in which the most intelligent are promoted. An example that fails is the following: "All Jews are inferior," inferring "Hence all Jews should be exterminated." This example brings us back to the naturalistic fallacy.

These examples help to illuminate how important *both* the functional (performative) and the logical aspects of the is–ought dichotomy are. And one gains an intuitive feel for why is and ought should be separated: 'what is, is (or should be) right' is an obnoxious shibboleth, connected in an uncomfortably close way with "might makes right." Still, a careful scrutiny of the various ways in which *is* is used seems to indicate that the separation simply is not perfect.

APPENDIX II

Economists Whose First Work or Works Deal with Methodological and/ or Philosophical Topics

Albert, Hans. 1954; 2d ed., 1972. *Ökonomische Ideologie und politische Theorie.*

Allais, Maurice. 1943. *A la recherche d'une discipline économique.*

Arrow, Kenneth. 1951. *Social Choice and Individual Values.*

Ayres, Clarence. 1938. *The Problem of Economic Order.*

Boland, Lawrence A. 1982. *The Foundations of Economic Methodology,* followed in 1986 by *Methodology for a New Microeconomics.*

Buchanan, James, and Gordon Tulloch. 1962. *The Calculus of Consent.*

Caldwell, Bruce J. 1982. *Beyond Positivism,* followed in 1984 by *Appraisal and Criticism in Economics: A Book of Readings.*

Clark, John Bates. 1886. *The Philosophy of Wealth.*

Hutchison, Terence W. 1938. *The Significance and Basic Postulates of Economic Theory.*

Keynes, John Neville. 1884. *Studies and Exercises in Formal Logic,* followed in 1891 by *The Scope and Method of Political Economy.*

Klein, Lawrence. 1947. *The Keynesian Revolution.*

Leijonhufvud, Axel. 1968. *On Keynesian Economics and the Economics of Keynes.*

Lipsey, Richard. 1963; 6th ed., 1983. *An Introduction to Positive Economics* (see the introduction).

Lowe, Adolph. 1915. *Economics and Sociology.*

McCullough, J. R. 1824. *A Discourse on the Rise, Progress, Peculiar Objects and Importance of Political Economy.*

Mill, J. S. 1836. *On the Definition of Political Economy; and on the Method of Investigation Proper to it,* followed in 1843 by *A System of Logic, Raciocinative and Inductive, Being a Connected View of the Principles of Evidence and the Methods of Scientific Investigation,* followed in 1844 by *Essays on Some Unsettled Questions on Political Economy.*

Morgenstern, Oskar. 1928. *Wirtschaftsprognose,* followed in 1934 by *Die Grenzen der Wirtschaftspolitik.*

Ramsey, Frank. 1931. *Foundations of Mathematics and Other Essays.*

Robbins, Lionel. 1932; 3d ed., 1983. *An Essay on the Nature and Significance of Economic Science.*

Roscher, W.G.F. 1843. *Grundriß zu Vorlesungen über die Staatswirtschaft nach geschichtlicher Methode.*

Samuelson, Paul. 1947. *Foundations of Economic Analysis* (see the introduction).

Schumpeter, Joseph A. 1908. *Das Wesen und der Hauptinhalt der theoretischen Nationalökonomie.*

Senior, Nassau. 1827. *Introductory Lecture on Political Economy,* followed in 1936 by *An Outline of the Science of Political Economy.*

Silberberg, Eugene. 1978. *The Structure of Economics: A Mathematical Analysis* (see the introduction).

Smith, Adam. ca. 1750. *The Principles Which Lead and Direct Philosophical Enquiries; Illustrated by the History of Astronomy.*

Steuart, James. 1767. *An Inquiry into the Principles of Political Economy.*

Tarascio, Vincent. 1968. *Pareto's Methodological Approach to Economics: A Study in the History of Some Scientific Aspects of Economic Thought.*

Veblen, Thorstein. 1899. *The Theory of the Leisure Class.*

Walras, Léon. 1860. *L'Economie politique et la justice: examen critique et réfutation des doctrines économique de M.P.-J. Proudhon.*

von Wieser, Friedrich. 1884. *Über den Ursprung und die Hauptgesetze des wirtschaftlichen Werthes* (see chapter 1).

BIBLIOGRAPHY

Abele, Hans. 1971. "From One-Dimensional to Multidimensional Economics: 'Paradigm' Lost." *Zeitschrift für Nationalökonomie* 31:45–62.

Achinstein, Peter. 1977. "History and Philosophy of Science: A Reply to Cohen." In Suppe, 1977b, pp. 350–60.

Achinstein, Peter, and Stephen F. Barker, eds. 1969. *The Legacy of Logical Positivism.* Baltimore, Md.: The John Hopkins University Press.

Ackermann, Robert. 1985. "Popper and German Social Philosophy." In *Popper and the Human Sciences,* edited by Gregory Currie and Alan Musgrave. Dordrecht, The Netherlands: Martinus Nijhoff.

———. 1983. "Methodology and Economics." *Philosophical Forum* 14, nos. 3–4 (Spring–Summer): 389–402.

———. 1976. *The Philosophy of Karl Popper.* Amherst: University of Massachusetts Press.

Addison, John T., John Burton, and Thomas S. Torrance. 1981. "Causation, Social Science and Sir John Hicks." *Oxford Economic Papers* 36, no. 1 (March): 1–11.

Adorno, Theodor W., et al. 1976. *The Positivist Dispute in German Sociology.* Translation of *Der Positivismusstreit in der deutschen Soziologie,* 1969, by Glyn Adey and David Frisby, New York: Harper and Row.

Aeppli, Roland. 1980. "Ökonomie als multiparadigmatische Wissenschaft." *Kyklos* 33, no. 4: 682–708.

Agassi, Joseph. 1986a. "Lakatos: An Exchange." *Philosophia* 16:209–38.

———. 1986b. "Refutation à la Popper: A Rejoinder." *Philosophia* 16:245–47.

———. 1979. "The Legacy of Lakatos." *Philosophy of the Social Sciences* 9, no. 3 (September): 316–26.

———. 1976. "The Lakatosian Revolution." In Cohen, Feyerabend, and Wartofsky, 1976, pp. 9–21.

———. 1975. "The Present State of the Philosophy of Science." *Philosophia* 15, no. 1: 5–20.

———. 1971a. "The Standard Misinterpretation of Skepticism." *Philosophical Studies* 22, no. 4 (June): 49–50.

———. 1971b. "Tautology and Testability in Economics." *Philosophy of the Social Sciences* 1:49–63.

———. 1967. "Science in Flux." In Cohen and Wartofsky, 1967, pp. 293–323.

————. 1963. *Towards an Historiography of Science. History and Theory,* Beiheft 2. Middletown, Conn.: Wesleyan University Press.

————. 1960. "Methodological Individualism." *British Journal of Sociology* 11:244–70.

Akerlof, George A. 1984. "A Theory of Social Custom, of Which Unemployment May Be One Consequence." In *An Economic Theorist's Book of Tales,* pp. 69–99. Cambridge: Cambridge University Press, 1984.

————. 1979. "The Case Against Conservative Macroeconomics: An Inaugural Lecture." *Economica* 46 (August): 219–37.

Albert, Hans. 1980. *Traktat über kritische Vernunft.* 4th rev. ed. Tübingen: J. B. C. Mohr.

————. 1979. "The Economic Tradition." In Brunner, 1979, pp. 1–27.

————. 1972a (1954). *Ökonomische Ideologie und politische Theorie. Das ökonomische Argument in der ordnungspolitischen Debatte.* 2d ed. Göttingen: Verlag Otto Schwartz.

————. 1972b. "Theorie und Prognose in den Sozialwissenschaften." In Topitsch, 1972, pp. 126–37.

————. 1972c. "Theorien in den Sozialwissenschaften." In his *Theorie und Realität,* pp. 3–25. 2d ed. Tübingen: J.B.C. Mohr, Paul Siebeck, 1972.

————. 1965. "Tradition und Kritik." *Club Voltaire: Jahrbuch für kritische Aufklärung II,* pp. 156–66. München: Szczesny Verlag.

————. 1964. "Die Idee der kritischen Vernunft." *Club Voltaire: Jahrbuch für kritische Aufklärung I,* pp. 17–30. München: Szczesny Verlag.

————. 1956. "Das Werturteilsproblem im Lichte der logischen Analyse." *Zeitschrift für die gesamte Staatswissenschaft* 112:410–39.

Alexander, Peter. 1967a. "Conventionalism." *Encyclopedia of Philosophy.* 1967 ed.

————. 1967b. "Duhem, Pierre Maurice Marie." *Encyclopedia of Philosophy.* 1967 ed.

————. 1967c. "Poincaré, Jules Henri." *Encyclopedia of Philosophy.* 1967 ed.

————. 1964. "The Philosophy of Science, 1850–1910." In *A Critical History of Western Philosophy,* edited by D. J. O'Connor, pp. 402–25. New York: The Free Press.

Allais, Maurice. 1968. "L'Economique en tant que science." *Revue d'économie politique* 78, no. 1 (January–February): 5–30.

Alter, Max. 1982. "Carl Menger and Homo Oeconomicus: Some Thoughts on Austrian Theory and Methodology." *Journal of Economics Issues* 16, no. 1 (March): 149–60.

Amsterdamski, Stephen. 1975. *Between Experience and Metaphysics.* Boston Studies in the Philosophy of Science, vol. 35. S. Cohen and Marx Wartofsky, Dordrecht, The Netherlands: D. Reidel.

Andersson, Gunnar. 1986. "Lakatos and Progress and Rationality in Science: A Reply to Agassi." *Philosophia* 16:239–43.

————, ed. 1984. *Rationality and Progress in Science and Politics.* Boston Studies in the Philosophy of Science, vol. 79. Dordrecht, The Netherlands: D. Reidel.

Archibald, G. C. 1979. "Method and Appraisal in Economics." *Philosophy of the Social Sciences* 9, no. 3 (September): 304–15.

————. 1967. "Refutation or Comparison?" *British Journal for the Philosophy of Science* 17, no. 4 (February): 279–96.

————. 1965. "The Qualitative Content of Maximizing Models." *Journal of Political Economy* 73 (February): 27–36.

————. 1959. "The State of Economic Science." *British Journal for the Philosophy of Science* 10, no. 37 (May): 58–69.

Armstrong, J. Scott. 1978. "Forecasting with Econometric Methods: Folklore Versus Fact." *Journal of Business* 51, no. 4: 549–64.

Arndt, Helmut. 1984. *Economic Theory v. Economic Reality.* Rev. English ed. of his 1979 volume. Translated by William A. Kirby. East Lansing: Michigan State University Press.

———. 1979. *Irrwege der politischen Ökonomie: Die Notwendigkeit einer wirtschaftstheoretischen Revolution.* München: C. H. Beck.

Arouh, Albert. 1987. "The Mumpsimus of Economists and the Role of Time and Uncertainty in the Progress of Economic Knowledge." *Journal of Post Keynesian Economics* 9, no. 3 (Spring): 395–423.

Arrow, Kenneth J. 1980. "Real and Nominal Values in Economics." *The Public Interest* (Special Issue: The Crisis in Economic Theory), pp. 139–50.

"The Art of Crunching Numbers." 1987. *The Economist,* 9 May, pp. 70–71.

Ash, Mitchell. 1980. "Academic Politics in the History of Science: Experimental Psychology in Germany, 1879–1941." *Central European History* 13, no. 3 (September): 255–86.

Ashby, W. R. 1964. "Logical Positivism." In *A Critical History of Western Philosophy,* edited by D. J. O'Connor, pp. 492–508. New York: The Free Press.

Ashley, W. J. 1893. "On the Study of Economic History." *Quarterly Journal of Economics* 7:115–36.

Asquith, Peter D., and Ronald N. Giere, eds. 1980. *PSA 1980,* vol. 1. East Lansing, Mich.: Philosophy of Science Association.

Asquith, Peter D., and P. Kitcher, eds. 1984. *PSA 1984,* vol. 2. East Lansing, Mich.: Philosophy of Science Association.

Asquith, Peter D., and Henry E. Kyburg, eds. 1979. *Current Research in Philosophy of Science.* East Lansing, Mich.: Philosophy of Science Association.

Axelsson, Runo. 1973. "The Economic Postulate of Rationality—Some Methodological Views." *Swedish Journal of Economics* 75 (September): 289–95.

Ayer, A. J., ed. 1959. *Logical Positivism.* New York: The Free Press.

Ayres, Charles E. 1935. "Moral Confusion in Economics." *Ethics* 45 (January): 170–99.

Baier, Kurt. 1977. "Rationality and Morality." *Erkenntnis* 11:197–223.

Baker, John R. 1979. "In the Cause of Freedom of Science." *New Scientist* 83, no. 1163 (12 July): 108–9.

Balogh, Thomas. 1982. *The Irrelevance of Conventional Economics.* London: Weidenfeld and Nicolson.

Baranzini, Mauro, and Roberto Scazzieri, eds. 1986a. *Foundations of Economics.* Oxford: Basil Blackwell.

———. 1986b. "Knowledge in Economics: A Framework." In their 1986a, pp. 2–87.

Barber, Bernard. 1968. "Science: The Sociology of Science." *International Encyclopedia of the Social Sciences.* 1968 ed.

Barber, William J. 1987. "Should the American Economic Association Have Toasted Simon Newcomb at Its 100th Birthday Party?" *Journal of Economic Perspectives* 1, no. 1 (Summer): 179–83.

Barnes, Barry. 1982. *T. S. Kuhn and Social Science.* New York: Columbia University Press.

Barnes, Barry, and David Bloor. 1982. "Relativism, Rationalism and the Sociology of Science." In *Rationality and Relativism,* edited by Martin Hollis and Steven Lukes, pp. 21–47. Oxford: Basil Blackwell, 1982.

Barnes, S. B. 1969. "Paradigms—Scientific and Social." *Man* 4, no. 1 (March): 94–102.

Barry, Norman P. 1981. "Re-stating the Liberal Order: Hayek's Philosophical Economics." In Shackleton and Locksley, 1981, pp. 87–107.

Bartley, William Warren, III. 1987a. "Alienation Alienated: The Economics of Knowledge Versus the Psychology and Sociology of Knowledge." In Radnitzky and Bartley, 1987, pp. 423–51.

———. 1987b. "A Refutation of the Alleged Refutation of Comprehensively Critical Rationalism." In Radnitzky and Bartley, 1987, pp. 314–41.

———. 1987c. "Theories of Rationality." In Radnitzky and Bartley, 1987, pp. 205–14.

———. 1984a. "Logical Strength and Demarcation." In Andersson, 1984, pp. 69–93.

———. 1984b (1962). *The Retreat to Commitment.* 2d rev., enl. ed. La Salle, Ill.: Open Court.

———. 1982a. "Critical Study: The Philosophy of Karl Popper Part III: Rationality, Criticism, and Logic." *Philosophia* 11, no. 1 (February): 121–221.

———. 1982b. "A Popperian Harvest." In *In Pursuit of Truth,* edited by Paul Levinson, pp. 249–89. Atlantic Highlands, N.J.: Humanities Press, 1982.

———. 1980. "Ein schwieriger Mensch: eine Porträtskizze von Sir Karl Popper." In *Physiognomien: Philosophen des 20. Jahrhunderts in Portraits,* edited by Eckhard Nordhofen, pp. 45–69. Königstein: Athenäum, 1980.

———. 1977. "Critical Study: The Philosophy of Karl Popper Part II: Consciousness and Physics." *Philosophia* 7, no. 2 (March): 675–716.

———. 1976a. "Critical Study: The Philosophy of Karl Popper Part I: Biology and Evolutionary Epistemology." *Philosophia* 6, nos. 3–4 (September–December): 463–94.

———. 1976b. "On Imre Lakatos." In Cohen, Feyerabend, and Wartofsky, 1976, pp. 37–38.

———. 1975. "Wissenschaft und Glaube: Die Notwendigkeit des Engagements." In *Neue Anthropologie,* vol. 7. Edited by Hans-Georg Godamer and Paul Vogler, pp. 64–102. Stuttgart: n.p., 1975.

———. 1974. "Theory of Language and Philosophy of Science as Instruments of Educational Reform: Wittgenstein and Popper as Austrian Schoolteachers." In Cohen and Wartofsky, 1974, pp. 307–37.

———. 1970. "Die österreichische Schulreform als die Wiege der modernen Philosophie." *Club Voltaire: Jahrbuch für kritische Aufklärung IV,* pp. 349–66. Reinbeck bei Hamburg: Rowohlt Verlag.

———. 1969a. "Approaches to Science and Skepticism." *Philosophical Forum* 1, no. 3 (Spring): 318–31.

———. 1969b. "Sprach- und Wissenschaftstheorie als Werkzeuge einer Schulreform: Wittgenstein und Popper als österreichische Schullehrer." *Conceptus* 3, no. 1: 6–22.

———. 1968. "Theories of Demarcation Between Science and Metaphysics." In Lakatos and Musgrave, 1968, pp. 40–64.

———. 1965. "Das Haus der Wissenschaft." In *Club Voltaire: Jahrbuch für kritische Aufklärung II,* pp. 118–34. München: Szczesny Verlag.

———. 1964. "Rationality versus the Theory of Rationality." In *The Critical Approach to Science and Philosophy,* edited by Mario Bunge, pp. 3–31. Glencoe, Ill. The Free Press.

———. 1962. "Achilles, the Tortoise, and Explanation in Science and History." *British Journal for the Philosophy of Science* 13 (May 1962–February 1963): 15–33.

Barucci, Piero. 1983. "The Scope and Method of Political Economy in the First Histories of Economics." In Coats, 1983c, pp. 125–36.

Basmann, R. L. 1975. "Modern Logic and the Suppositious Weakness of the Empirical Foundations of Economic Science." *Schweizerische Zeitschrift für Volkswirtschaft und Statistik* 111, no. 2 (April): 153–76.

Bauer, P. T., and A. A. Walters. 1975. "The State of Economics." *Journal of Law and Economics* 8 (April): 1–23.

Baumberger, Jörg. 1977. "No Kuhnian Revolutions in Economics." *Journal of Economic Issues* 11 (March): 1–20. Repr. in Samuels, 1980, pp. 322–41.

Baumol, William J. 1985. "On Method in U.S. Economics a Century Earlier." *American Economic Review: Special Issue* 75, no. 6 (December): 1–12.

Bear, D.V.T., and Daniel Orr. 1967. "Logic and Expediency in Economic Theorizing." *Journal of Political Economy* 75, no. 2 (April): 188–96.

Becker, Gary. 1962. "Irrational Behavior and Economic Theory." *Journal of Political Economy* 70, no. 1 (February): 1–13.

Bell, Daniel. 1980. "Models and Reality in Economic Discourse." *The Public Interest* (Special Issue: The Crisis in Economic Theory), pp. 46–80.

Ben-David, Joseph. 1964. "Scientific Growth: A Sociological View." *Minerva* 2 no. 2 (Summer): 455–76.

Benn, S. I., and G. W. Mortimore, eds. 1976. *Rationality and the Social Sciences.* London: Routledge and Kegan Paul.

Bennett, James T., and Manuel H. Johnson. 1979. "Mathematics and Quantification in Economic Literature: Paradigm or Paradox?" *Journal of Economic Education* (Fall): 40–42.

Bensusan-Butt, D. M. 1980a. "Keynes's General Theory: Then and Now." In his 1980b, pp. 25–40.

———. 1980b. *On Economic Knowledge: A Sceptical Miscellany.* Canberra: Australian National University.

———. 1980c. "Thoughts on a Horrible Remark." In his 1980b, pp. 11–23.

Berkson, William, and John Wettersten. 1982. *Lernen aus dem Irrtum: die Bedeutung von Karl Poppers Lerntheorie für die Psychologie und Philosophie der Wissenschaft.* Hamburg: Hoffman und Campe. Published in English as *Learning from Error.* La Salle, Ill.: Open Court, 1984.

Berlin, Sir Isaiah. 1978. *Concepts and Categories.* London: The Hogarth Press.

———. 1966. "The Concept of Scientific History." In Dray, 1966, pp. 5–53.

Bernadelli, Harro, 1961. "The Origins of Modern Economic Theory." *Economic Record* 37 (September): 320–38.

Bhaskar, Roy. 1975. "Feyerabend and Bachelard: Two Philosophies of Science." *New Left Review* 94 (November–December): 31–55.

Black, Max. 1967. "Induction." *Encyclopedia of Philosophy.* 1967 ed.

Black, R. D. Collison. 1983. "The Present Position and Prospects of Political Economy." In Coats, 1983c, pp. 55–70.

Black, R. D. Collison, A. W. Coats, and Craufurd D. W. Goodwin, eds. 1973. *The Marginal Revolution in Economics: Interpretation and Evaluation.* Papers presented at a conference held at the Villa Serbelloni, Bellagio, Italy, August 22–28, 1971. Durham: Duke University Press.

Blackman, James H. 1971. "The Outlook for Economics." *Southern Economic Journal* 37, no. 4 (April): 385–95.

Blair, Douglas H., and Robert A. Pollack. 1983. "Rational Collective Choice." *Scientific American* 249, no. 2 (August): 76–83.

Blaug, Mark. 1985. "Comment on D. Wade Hands, 'Karl Popper and Economic Methodology: A New Look.' " *Economic and Philosophy* 1:286–88.

———. 1983. "A Methodological Appraisal of Radical Economics." In Coats, 1983c, pp. 211–46.

———. 1980a. *A Methodological Appraisal of Marxian Economics.* Amsterdam: North-Holland.

———. 1980b. *The Methodology of Economics or How Economists Explain.* Cambridge: Cambridge University Press.

———. 1978. *Economic Theory in Retrospect,* 3d ed. Cambridge: Cambridge University Press.

———. 1976. "Kuhn Versus Lakatos *Or* Paradigms Versus Research Programmes in the History of Economics." In Latsis, 1976a, pp. 149–80. Repr. from *History of Political Economy* 7, no. 4 (Winter 1975): 399–433. Repr. in Gutting, 1980, pp. 137–59.

———. 1973. "Was There a Marginal Revolution?" In Black, Coats, and Goodwin, 1973, pp. 3–14.

Blegvad, Mogens. 1977. "Competing and Complementary Patterns of Explanation in Social Science." In Butts and Hintikka, 1977, pp. 127–58.

Bliss, Christopher. 1986. "Progress and Anti-Progress in Economic Science." In Baranzini and Scazzieri, 1986a, pp. 363–76.

Bloor, David. 1974. "Popper's Mystification of Objective Knowledge." *Science Studies* 4, no. 1 (January): 65–76.

Böhler, Dietrich. 1972. "Paradigmawechsel in analytischer Wissenschaftstheorie." *Zeitschrift für allgemeine Wissenschaftstheorie* 3, no. 2: 219–42.

Bohm, David. 1961. "On the Relationship Between the Methodology in Scientific Research and the Content of Scientific Knowledge." *British Journal for the Philosophy of Science* 12, no. 46 (August): 103–16.

Boland, Lawrence A. 1986. *Methodology for a New Microeconomics.* Boston: Allen and Unwin.

———. 1985. "Reflections on Blaug's *Methodology of Economics:* Suggestions for a Revised Edition." *Eastern Economic Journal* 11, no. 4 (October–December): 450–54.

———. 1983a. "The Neoclassical Maximization Hypothesis: Reply." *American Economic Review* 73, no. 4 (September): 828–30.

———. 1983b. "On the Best Strategy for Doing Philosophy of Economics." *British Journal for the Philosophy of Science* 34 (December): 387–92.

———. 1982. *The Foundations of Economic Method.* London: Allen and Unwin.

———. 1981a. "On the Futility of Criticizing the Neoclassical Maximization Hypothesis." *American Economic Review* 71, no. 5 (December): 1031–36.

———. 1981b. "Satisficing in Methodology: A Reply to Fels." *Journal of Economic Literature* 19, no. 1 (March): 84–86.

———. 1980. "Friedman's Methodology vs. Conventional Empiricism: A Reply to Rotwein." *Journal of Economic Literature* 18 (December): 1555–57.

———. 1979a. "A Critique of Friedman's Critics." *Journal of Economic Literature* 17, no. 2 (June): 503–22.

———. 1979b. "Knowledge and the Role of Institutions in Economic Theory." *Journal of Economic Issues* 13 (December): 957–72.

———. 1978. "Time in Economics vs. Economics in Time: The 'Hayek Problem.' " *Canadian Journal of Economics* 11, no. 2 (May): 240–62.

———. 1977a. "Model Specification and Stochasticism in Economic Methodology." *South African Journal of Economics* 45 (June): 182–89.

———. 1977b. "Testability in Economic Science." *South African Journal of Economics* 45 (March): 93–105.

————. 1971. "Discussion: Methodology as an Exercise in Economic Analysis." *Philosophy of Science* 38 (March): 105–17.

————. 1970a. "Axiomatic Analysis and Economic Understanding." *Australian Economic Papers* 9, no. 14 (June): 62–75.

————. 1970b. "Conventionalism and Economic Theory." *Philosophy of Science* 37, no. 2 (June): 239–48.

————. 1969. "Economic Understanding and Understanding Economics." *South African Journal of Economics* 37 (June): 144–60.

————. 1968. "The Identification Problem and the Validity of Economic Models." *South African Journal of Economics* 36 (September): 236–40.

Bostaph, Samuel. 1978. "The Methodological Debate Between Carl Menger and the German Historicists." *Atlantic Economic Journal* 6, no. 3 (September): 3–16.

Boulding, Kenneth E. 1986. "What Went Wrong with Economics?" *American Economist* 30, no. 1 (Spring): 5–12.

————. 1978. "Do the Values of Science Lead to a Science of Value?" *Social Science Quarterly* 11 (March): 548–50.

————. 1971a. "After Samuelson, Who Needs Adam Smith?" *History of Political Economy* 3, no. 2 (Fall): 225–37.

————. 1971b. "The Misallocation of Intellectual Resources in Economics." In *The Use and Abuse of Social Science,* edited by Irving Louis Horowitz. New Brunswick, N.J.: Transaction Books, pp. 34–51.

————. 1970. *Economics as a Science.* New York: McGraw-Hill.

————. 1969. "Economics as a Moral Science." *American Economic Review* 59, no. 1 (March): 1–12.

————. 1966. "The Economics of Knowledge and the Knowledge of Economics." *American Economic Review: Supplement* 56, no. 2 (May): 1–13.

Bowley, Marion. 1972. "The Predecessors of Jevons–The Revolution That Wasn't." *Manchester School of Economic and Social Studies* 40, no. 1 (March): 9–29.

Braun, Gunther E. 1977. "Von Popper zu Lakatos: das Abgrenzungsproblem zwischen Wissenschaft und Pseudo-Wissenschaft." *Conceptus* 11:217–42.

Bray, Jeremy. 1977. "The Logic of Scientific Method in Economics." *Journal of Economic Studies* 4, no. 1 (May): 1–28.

Brems, Hans. 1975. "Marshall on Mathematics." *Journal of Law and Economics* 18, no. 2 (October): 583–85.

Brennan, Timothy. 1979. "Explanation and Value in Economics." *Journal of Economic Issues* 13, no. 4 (December): 911–32.

Brinkmann, Carl. 1956. "Historische Schule." *Handwörterbuch der Sozialwissenschaften,* vol. 5. 1956 ed.

Broad, William J. 1979. "Paul Feyerabend: Science and the Anarchist." *Science* 209, no. 4418 (2 November): 534–37.

Brodbeck, May. 1966. "Methodological Individualism: Definition and Reduction." In Dray, 1966, pp. 297–329.

Bronfenbrenner, Martin. 1971. "The 'Structure of Revolutions' in Economic Thought." *History of Political Economy* 3, no. 1 (Spring): 136–51.

————. 1966. "Trends, Cycles, and Fads in Economic Writing." *American Economic Review: Supplement* 56, no. 2 (May): 538–52.

Brooks, Harvey. 1984. "Sponsorship and Social Science Research." *Society* 21, no. 4 (May/June): 81–83.

Brooks, John Graham. 1893. "Philosophy and Political Economy." *Quarterly Journal of Economics* 8 (October): 93–97.

Brown, E. Cary, and Robert M. Solow, eds. 1983. *Paul Samuelson and Modern Economic Theory.* New York: McGraw-Hill.

Brown, Elba K. 1981. "The Neoclassical and Post-Keynesian Research Programmes: The Methodological Issues." *Review of Social Economy* 39, no. 2 (October): 111–32.

Brown, Harold I. 1983. "Incommensurability." *Inquiry* 26, no. 1 (March): 3–30.

———. 1977. *Perception, Theory and Commitment: The New Philosophy of Science.* Chicago: University of Chicago Press.

Brown, James Robert. 1980. "History and the Norms of Science." In Asquith and Giere, 1980, pp. 236–48.

Brunner, Karl, ed. 1979. *Economics and Social Institutions: Insights from the Conferences on Analysis and Ideology.* Boston: Martinus Nijhoff.

———. 1969. " 'Assumptions' and the Cognitive Quality of Theories." *Synthese* 20, no. 4 (December): 501–25.

Brush, Stephen G. 1974. "Should the History of Science Be Rated X?" *Science* 183, no. 4130 (22 April): 1164–72.

Buchanan, James M. 1979. *What Should Economists Do?* Indianapolis: Liberty Press.

———. 1970. "Is Economics the Science of Choice?" In Streissler, 1970 [1969], pp. 47–64.

Buck, Roger C., and Robert S. Cohen, eds. 1971. *PSA 1970 in Memory of Rudolf Carnap.* Boston Studies in the Philosophy of Science, vol. 8. Dordrecht, The Netherlands: D. Reidel.

Burian, Richard M. 1977. "More than a Marriage of Convenience: On the Inextricability of History and Philosophy of Science." *Philosophy of Science* 44, no. 1 (March): 1–42.

Butts, Robert E., and Jaako Hintikka, eds. 1977. *Historical and Philosophical Dimensions of Logic, Methodology, and Philosophy of Science.* University of Western Ontario Series in Philosophy of Science, vol. 12. Dordrecht, The Netherlands: D. Reidel.

Cairnes, J. E. 1965 (1888). *The Character and Logical Method of Political Economy.* 2d ed. London: Macmillan.

Caldwell, Bruce J., ed. 1984a. *Appraisal and Criticism in Economics: A Book of Readings.* Boston: Allen and Unwin.

———. 1984b. "Praxeology and Its Critics: An Appraisal." *History of Political Economy* 16, no. 3 (Fall): 363–79.

———. 1984c. "Some Problems with Falsification in Economics." *Philosophy of the Social Sciences* 14:489–95.

———. 1983. "The Neoclassical Maximization Hypothesis: Comment." *American Economic Review* 73 (September): 824–27.

———. 1982. *Beyond Positivism.* London: Allen and Unwin.

———. 1980a. "A Critique of Friedman's Methodological Instrumentalism." *Southern Economic Journal* 47, no. 2 (October): 366–74.

———. 1980b. "Positivist Philosophy of Science and the Methodology of Economics." *Journal of Economics* 14 (March): 53–76.

———. 1979a. "Positivism and the Methodology of Economics: The End of an Epoch." Working Paper Series no. 790301. Center for Applied Research, University of North Carolina, Greensboro, March.

———. 1979b. "Two Suggestions for the Improvement of Methodological Work in Economics." *American Economist* 23 (Fall): 56–61.

Caldwell, Bruce, and A. W. Coats. 1984. "The Rhetoric of Economists: A Comment on McCloskey." *Journal of Economic Literature* 22 (June): 575–78.

Canterbery, E. Ray. 1980. *The Making of Economics.* 2d ed. Belmont, Calif.: Wadsworth.

Canterbery, E. Ray, and Robert J. Burkhardt. 1983. "What Do We Mean by Asking Whether Economics Is a Science?" In Eichner, 1983, pp. 15–40.

Carnap, Rudolf. 1966. *The Philosophy of Science.* Edited by Martin Gardner. New York: Basic Books.

———. 1963. "Replies and Systematic Expositions." In Schilpp, 1963, pp. 859–1013.

Cartwright, Nancy, 1983. *How the Laws of Physics Lie.* Oxford: Oxford University Press.

Caws, Peter. 1967. "Scientific Method." *Encyclopedia of Philosophy.* 1967 ed.

Cesarano, Filippo. 1983. "On the Role of the History of Economic Analysis." *History of Political Economy* 15, no. 1 (Spring): 63–82.

Chalk, Alfred. 1970. "Concepts of Change and the Role of Predictability in Economics." *History of Political Economy* 2:97–117.

Chalmers, A. F. 1982. *What Is This Thing Called Science?.* 2d ed. St. Lucia: University of Queensland Press.

———. 1973. "On Learning from Our Mistakes." *British Journal for the Philosophy of Science* 24, no. 2 (June): 164–73.

Chase, Richard X. 1983. "Adolf Lowe's Paradigm Shift for a Scientific Economics: An Interpretative Perspective." *American Journal of Economics and Sociology* 42, no. 2 (April): 167–77.

Checkland, S. G. 1951. "The Advent of Academic Economics in England." *Manchester School of Economic and Social Studies* 19, no. 1 (January): 43–70.

Cheung, Steven N. S. 1973. "The Fable of the Bees: An Economic Investigation." *Journal of Law and Economics* 16, no. 1 (April): 11–33.

Clapham, J. H. 1922. "Of Empty Economic Boxes." *Economic Journal* 32 (September): 305–14.

Clark, A. F. 1977. "Testability in Economic Science: A Comment." *South African Journal of Economics* 45: 106–7.

Clower, Robert. 1975. "Reflections on the Keynesian Perplex." *Zeitschrift für Nationalökonomie* 35:1–24.

———. 1974. "Reflections on Science . . . and Economics." *Intermountain Economic Review* 5:1–12.

Clower, Robert, and Axel Leijonhufvud. 1973. "Say's Principle, What It Means and Doesn't Mean: Part I." *Intermountain Economic Review* 4, no. 2 (Fall): 1–16. (Part II never appeared.)

Coase, R. H. 1975. "Marshall on Method." *Journal of Law and Economics* 18 (April): 1–23.

Coats, A. W. 1983a. "The First Decade of HOPE (1969–79)." *History of Political Economy* 15, no. 3 (Fall): 303–19.

———. 1983b. "Half a Century of Methodological Controversy in Economics: As Reflected in the Writings of T. W. Hutchison." In Coats, 1983c, pp. 1–42.

———, ed. 1983c. *Methodological Controversy in Economics: Historical Essays in Honor of T. W. Hutchison.* Greenwich, Conn.: Jai Press.

———. 1983d. "The Revival of Subjectivism in Economics." In Wiseman, 1983a, pp. 87–103.

———. 1982. "The Methodology of Economics: Some Recent Contributions." *Kyklos* 35:310–21.

———. 1980a. "The Culture and the Economists: Some Reflections on Anglo-American Differences." *History of Political Economy* 12, no. 4 (Winter): 588–609.

———. 1980b. "The Historical Context of the 'New' Economic History." *Journal of Economic History* 9, no. 1 (Spring): 185–207.

———. 1978a. "Methodology in Economics: A Subordinate Theme in Machlup's Writings." In Dreyer, 1978, pp. 23–35.

———. 1978b. "Reflections on the Role of the History of Economics in the Training of Economics." In *Studi in Memoria di Federigo Melis* (editor unknown), vol. 5. Rome: Giannini.

———. 1977. "The Current 'Crisis' in Economics in Historical Perspective." *Nebraska Journal of Economics and Business* 16, no. 3 (Summer): 3–16.

———. 1976. "Economics and Psychology: The Death and Resurrection of a Research Programme." In Latsis, 1976a, pp. 43–64.

———. 1975. "The Development of the Economics Profession in England." In *International Congress of Economic History and History of Economic Theories in Piraeus,* edited by L. Th. Houmanidis, pp. 277–90. Piraeus: The Piraeus School of Industrial Studies, 1975.

———. 1972. "Situational Determinism in Economics: The Implications of Latsis' Argument for the Historian of Economics." *British Journal for the Philosophy of Science* 23 (1972): 285–88.

———. 1971. "The Role of Scholarly Journals in the History of Economics: An Essay." *Journal of Economic Literature* 9:29–44.

———. 1969. "Is There a 'Structure of Scientific Revolutions' in Economics?" *Kyklos* 22, no. 2:289–96.

———. 1968. "The Origins and Early Development of the Royal Economic Society." *Economic Journal* 78, no. 310 (June): 349–71.

———. 1967a. "Alfred Marshall and the Early Development of the London School of Economics: Some Unpublished Letters." *Economica* 34, no. 136 (November): 408–17.

———. 1967b. "Sociological Aspects of British Economic Thought." *Journal of Political Economy* 75, no. 5 (October): 706–29.

———. 1964. "The American Economic Association, 1904–29." *American Economic Review* 54, no. 4 (June): 261–85.

———. 1963. "The Origins of the 'Chicago School(s)'?" *Journal of Political Economy* 71 (October): 487–93.

———. 1961. "Alfred Marshall and Richard T. Ely: Some Unpublished Letters." *Economica* 28 n.s. (May): 191–94.

———. 1960. "The First Two Decades of the American Economic Association." *American Economic Review* 50, no. 4 (September): 555–74.

Coats, A. W., and S. E. Coats. 1973. "The Changing Social Composition of the Royal Economic Society 1890–1960 and the Professionalization of British Economics." *British Journal of Sociology* 24, no. 2 (June): 165–87.

Coddington, Alan. 1983. "Economists and Policy." In Marr and Raj, 1983, pp. 229–38.

———. 1979. "Friedman's Contribution to Methodological Controversy." *British Review of Economic Issues* 2, no. 4 (May): 1–13.

———. 1972. "Positive Economics." *Canadian Journal of Economics* 5, no. 1 (February): 1–15.

Cohen, I. Bernard. 1985. *Revolution in Science.* Cambridge: Harvard University Press, Belknap Press.

——. 1977. "History and the Philosophy of Science." In Suppe, 1977b, pp. 308–49.

——. 1976. "The Eighteenth-Century Origins of the Concept of Scientific Revolution." *Journal of the History of Ideas* 37:257–88.

Cohen, Robert S., Paul K. Feyerabend, and Marx W. Wartofsky, eds. 1976. *Essays in Memory of Imre Lakatos*. Boston Studies in the Philosophy of Science, vol. 39. Dordrecht, The Netherlands, D. Reidel.

Cohen, Robert S., and Marx W. Wartofsky, eds. 1983. *Epistemology, Methodology, and the Social Sciences*. Boston Studies in the Philosophy of Science, vol. 71. Dordrecht, The Netherlands: D. Reidel.

——, eds. 1974. *Methodological and Historical Essays in the Natural and Social Sciences*. Boston Studies in the Philosophy of Science, vol. 14. Dordrecht, The Netherlands: D. Reidel.

——, eds. 1967. *In Memory of Norwood Russell Hanson*. Boston Studies in the Philosophy of Science, vol. 3. Dordrecht, The Netherlands: D. Reidel.

Colander, David, and Arjo Klamer. 1987. "The Making of an Economist." *Journal of Economic Perspectives* 1, no. 2 (Fall): 95–111.

Collard, D. A. 1964. "Swans, Falling Bodies, and Five-Legged Dogs." *Quarterly Journal of Economics* 78 (November): 645–46.

Colodny, Robert G., ed. 1965. *Beyond the Edge of Certainty*. University of Pittsburgh Series in the Philosophy of Science, vol. 2. Englewood Cliffs, N.J.: Prentice-Hall.

Corry, Bernard A. 1975. "Should Economists Abandon HOPE?" *History of Political Economy* 7, no. 2: 252–60.

Coser, Lewis A. 1984. *Refugee Scholars in America: Their Impact and Their Experiences*. New Haven: Yale University Press.

Craver, Earlene, and Axel Leijonhufvud. 1987. "Economics in America: The Continental Influence." *History of Political Economy* 19, no. 2: 173–82.

Crittenden, P. J. 1983. "Anarchistic Epistemology and Education." *Methodology and Science* 16:211–29.

Cross, Rod. 1984. "Monetarism and Duhem Thesis." In Wiles and Routh, 1984, pp. 78–99. With comments by Brian McCormick (pp. 100–4) and Andrew Hartropp and David Heathfield (pp. 105–10), and Cross's reply (pp. 111–20).

——. 1982. "The Duhem-Quine Thesis, Lakatos and the Appraisal of Theories." *Economic Record* 92 (June): 320–40.

Dacey, Raymond. 1981. "Some Implications of 'Theory Absorption' for Economic Theory and the Economics of Information." In Pitt, 1981, pp. 111–36.

——. 1976. "Theory Absorption and the Testability of Economic Theory." *Zeitschrift für Nationalökonomie* 36:247–67.

Dancy, R. M. 1978. "Model Behavior." *Journal of Philosophy* 75, no. 11 (November): 677–79.

Danto, Arthur C. 1967. "Philosophy of Science, Problems of." *Encyclopedia of Philosophy*. 1967 ed.

——. 1964. "The Historical Individual." In Dray, 1966, pp. 265–96.

Dasgupta, A. K. 1985. *Epochs of Economic Theory*. Oxford: Basil Blackwell.

Davidson, Paul. 1980. "Post Keynesian Economics." *The Public Interest* (Special Issue: The Crisis in Economic Theory), pp. 151–73.

de Alessi, Louis. 1965. "Economic Theory as a Language." *Quarterly Journal of Economics* 79, no. 3 (August): 472–77.

——. 1971. "Reversals of Assumptions and Implications." *Journal of Political Economy* 79, no. 4 (August): 867–77.

Dean, James W. 1980. "The Dissolution of the Keynesian Consensus." *The Public Interest* (Special Issue: The Crisis in Economic Theory), pp. 19–34.

Deane, Phyllis. 1983. "The Scope and Method of Economic Science." *Economic Journal* 93, no. 369 (March): 1–12.

de Marchi, Neil, ed. 1988. *The Popperian Legacy in Economics.* Papers presented at a symposium in Amsterdam, December 1985. Cambridge: Cambridge University Press.

———. 1983. "The Case for James Mill." In Coats, 1983c, pp. 155–84.

———. 1976. "Anomaly and the Development of Economics: The Case of the Leontief Paradox." In Latsis, 1976a, pp. 109–27.

———. 1973. "The Noxious Influence: A Correction of Jevons' Charge." *Journal of Law and Economics* 16, no. 1 (April): 179–90.

de Marchi, Neil, and John Lodewijks. 1983. "HOPE and the Journal Literature in the History of Economic Thought." *History of Political Economy* 15, no. 3 (Fall): 321–43.

de Vries, G. J. 1952. *Antistenes Redivivus: Popper's Attack on Plato.* Amsterdam: North-Holland.

Dewey, John. 1939. *Theory of Valuation,* Foundations of the Unity of Science, vol. 2, no. 4. Chicago: University of Chicago Press.

Diederich, Werner. 1986. "What Is Revolutionary about the Copernican Revolution?" In *Essays on Creativity and Science.* Papers delivered at a conference, Honolulu, Hawaii, 23–24 March, 1985, pp. 31–42. Hawaii Council of Teachers of English, 1986.

———, ed. 1974. *Theorien der Wissenschaftsgeschichte: Beiträge zur diachronischen Wissenschaftstheorie.* Frankfurt am Main: Suhrkamp.

Dillard, Dudley. 1978. "Revolutions in Economic Theory." *Southern Economic Journal* 44, no. 4 (April): 705–24.

Dolan, Edwin G. 1976a. "Austrian Economics as Extraordinary Science." In his 1976b, pp. 3–15.

———, ed. 1976b. *The Foundations of Modern Austrian Economics.* Kansas City, Kans.: Sheed and Ward.

Donagan, Alan. 1974. "Popper's Examination of Historicism." In Schilpp, 1974, pp. 905–24.

Dorfman, Joseph. 1969a (1949). *The Economic Mind in American Civilization: 1865– 1918,* vol. 3. New York: Kelley.

———. 1969b (1959). *The Economic Mind in American Civilization: 1918–1933,* vols. 4 and 5. New York: Kelley.

Dow, Sheila C. 1982–1983. "Substantive Mountains and Methodological Molehills: A Rejoinder." *Journal of Post Keynesian Economics* 5 (Winter): 304–8.

———. 1982. "Neoclassical Tautologies and the Cambridge Controversies: Reply to Salanti." *Journal of Post Keynesian Economics* 5, no. 1 (Fall): 132–34.

———. 1981. "Weintraub and Wiles: the Methodological Basis of Policy Conflict." *Journal of Post Keynesian Economics* 3, no. 3 (Spring): 325–39.

———. 1980. "Methodological Morality in the Cambridge Controversies." *Journal of Post Keynesian Economics* 2, no. 3 (Spring): 368–80.

Drake, S. 1967. "Galileo Galilei." *Encyclopedia of Philosophy.* 1967 ed.

Dray, William H., ed. 1966. *Philosophical Analysis and History.* Englewood Cliffs, N.J.: Prentice-Hall.

———. 1964. *The Philosophy of History.* Englewood Cliffs, N.J.: Prentice-Hall.

Dreyer, Jacob S., ed. 1978. *Breadth and Depth in Economics: Fritz Machlup—the Man and His Ideas*. Lexington, Mass.: Lexington Books.

Drucker, Peter F. 1980. "Toward the Next Economics." *The Public Interest* (Special Issue: The Crisis in Economic Theory), pp. 4–18.

Duerr, Hans Peter, ed. 1980, 1981. *Versuchungen, Aufsätze zur Philosophie Paul Feyerabends*. 2 vols. Frankfurt am Main: Suhrkamp.

Dugger, William M. 1983. "Two Twists in Economic Methodology: Positivism and Subjectivism." *American Journal of Economics and Sociology* 42, no. 1 (January): 75–91.

———. 1979. "Methodological Differences Between Institutional and Neoclassical Economics." *Journal of Economic Issues* 13 (December): 899–909.

Duhem, Pierre. 1962. *The Aim and Structure of Physical Theory*. Translated by Philip P. Wiener. New York: Atheneum.

Dunbar, Charles. 1891. "The Academic Study of Political Economy." *Quarterly Journal of Economics* (July): 397–416.

Dwyer, Larry. 1983. " 'Value Freedom' and the Scope of Economic Inquiry, II: The Fact/Value Continuum and the Basis for Scientific and Humanistic Policy." *American Journal of Economics and Sociology* 42, no. 3 (July): 353–68.

———. 1982a. "The Alleged Value Neutrality of Economics: An Alternative View." *Journal of Economic Issues* 16 (March): 75–106.

———. 1982b. " 'Value Freedom' and the Scope of Economic Inquiry, I: Positivism's Standard View and the Political Economists." *American Journal of Economics and Sociology* 41, no. 2 (April): 159–68.

Dyke, Charles. 1981. *Philosophy of Economics*. Englewood Cliffs, N. J.: Prentice-Hall.

Dziech, Billie Wright, and Linda Weiner. 1984. *The Lecherous Professor: Sexual Harrassment on Campus*. Boston: Beacon Press.

Eagley, Robert V. 1974. "Contemporary Profile of Conventional Economists." *History of Political Economy* 6, no. 1: 76–91.

———, ed. 1968. *Events, Ideology and Economic Theory*. Detroit: Wayne State University Press.

Earl, Peter. 1983. "The Consumer in His/Her Social Setting—Subjectivist View." In Wiseman, 1983a, pp. 176–91.

Eichner, Alfred S., ed. 1983. *Why Economics Is Not Yet a Science*. London: Macmillan.

Eichner, Alfred S., and J. A. Kregel. 1975. "An Essay on Post-Keynesian Theory: A New Paradigm in Economics." *Journal of Economic Literature* 13, no. 4 (December): 1293–1314.

Eisermann, Gottfried. 1956. *Die Grundlagen des Historismus in der deutschen Nationalökonomie*. Stuttgart: Ferdinand Enke Verlag.

Eisner, Robert. 1978. "Machlup on Academic Freedom." In Dreyer, 1978, pp. 3–12.

Ekelund, Robert B., Jr., and Emile S. Olsen. 1973. "Comte, Mill, Cairnes: The Positivist–Empiricist Interlude in Late Classical Economics." *Journal of Economic Issues* 7, no. 3 (September): 383–416.

Eysenck, H. J. 1979. "Astrology—Science Or Superstition?" *Encounter* 53, no. 6 (December): 85–90.

Faber, Karl-Georg. 1977. "Review of Wehler's *Geschichte und Soziologie, Geschichte und Ökonomie;* Ludz's *Soziologie und Sozialgeschichte;* Schulz's *Geschichte Heute.*" *History and Theory* 16:51–66.

Fabian, Robert G. 1971. "Cultural Influences on Economic Theory." *Journal of Economic Issues* 5, no. 3 (September): 46–59.

Fallon, Daniel. 1980. *The German University: A Heroic Ideal in Conflict with the Modern World.* Boulder: Colorado Associated University Press.

Farr, James. 1983. "Popper's Hermeneutics." *Philosophy of the Social Sciences* 13:157–76.

Fearnside, W. Ward, and William B. Holther. 1959. *Fallacy: The Counterfeit of Argument.* Englewood Cliffs, N. J.: Prentice-Hall.

Feige, Edgar L. 1975. "The Consequences of Journal Editorial Policies and a Suggestion for Revision" (with "Editorial Comment"). *Journal of Political Economy* 83, no. 6: 1291–96.

Feigl, Herbert. 1981a. *Inquiries and Provocations: Selected Writings, 1929–74,* Vienna Circle Collection, edited by Robert S. Cohen, vol. 14. Dordrecht, The Netherlands: D. Reidel.

———. 1981b. "The Origin and Spirit of Logical Positivism." In his 1981a, pp. 21–37.

———. 1981c. "The Power of Positivistic Thinking." In his 1981a, pp. 38–55.

———. 1981d. "The *Wiener Kreis* in America." In his 1981a, pp. 57–94.

———. 1971. "Research Programmes and Induction." In Buck and Cohen, 1971, pp. 147–82.

———. 1970. "The 'Orthodox' View of Theories: Remarks in Defense as Well as Critique." In Radner and Winokur, 1970, pp. 3–16.

Feigl, Herbert, and Grover Maxwell, eds. 1962. *Scientific Explanation, Space and Time,* Minnesota Studies in the Philosophy of Science, vol. 3. Minneapolis: University of Minnesota Press.

———, eds. 1961. *Current Issues in the Philosophy of Science.* New York: Holt, Rinehart and Winston.

Fels, Rendigs. 1981. "Boland Ignores Simon: A Comment." *Journal of Economic Literature* 19 (March): 83–84.

Fetter, Frank W. 1965. "The Relation of the History of Economic Thought to Economic History." *American Economic Review: Supplement* 55, no. 2 (May): 136–42; discussion, 143–49.

Feyerabend, Paul K. 1983 (1976). *Wider den Methodenzwang. Skizze einer anarchistischen Erkenntnistheorie* (teils gekürzt, teils ergänzt, teils umgeschrieben). Frankfurt am Main: Suhrkamp.

———. 1982. "Academic Ratiofascism: Comments on Tibor Machan's Review." *Philosophy of the Social Sciences* 12, no. 2 (June): 191–95.

———. 1981a. *Problems of Empiricism.* Vol. 2 of his *Philosophical Papers.* Cambridge: Cambridge University Press.

———. 1981b. *Realism, Rationalism, and Scientific Method.* Vol. 1 of his *Philosophical Papers.* Cambridge: Cambridge University Press.

———. 1980. "Democracy, Elitism, and Scientific Method." *Inquiry* 23, no. 1 (March): 3–18.

———. 1979. *Erkenntnis für freie Menschen.* Frankfurt am Main: Suhrkamp.

———. 1978a (1975). *Against Method.* London: Verso.

———. 1978b. *Science in a Free Society.* London: New Left Books.

———. 1978c. *Der wissenschaftstheoretische Realismus und die Autorität der Wissenschaften.* Braunschweig: Vieweg.

———. 1977a. "Changing Patterns of Reconstruction." *British Journal for the Philosophy of Science* 28:351–82.

———. 1977b. "Rationalism, Relativism and Scientific Method." *Philosophy in Context* 6:7–19.

————. 1976. "On the Critique of Scientific Reason." In Cohen, Feyerabend, and Wartofsky, 1976, pp. 109–43.

————. 1975a. "How to Defend Society Against Science." *Radical Philosophy* 11:3–8. Repr. in Hacking, 1981b, pp. 156–67.

————. 1975b. "Imre Lakatos." *British Journal for the Philosophy of Science* 26, no. 1 (March): 1–18.

————. 1975c. " 'Science.' The Myth and Its Role in Society." *Inquiry* 18, no. 2 (Summer): 167–81.

————. 1974. "Popper's *Objective Knowledge.*" *Inquiry* 17, no. 4 (Winter): 475–507.

————. 1972. "Von der beschränkten Gültigkeit methodologischer Regeln." *Neue Hefte für Philosophie* 2/3:124–71.

————. 1970a. "Consolations for the Specialist." In Lakatos and Musgrave, 1970, pp. 197–230.

————. 1970b. "Philosophy of Science: A Subject with a Great Past." In Stuewer, 1970, pp. 172–83.

————. 1969. "Science Without Experience." *Journal of Philosophy* 66, no. 22 (November): 791–94.

————. 1968. "Science, Freedom, and the Good Life." *Philosophical Forum* 1, no. 2 (Winter): 127–35.

————. 1967. "On the Improvement of the Sciences and the Arts, and the Possible Identity of the Two." In Cohen and Wartofsky, 1967, pp. 307–415.

————. 1965. "Problems of Empiricism." In Colodny, 1965, pp. 145–260.

————. 1964. "Realism and Instrumentalism." In *The Critical Approach to Science and Philosophy.* Edited by Mario Bunge. New York: The Free Press, pp. 280–308.

————. 1962. "Explanation, Reduction, and Empiricism." In Feigl and Maxwell, 1962, pp. 28–97.

————. 1961a. "Comments on Hanson's 'Is There a Logic of Scientific Discovery?' " In Feigl and Maxwell, 1961, pp. 35–39.

————. 1961b. "Knowledge Without Foundations." The Nellie Heldt Lectures, 8. Oberlin College, Oberlin, Ohio.

Feyerabend, Paul K., and Martin Gardner. 1983. "Feyerabend vs. Gardner: Science, Church or Instrument of Research?" *Free Inquiry* 3 (Summer): 58–60.

Finger, J. M. 1971. "Is Equilibrium an Operational Concept?" *Economic Journal* 81 (September): 609–12.

Finley, M. I. 1977. " 'Progress' in Historiography." *Daedalus* 106, no. 3 (Summer): 125–42.

Fire, Arthur, and Peter Machamer, eds. 1986. *PSA 1986,* vol. 1. East Lansing, Mich.: Philosophy of Science Association.

Fischer, David Hackett. 1970. *Historians' Fallacies: Toward a Logic of Historical Thought.* New York: Harper and Row.

Fisher, Robert M. 1986. *The Logic of Economic Discovery: Neoclassical Economics and the Marginal Revolution.* Brighton, Eng.: Wheatsheaf.

Fleck, Ludwik. 1980 (1935). *Entstehung und Entwicklung einer wissenschaftlichen Tatsache.* Edited by Lothar Schäfer and Thomas Schnelle. Frankfurt am Main: Suhrkamp.

Fogel, Robert William. 1965. "The Unification of Economic History with Economic Theory." *American Economic Association Papers and Proceedings* 60, no. 2 (May): 92–98.

Fogel, Robert William, and Stanley Engerman. 1974. *Time on the Cross.* Boston: Little, Brown.

Fontana, Biancamaria. 1986. "Democracy and Civilization: John Stuart Mill and the Critique of Political Economy." *Economies et sociétés* 20, no. 3 (March): 3–24.

Frazer, W., and Lawrence Boland. 1983. "An Essay on the Foundations of Friedman's Methodology." *American Economic Review* 73, no. 1 (March): 129–44.

Freese, Lee. 1972. "Cumulative Sociological Knowledge." *American Sociological Review* 37, no. 4 (August): 472–82; with discussion, 483–87.

Frey, Bruno S., et al. 1984. "Consensus and Dissension Among Economists: An Empirical Inquiry." *American Economic Review* 74, no. 5 (December): 986–94.

Friedman, Milton. 1953. "The Methodology of Positive Economics." In his *Essays in Positive Economics,* pp. 3–43. Chicago: University of Chicago Press, 1953.

Frisby, David. 1972. "The Popper-Adorno Controversy: The Methodological Dispute in German Sociology." *Philosophy of the Social Sciences* 2:105–19.

Fulton, G. 1984. "Research Programmes in Economics." *History of Political Economy* 16, no. 2 (Summer): 187–205.

Furner, Mary O. 1975. *Advocacy and Objectivity: A Crisis in the Professionalization of American Social Science, 1865–1905.* Lexington: University of Kentucky Press.

"The Furniture Movers." 1978. *Business Week,* 16 January, p. 120.

Fusfeld, Daniel R. 1980. "The Conceptual Framework of Modern Economics." *Journal of Economic Issues* 14, no. 1 (March): 1–52.

Gäfgen, Gérard. 1974. "On the Methodology and Political Economy of Galbraithian Economics." *Kyklos* 27, no. 4:705–31.

Gähde, Ulrich. Forthcoming. "Bridge Structures and the Borderline Between the Internal and External History of Science," Boston Studies in the Philosophy of Science, special issue.

Galbraith, John K. 1978a. "On Post Keynesian Economics." *Journal of Post Keynesian Economics* 1, no. 1 (Fall): 8–11.

———. 1978b. "Writing, Typing, and Economics." *Atlantic Monthly* 241 (March): 102–5.

———. 1973. "Power and the Useful Economist." *American Economic Review* 63, no. 1 (March): 1–11.

Ganslandt, Herbert R. 1984. "Neopositivismus." *Enzyklopädie Philosophie und Wissenschaftstheorie.* 1984 ed.

Garb, Gerald. 1964. "The Problem of Causality in Economics." *Kyklos* 17, no. 4:594–609.

Gardner, John W. 1961. *Excellence: Can We Be Equal and Excellent Too?* New York: Harper and Brothers.

Gardner, Martin. 1982/83. "Anti-Science: The Strange Case of Paul Feyerabend." *Free Inquiry* 3 (Winter): 32–35.

———. 1976. "Mathematical Games: On the Fabric of Inductive Logic, and Some Probability Paradoxes." *Scientific American* 234, no. 3 (March): 119–24.

Georgescu-Roegen, Nicholas. 1979. "Methods in Economics Science." *Journal of Economic Issues* 13 (June): 317–28.

———. 1971. *The Entropy Law and the Economic Process.* Cambridge: Harvard University Press.

Gerschenkron, Alexander. 1969. "History of Economic Doctrines and Economic History." *American Economic Association Papers and Proceedings* 59, no. 2 (May): 1–17.

Gethmann, Carl F. 1984a. "Feyerabend." *Enzyklopädie Philosophie und Wissenschaftstheorie.* 1984 ed.

———. 1984b. "Kuhn." *Enzyklopädie Philosophie und Wissenschaftstheorie.* 1984 ed.

————. 1984c. "Lakatos." *Enzyklopädie Philosophie und Wissenschaftstheorie.* 1984 ed.

Gibbard, Allan, and Hal R. Varion. 1978. "Economic Models." *Journal of Philosophy* 75, no. 11 (November): 664–77.

Giere, Ronald N. 1984. *Understanding Scientific Reasoning.* New York: Holt, Rinehart and Winston.

————. 1973. "History and Philosophy of Science: Intimate Relationship or Marriage of Convenience?" *British Journal for the Philosophy of Science* 24, no. 3 (September): 282–97.

Glymour, Clark. 1985. "Interpreting Leamer." *Economics and Philosophy* 1:290–94.

————. 1980. *Theory and Evidence.* Princeton: Princeton University Press.

Godelier, Maurice. 1972. *Rationality and Irrationality in Economics.* Translated by Brian Pearce from the French *Rationalité et irrationalité en économie* (1966). London: New Left Books.

Good, I. J. 1981. "Some Logic and History of Hypothesis Testing." In Pitt, 1981, pp. 149–74.

Goodwin, Craufurd. 1980. "Toward a Theory of the History of Economics." *History of Political Economy* 12, no. 4 (Winter): 610–19.

Gordon, Donald F. 1965. "The Role of the History of Economic Thought in the Understanding of Modern Economic Theory." *American Economic Review: Supplement* 55, no. 2 (May): 119–27.

————. 1955. "Operational Propositions in Economic Theory." *Journal of Political Economy* 63 (April): 150–62.

Gordon, Robert Aaron. 1976. "Rigor and Relevance in a Changing Institutional Setting." *American Economic Review* 66, no. 1 (March): 1–14.

Gordon, Scott. 1978. "Should Economists Pay Attention to Philosophers?" *Journal of Political Economy* 86, no. 4 (August): 717–28.

————. 1977. "Social Science and Value Judgements." *Canadian Journal of Economics* 10, no. 4 (November): 529–46.

Gorman, W. M. 1984. "Towards a Better Economic Methodology?" In Wiles and Routh, 1984, pp. 260–88; with reply by Wiles, 289–92.

Grammp, William D. 1983. "An Episode in the History of Thought and Policy." In Coats, 1983c, pp. 137–54.

————. 1973. "Classical Economics and Its Moral Critics." *History of Political Economy* 5:359–74.

————. 1965. "On the History of Thought and Policy." *American Economic Review: Supplement* 55, no. 2 (May): 128–35.

Green, Edward J. 1981. "On the Role of Fundamental Theory in Positive Economics." In Pitt, 1981, pp. 5–15.

Grofman, Bernard. 1974. "Rational Choice Models and Self-Defeating Prophecies." In Leinfellner and Köhler, 1974, pp. 381–83.

Grubel, Herbert G., and Anthony D. Scott. 1967. "The Characteristics of Foreigners in the U.S. Economics Profession." *American Economic Review* 57, no. 1 (March): 131–45.

Grünbaum, Adolf. 1976. "Is Falsifiability the Touchstone of Scientific Rationality? Karl Popper Versus Inductivism." In Cohen, Feyerabend, and Wartofsky, 1976, pp. 213–52.

Grunberg, Emile. 1957. "Notes on the Verifiability of Economic Laws." *Philosophy of Science* 24:337–48.

Grünfeld, Joseph. 1984. "Feyerabend's Irrational Science." *Logique et analyse* 27, no. 106 (June): 221–31.

———. 1979. "Progress in Science." *Logique et analyse* 85–86 (May–June): 208–21.

Gurley, John. 1971. "The State of Political Economics." *American Economic Association Papers and Proceedings* 61, no. 2 (May): 53–68.

Guthrie, William G. 1982. "The Methodological and the Ethical Context of Positive Economics: A Comment on McKenzie." *Journal of Economic Issues* 16 (December): 1109–16.

Gutting, Gary, ed. 1980. *Paradigms and Revolutions: Appraisals and Applications of Thomas Kuhn's Philosophy of Science.* Notre Dame: University of Notre Dame Press.

———. 1979. "Continental Philosophy of Science." In Asquith and Kyburg, 1979, pp. 94–117.

Haavelmo, Trygve. 1944. "The Probability Approach in Econometrics." *Econometrica Supplement* 12 (July): 1–115.

Haberler, Gottfried. 1969. "Joseph Alois Schumpeter, 1883–1950." In S. Harris, 1969, pp. 24–47.

Hacking, Ian. 1981a. "Lakatos's Philosophy of Science." In his 1981b, pp. 128–43.

———, ed. 1981b. *Scientific Revolutions.* Oxford: Oxford University Press.

Händler, Ernst W. 1982. "The Evolution of Economic Theories: A Formal Approach." *Erkenntnis* 18, no. 1 (July): 65–96.

———. 1980. "The Role of Utility and of Statistical Concepts in Empirical Economic Theories: The Empirical Claims of the Systems of Aggregate Market Supply and Demand Functions Approach." *Erkenntnis* 15, no. 2 (July): 129–57.

Hahn, Frank H. 1983. "On General Equilibrium and Stability." In E. C. Brown and Solow, 1983, pp. 31–55.

———. 1980. "General Equilibrium Theory." *The Public Interest* (Special Issue: The Crisis in Economic Theory), pp. 123–38.

———. 1973. "The Winter of Our Discontent." *Economica* 40, no. 159 (August): 322–30.

Hahn, Frank, and Martin Hollis. 1979. *Philosophy and Economic Theory.* Oxford: Oxford University Press.

Hall, Richard J. 1971. "Can We Use the History of Science to Decide Between Competing Methodologies?" In Buck and Cohen, 1971, pp. 151–59.

Hall, Sir Robert. 1959. "Reflections on the Practical Application of Economics." *Economic Journal* 69 (December): 639–52.

Hamminga, Bert. 1983. *Neoclassical Theory Structure and Theory Development: An Empirical–Philosophical Case Study Concerning the Theory of International Trade.* Berlin: Springer-Verlag.

Hamminga, Bert, and W. Balzer. 1986. "The Basic Structure of Neoclassical General Equilibrium Theory." *Erkenntnis* 25, no. 1 (July): 31–46.

Hands, Douglas W. 1985a. "Karl Popper and Economic Methodology." *Economics and Philosophy* 1:83–99.

———. 1985b. "Second Thoughts on Lakatos." *History of Political Economy* 17, no. 1: 1–16.

———. 1985c. "The Structuralist View of Economic Theories: A Review Essay." *Economics and Philosophy* 1:303–35.

———. 1984a. "Blaug's Economic Methodology." *Philosophy of the Social Sciences* 14, no. 1: 115–25.

———. 1984b. "The Role of Crucial Counter-Examples in the Growth of Economic Knowledge: Two Case Studies in the Recent History of Economic Thought." *History of Political Economy* 16, no. 1 (Spring): 59–67.

204   *Bibliography*

———. 1984c. "What Economics Is Not: An Economist's Response to Rosenberg."
  Philosophy of Science 51, no. 3 (September): 495–503.
———. 1979. "The Methodology of Economic Research Programmes." *Philosophy of
  the Social Sciences* 9, no. 3 (September): 293–303.
Hanson, Norwood Russell. 1971a. "The Irrelevance of History of Science to Philosophy of Science." In his 1971c, pp. 274–87.
———. 1971b. *Observation and Explanation: A Guide to Philosophy of Science.* New
  York: Harper and Row.
———. 1971c. *What I Do Not Believe, and Other Essays.* Edited by Stephen Toulmin
  and Harry Woolf. Dordrecht, The Netherlands: D. Reidel.
———. 1969a. "Logical Positivism and the Interpretation of Scientific Theories." In
  Achinstein and Barker, 1969, pp. 57–84.
———. 1969b. *Perception and Discovery.* San Francisco: Freeman, Cooper.
———. 1967. "Nicolas Copernicus." *Encyclopedia of Philosophy.* 1967 ed.
———. 1961a. "Is There a Logic of Discovery?" In Feigl and Maxwell, 1961, pp.
  20–35.
———. 1961b. "Rejoinder to Feyerabend." In Feigl and Maxwell, 1961, pp. 40–42.
———. 1960. "More on 'The Logic of Discovery.' " *Journal of Philosophy* 57, no. 6
  (March): 182–88.
———. 1958a. "The Logic of Discovery." *Journal of Philosophy* 60, no. 25 (December): 1073–89.
———. 1958b. *Patterns of Discovery.* Cambridge: Cambridge University Press.
Harding, Sandra G., ed. 1976a. *Can Theories Be Refuted? Essays on the Duhem-Quine
  Thesis.* Dordrecht, The Netherlands: D. Reidel.
———. 1976b. "Introduction." In her 1976a, pp. ix–xxi.
Harré, Rom. 1967. "Philosophy of Science, History of." *Encyclopedia of Philosophy.*
  1967 ed.
Harris, Errol E. 1972. "Epicyclic Popperism." *British Journal for the Philosophy of
  Science* 23:55–67.
Harris, Seymour Edwin, ed. 1969. *Schumpeter: Social Scientist.* Freeport, N.Y.: Books
  for Libraries Press.
Harrison, Jonathan. 1967. "Ethical Naturalism." *Encyclopedia of Philosophy.* 1967 ed.
Harrison, Royden. 1963. "Two Early Articles by Alfred Marshall." *Economic Journal*
  73 (September): 422–30.
Harrod, Roy R. 1938. "Scope and Method of Economics." *Economic Journal* 48
  (September): 383–412.
Hart, H. L., and A. M. Honoré. 1966. "Causal Judgement in History and in the Law."
  In Dray, 1966, pp. 213–37.
Harvey-Phillips, M. B. 1983. "T. R. Malthus on the 'Metaphysics of Political Economy': Ricardo's Critical Ally." In Coats, 1983c, pp. 185–210.
Hausman, Daniel M. 1989. "Economic Methodology in a Nutshell." *Journal of Economic Perspectives* 3, no. 2 (Spring): 115–27.
———. 1984a. "Philosophy and Economic Methodology." In Asquith and Kitcher,
  1984, pp. 231–49.
———. 1984b. *The Philosophy of Economics: An Anthology.* Cambridge: Cambridge
  University Press.
———. 1981a. "Are General Equilibrium Theories Explanatory?" In Pitt, 1981, pp.
  17–32.
———. 1981b. *Capital, Profits, and Prices: An Essay in the Philosophy of Economics.*
  New York: Columbia University Press.

————. 1981c. "John Stuart Mill's Philosophy of Economics." *Philosophy of Science* 48, no. 3 (September): 363–85.

————. 1980. "How to Do Philosophy of Economics." In Asquith and Giere, 1980, pp. 353–62.

Hayek, Friedrich A. von. 1979. *The Counter-Revolution of Science: Studies in the Abuse of Reason.* 2d ed. Indianapolis: Liberty Press.

————. 1978. *New Studies in Philosophy, Politics, Economics and the History of Ideas.* London: Routledge and Kegan Paul.

————. 1973. "The Place of Menger's *Grundsätze* in the History of Economic Thought." In Hicks and Weber, 1973, pp. 1–14.

————. 1937. "Economics and Knowledge." *Economica* 4, nos. 13–16 (February): 33–54.

Heelan, Patrick A. 1979. "Continental Philosophy and the Philosophy of Science." In Asquith and Kyburg, 1979, pp. 84–93.

Heilbroner, Robert. 1979. "Modern Economics as a Chapter in the History of Economic Thought." *History of Political Economy* 11, no. 2 (Summer): 192–98.

————. 1970. "On the Limits of Economic Prediction." *Diogenes* 70 (Summer): 27–40.

Heiner, Ronald A. 1983. "The Origin of Predictable Behavior." *American Economic Review* 73, no. 4 (September): 560–95.

Heller, Walter W. 1975. "What's Right with Economics?" *American Economic Review* 65, no. 1 (March): 1–26.

Hempel, C. G. 1973. "The Meaning of Theoretical Terms: A Critique of a Critique of the Standard Empiricist Construal." In Suppes et al., 1973, pp. 367–78.

————. 1966. "Explanation in Science and History." In Dray, 1966, pp. 95–126.

Hendry, David F. 1980. "Econometrics: Alchemy or Science?" *Economica* 47 (November): 387–406.

Hennings, Klaus H. 1986. "The Exchange Paradigm and the Theory of Production and Distribution." In Baranzini and Scazzieri, 1986, pp. 221–43.

Hesse, Mary. 1980. *Revolutions and Reconstructions in the Philosophy of Science.* Brighton, Eng.: Wheatsheaf Books.

Hey, John D. 1983. "Towards the Double Negative Economics." In Wiseman, 1983a, pp. 160–75.

Hicks, Sir John. 1986. "Is Economics a Science?" In Baranzini and Scazzieri, 1986, pp. 91–101.

————. 1984. "The 'New Causality': An Explanation." *Oxford Economic Papers* 36, no. 1 (March): 12–15.

————. 1983a. *Classics and Moderns.* Vol. 3 of his *Collected Essays on Economic Theory.* Oxford: Basil Blackwell.

————. 1983b. "A Discipline Not a Science." In his 1983b, pp. 365–75.

————. 1980–1981. "IS–LM: An Explanation." *Journal of Post Keynesian Economics* 3, no. 2 (Winter): 139–54.

————. 1979a. *Causality in Economics.* Oxford: Basil Blackwell.

————. 1979b. "The Formation of an Economist." *Banca Nationale del Lavoro Quarterly Review* 32, no. 130: 195–204. Repr. in his 1983a, pp. 355–64.

————. 1976. " 'Revolutions' in Economics." In Latsis, 1976a, pp. 207–18. Repr. in his 1983a, pp. 3–16.

————. 1974. *The Crisis in Keynesian Economis.* Oxford: Basil Blackwell.

————. 1937. "Mr. Keynes and the Classics: A Suggested Interpretation." *Econometrica* 5:147–59.

Hicks, Sir John, and W. Weber, eds. 1973. *Carl Menger and the Austrian School of Economics*. Oxford: Oxford University Press.

Hill, Lewis E. 1979. "The Metaphysical Preconceptions of the Economic Science." *Review of Social Economy* 37 (October): 189–97.

Hindess, Barry. 1977. *Philosophy and Methodology in the Social Sciences*. Atlantic Highlands, N.J.: Humanities Press.

Hirsch, Abraham. 1980. "The 'Assumptions' Controversy in Historical Perspective." *Journal of Economic Issues* 14, no. 1 (March): 99–118.

Hirsch, Abraham, and Neil de Marchi. 1986. "Making a Case When Theory Is Unfalsifiable: Friedman's Monetary History." *Economics and Philosophy* 2, no. 1 (April): 1–21.

Hirsch, Abraham, and Eva Hirsch. 1975. "The Heterodox Methodology of Two Chicago Economists." *Journal of Economic Studies* 9 (December): 645–64.

Hirshleifer, J. 1977. "Economics from a Biological Viewpoint." *Journal of Law and Economics* 20, no. 1 (April): 1–52.

Hollander, Samuel. 1983. "William Whewell and John Stuart Mill on the Methodology of Political Economy." *Studies in History and Philosophy of Science* 14, no. 2 (June): 127–68.

———. 1977. "Adam Smith and the Self-Interest Axiom." *Journal of Law and Economics* 20, no. 1 (April): 133–52.

Hollinger, David A. 1973. "T. S. Kuhn's Theory of Science and Its Implications for History." *American Historical Review* 78, no. 2 (April): 370–93.

Hollis, Martin, and Edward J. Nell. 1975. *Rational Economic Man*. London: Cambridge University Press.

Holmstrom, Engin Inel, and Robert W. Holmstrom. 1974. "The Plight of the Woman Doctoral Student." *American Educational Research Journal* 11, no. 1 (Winter): 1–17.

Holton, Gerald. 1974. "On Being Caught Between Dionysians and Apollonians." *Daedalus* 103, no. 3 (Summer): 65–81.

———. 1973. *Thematic Origins of Scientific Thought*. Cambridge: Harvard University Press.

Homans, George C. 1978. "What Kind of Myth Is the Myth of a Value-Free Social Science?" *Social Science Quarterly* 58, no. 4 (March): 530–41.

Hook, Sidney, Paul Kurtz, and Miro Todorovich, eds. 1974. *The Idea of a Modern University*. Buffalo: Prometheus.

Horowitz, Daniel. 1974. "Historians and Economists: Perspectives on the Development of American Economic Thought." *History of Political Economy* 6, no. 4: 454–62.

Howard, V. A. 1968. "Do Anthropologists Become Moral Relativists by Mistake?" *Inquiry* 11:175–89.

Hübner, Kurt. 1974. "Zur Frage des Relativismus und des Fortschritts in den Wissenschaften." *Zeitschrift für allgemeine Wissenschaftstheorie* 5, no. 2: 285–303.

Hudson, W. D., ed. 1969. *The Is–Ought Question: A Collection of Papers on the Central Problem in Moral Philosophy*. London: Macmillan.

Huff, Darrell. 1954. *How to Lie with Statistics*. New York: Norton.

Huff, Toby E. 1981. "On the Methodology of the Social Sciences: A Review Essay Part I." *Philosophy of the Social Sciences* 11:461–75.

Hume, L.: J. 1956. "The Definition of Economics." *Economic Record* 32:152–57.

Hutchison, T. W. 1984. "Our Methodological Crisis." In Wiles and Routh, 1984, pp. 1–21; with comments by Hollis, pp. 22–29, and Blaug, pp. 30–36.

———. 1983. "From 'Dismal Science' to 'Positive Economics'—A Century-and-a-Half of Progress?" In Wiseman, 1983a, pp. 192–211.

———. 1981. *The Politics and Philosophy of Economics: Marxians, Keynesians and Austrians.* New York: New York University Press.

———. 1979. "Die Natur- und die Sozialwissenschaften und die Entwicklung und Unterentwicklung der Ökonomik: Methodologische Vorschriften für weniger entwickelte Wissenschaften." In *Theorie und Erfahrung.* Edited by Hans Albert and Kurt Stapf, pp. 245–68. Stuttgart: Klett, 1979.

———. 1978. *On Revolutions and Progress in Economic Knowledge.* Cambridge: Cambridge University Press.

———. 1977a. *Keynes Versus the 'Keynesians' . . . ?: An Essay in the Thinking of J. M. Keynes and the Accuracy of Its Interpretation by His Followers.* With Commentaries by Lord Kahn and Sir Austin Robinson. Hobart Paper no. 11. London: The Institute of Economic Affairs. Revised and without commentaries as "Keynes *Versus* the Keynesians," in Hutchison, 1981, pp. 108–54.

———. 1977b. *Knowledge and Ignorance in Economics.* Chicago: University of Chicago Press.

———. 1977c. "On the History and Philosophy of Science and Economics." In his 1977b, pp. 34–61. A slightly amended version appears in Latsis, 1976, pp. 181–205.

———. 1976. "Adam Smith and the Wealth of Nations." *Journal of Law and Economics* 19, no. 3 (October): 507–28.

———. 1973. "Some Themes from *Investigations into Method.*" In Hicks and Weber, 1973, pp. 15–37.

———. 1972. "The 'Marginal Revolution' and the Decline and Fall of English Classical Political Economy." *HOPE* 4 (Fall): 442–68. Repr. in Black, Coats, and Goodwin, 1973, pp. 176–202.

———. 1970. "Economic Thought and Policy: Generalizations and Ambiguities." In *International Congress on Economic History,* 5th ed., edited by H. van der Wee, V. A. Vinogradov, and G. G. Kotovsky, pp. 119–33. The Hague: Mouton, 1970.

———. 1966. "Testing Economic Assumptions: A Comment." *Journal of Political Economy* 74, no. 1 (February): 81–83.

———. 1965 (1938). *The Significance and Basic Postulates of Economic Theory.* New York: Kelley.

———. 1964. *'Positive' Economics and Policy Objectives.* London: George Allen and Unwin.

———. 1956. "Professor Machlup on Verification in Economics." *Southern Economic Journal* 22, no. 4 (April): 476–83.

———. 1941. "*The Significance and Basic Postulates of Economic Theory:* A Reply to Professor Knight." *Journal of Political Economy* 49, no. 5 (October): 732–50.

———. 1937. "Theoretische Ökonomie als Sprachsystem." *Zeitschrift für Nationalökonomie* 8, no. 1: 78–90.

Iggers, Georg G. 1973. "Historicism." *Dictionary of the History of Ideas.* 1973 ed.

———. 1968. *The German Conception of History.* Middletown, Conn.: Wesleyan University Press.

Irzik, Gürol. 1985. "Popper's Piecemeal Engineering: What Is Good for Science Is Not Always Good for Society." *British Journal for the Philosophy of Science* 36:1–10.

Jalladeau, Joel. 1978. "Research Program versus Paradigm in the Development of Economics." *Journal of Economic Issues* 12, no. 3 (September): 583–608.

Janik, Allan, and Stephen Toulmin. 1973. *Wittgenstein's Vienna.* New York: Simon and Schuster.

Jaschik, Scott. 1989. "Misconduct in Science: Congress's Interest in Ferreting Out Fraud Is Misinformed, Harmful Say Researchers." *Chronicle of Higher Education* 35, no. 39 (7 June): A1, A25.

Jefferson, Michael. 1983. "Economic Uncertainty and Business Decision-Making." In Wiseman, 1983a, pp. 122–59.

Jochimsen, Reimut, and Helmut Knobel. 1971. *Gegenstand und Methoden der Nationalökonomie*. Köln: Kiepenheuer und Witsch.

Joergensen, Joergen. 1970 (1951). *The Development of Logical Empiricism*. Vol. 2, no. 9, of International Encyclopedia of Unified Science. Chicago: University of Chicago Press.

Johansen, Leif. 1983. "Mechanistic and Organistic Analogies in Economics: The Place of Game Theory." *Kyklos* 36, no. 2: 304–7.

Johansson, Ingvar. 1975. *A Critique of Karl Popper's Methodology*. Stockholm: Scandanavian University Books.

Johnson, Elizabeth, and Harry G. Johnson. 1978. *The Shadow of Keynes: Understanding Keynes, Cambridge and Keynesian Economics*. Oxford: Basil Blackwell.

Johnson, Harry G. 1977. "The American Tradition in Economics." *Nebraska Journal of Economics* 16, no. 3 (Summer): 17–26.

———. 1973. "National Styles in Economic Research: The United States, the United Kingdom, Canada, and Various European Countries." *Daedalus* 102, no. 2 (Spring): 65–74.

———. 1971. "Revolution and Counter-Revolution in Economics." *Encounter* 36, no. 4 (April): 23–33.

———. 1968a. "A Catarrh of Economists?" *Encounter* 30 (May): 50–54.

———. 1968b. "The Economic Approach to Social Questions." *Economica* 35, no. 137 (February): 1–21.

Johnson, L. E. 1983. "Economic Paradigms: A Missing Dimension." *Journal of Economic Issues* 17, no. 4 (December): 1097–1111.

———. 1980. "A Neo-Paradigmatic Model for Studying the Development of Economic Reasoning." *Atlantic Economic Journal* 8, no. 4 (December): 52–61.

Jones, Evan. 1977. "Positive Economics or What?" *Economic Record* 53 (September): 350–63.

Juhos, Béla. 1971. "Formen des Positivismus." *Zeitschrift für allgemeine Wissenschaftstheorie* 2:27–65.

Kant, Immanuel. 1983 (1781). *Kant's Critique of Pure Reason: An Introductory Text*. Edited by Humphrey Palmer. Cardiff: University College Cardiff Press.

Katouzian, Homa. 1983. "Towards the Progress of Economic Knowledge." In Wiseman, 1983a, pp. 50–64.

———. 1980. *Ideology and Method in Economics*. New York: New York University Press.

———. 1974. "Scientific Method and Positive Economics." *Scottish Journal of Economics* 21, no. 3 (November): 279–86.

Katzner, Donald W. 1978–1979. "On Not Quantifying the Non-Quantifiable." *Journal of Post Keynesian Economics* 1, no. 2 (Winter): 113–28.

Kaufmann, Felix. 1937. "Do Synthetic Propositions a priori Exist in Economics?" *Economica*, n.s. 4 (August): 337–42.

Kaysen, Carl. 1945–1946. "A Revolution in Economic Theory." *Review of Economic Studies* 14:1–15.

Kearl, J. R., et al. 1979. "What Economists Think." *American Economic Association Papers and Proceedings* 69, no. 2 (May): 28–37.

Keller, Evelyn Fox. 1985. *Reflections on Gender and Science.* New Haven: Yale University Press.

Kendry, Adrian. 1981. "Paul Samuelson and the Scientific Awakening of Economics." In Shackleton and Locksley, 1981, pp. 219–39.

Keynes, John Maynard. 1972a (1924). "Alfred Marshall." In *Essays in Biography* (pp. 161–231). Vol. 10 of *The Collected Writings of John Maynard Keynes.* London: Macmillan for The Royal Economic Society, 1972.

———. 1972b (ca. 1930). "Economic Possibilities for Our Grandchildren." In *Essays in Persuasion* (pp. 321–32). Vol. 9 of *The Collected Writings of John Maynard Keynes.* London: Macmillan for The Royal Economic Society, 1972.

———. 1972c (1944). "Mary Paley Marshall." In *Essays in Biography* (pp. 232–50). Vol. 10 of *The Collected Writings of John Maynard Keynes.* London: Macmillan for The Royal Economic Society, 1972.

Keynes, John Neville. 1973 (1917). *The Scope and Method of Political Economy.* 4th ed. New York: Kelley.

King, M. D. 1971. "Reason, Tradition, and the Progressiveness of Science." *History and Theory* 10, no. 1: 3–32.

King-Farlow, John, and Wesley E. Cooper. 1983. "Comments on Farr's Paper (I) Sir Karl Popper: Tributes and Adjustments." *Philosophy of the Social Sciences* 13:177–82.

Kirzner, Israel M. 1980. "The 'Austrian' Perspective." *The Public Interest* (Special Issue: The Crisis in Economic Theory), pp. 111–22.

———. 1976a. "Equilibrium Versus Market Process." In Dolan, 1976b, pp. 115–25.

———. 1976b. "On the Method of Austrian Economics." In Dolan, 1976b, pp. 40–51.

———. 1976c. "Philosophical and Ethical Implications of Austrian Economics." In Dolan 1976b, pp. 75–88.

———. 1976d. "The Theory of Capital." In Dolan, 1976d, pp. 139–44.

———. 1962. "Rational Action and Economic Theory." *Journal of Political Economy* 70 (August): 380–85.

Kisiel, Theodore, with Galen Johnson. 1974. "New Philosophies of Science in the USA; A Selective Survey." *Zeitschrift für allgemeine Wissenschaftstheorie* 5, no. 1: 138–91.

Klamer, Arjo. 1985. "Economics as Discourse." Paper presented at a conference in honor of J. J. Klant, Amsterdam, December 1985.

———. 1984a. "Levels of Discourse in New Classical Economics." *History of Political Economy* 16, no. 2: 263–90.

———. 1984b. *The New Classical Macroeconomics: Conversations with the New Classical Economists and Their Opponents.* Brighton, Eng.: Wheatsheaf Books.

Klant, J. J. 1985. "The Slippery Transition." In *Keynes' Economics.* Edited by Tony Lawson and Hashem Pesaran, pp. 80–98. London: Croom Helm, 1985.

———. 1984. *The Rules of the Game: The Logical Structure of Economic Theories.* Translated by Ina Swart. Cambridge: Cambridge University Press.

———. 1982. "The Natural Order." Expanded and revised version of "Idealisatie: Idee en ideaal." In *Economie en ideaal.* Edited by Gaay Fortman. Samson, The Netherlands: Alphen aan de Rijn.

———. 1974. "Realism of Assumptions in Economic Theory." *Methodology and Science* 7:141–55.

Klappholz, Kurt. 1964. "Value Judgements and Economics." *British Journal for the Philosophy of Science* 15, no. 58 (August): 97–114.

Klappholz, Kurt, and Joseph Agassi. 1959. "Methodological Prescriptions in Economics." *Economica* 26, no. 101 (February): 60–74.

Klein, Lawrence. 1985. *Economic Theory and Econometrics*. Edited by Jaime Marquez. Oxford: Basil Blackwell.

———. 1968 (1966). *The Keynesian Revolution*. 2d ed. London: Macmillan.

Klein, Philip A. 1974. "Economics: Allocation or Valuation." *Journal of Economic Issues* 8, no. 4 (December): 785–811.

Knapp, Peter. 1984. "Can Social Theory Escape from History?" *History and Theory* 23, no. 1: 34–52.

Knight, Frank. 1956. *On the History and Method of Economics*. Chicago: University of Chicago Press.

———. 1941. "A Rejoinder." *Journal of Political Economy* 49, no. 5 (October): 750–53.

———. 1940. " 'What Is Truth' in Economics?" *Journal of Political Economy* 48, no. 1 (February): 1–32.

———. 1934. "Economic Science in Recent Discussion." *American Economic Review* 24, no. 2 (June): 225–38.

Kockelmans, Joseph J. 1972. "On the Meaning of Scientific Revolutions." *Philosophy Forum* 11 (September): 243–64.

Koertge, Noretta. 1979. "Braucht die Sozialwissenschaft wirklich Metaphysik?" *In Theorie und Erfahrung*, edited by Hans Albert and Kurt H. Stapf, pp. 55–81. Stuttgart: Klett, 1979.

———. 1978. "Towards a New Theory of Scientific Inquiry." In Radnitzky and Andersson, 1978, pp. 253–78.

———. 1976. "Rational Reconstructions." In Cohen, Feyerabend, and Wartofsky, pp. 359–69. Dordrecht, The Netherlands: D. Reidel, 1976.

———. 1975. "Popper's Metaphysical Research Program for the Human Sciences." *Inquiry* 18:437–62.

———. 1974. "Bartley's Theory of Rationality." *Philosophy of the Social Sciences* 4:75–81.

———. 1971. "Inter-Theoretic Criticism and the Growth of Science." In Buck and Cohen, 1971, pp. 160–73.

Kötter, Rudolf. 1980. "Empirismus und Rationalismus in der Ökonomie." Ph.D., diss., Friedrich–Alexander–Universität Erlangen–Nürnberg.

Kolakowski, Leszak. 1968. *The Alienation of Reason*. Translated by Norbert Guterman. New York: Doubleday.

Koopmans, Tjalling C. 1979. "Economics Among the Sciences." *American Economic Review* 69, no. 1 (March): 1–13.

———. 1957. *Three Essays on the State of Economic Science*. New York: McGraw-Hill.

———. 1974. "Measurement Without Theory." *Review of Economic Statistics* 29 (August): 161–72.

Koot, Gerard. 1980. "English Historical Economics and the Emergence of Economic History in England." *History of Political Economy* 12, no. 2: 174–205.

Kornai, Janos. 1983. "The Health of Nations: Reflections on the Analogy Between the Medical Science and Economics." *Kyklos* 36, no. 2:191–212.

Koselleck, Reinhart. 1977. "Standortbindung und Zeitlichkeit. Ein Beitrag zur historiographischen Erschließung der geschichtlichen Welt." In *Objektivität und Parteilichkeit*. Edited by R. Koselleck, Wolfgang J. Mommsen, and Jörn Rüsen, pp. 17–46. München: Deutscher Taschenbuch Verlag, 1977.

Koyré, Alexandre. 1961. "Influence of Philosophic Trends on the Formation of Scientific Theories." In *The Validation of Scientific Theories*. Edited by Philipp Frank, pp. 177–87. New York: Collier, 1961.

Kraft, Victor. 1974. "Popper and the Vienna Circle." In Schilpp, 1974, pp. 185–204.

Kristol, Irving. 1980. "Rationalism in Economics." *The Public Interest* (Special Issue: The Crisis in Economic Theory), pp. 201–18.

Krohn, Claus-Dieter. 1987. *Wissenschaft im Exil. Deutsche Sozial- und Wirtschaftswissenschaftler in den USA und die New School for Social Research*. Frankfurt: Campus Verlag.

———. 1985. "Die Krise der Wirtschaftswissenschaft in Deutschland im Vorfeld des Nationalisozialismus." *Leviathan: Zeitschrift für Sozialwissenschaft* 13:311–33.

Krüger, Lorenz. 1974. "Die systematische Bedeutung wissenschaftlicher Revolutionen, pro und contra Thomas Kuhn." In Diederich, 1974, pp. 210–46.

Krupp, Sherman R., ed. 1966. *The Structure of Economic Science*. Englewood Cliffs, N.J.: Prentice-Hall.

———. 1963. "Analytic Economics and the Logic of External Effects." *American Economic Review: Supplement* 53, no. 2 (May): 220–26; with discussion, 227–36.

Kuhn, Thomas S. 1977a. "Concepts of Cause in the Development of Physics." In his 1977b, pp. 21–30.

———. 1977b. *The Essential Tension*. Chicago: University of Chicago Press.

———. 1977c. "The Essential Tension: Tradition and Innovation in Scientific Research." In his 1977b, pp. 225–39.

———. 1977d. "A Function for Thought Experiments." In his 1977b, pp. 240–65. Repr. in Hacking 1981b, pp. 6–27.

———. 1977e. "The Function of Measurement in Modern Physical Science." In his 1977b, pp. 178–224.

———. 1977f. "The Historical Structure of Scientific Discovery." In his 1977b, pp. 165–77.

———. 1977g. "The History of Science." In his 1977b, pp. 105–26. Repr. from the *International Encyclopedia of the Social Sciences*. 1968 ed.

———. 1977h. "Mathematical Versus Experimental Traditions in the Development of Physical Science." In his 1977b, pp. 31–65.

———. 1977i. "Objectivity, Value Judgement, and Theory Choice." In his 1977b, pp. 320–39.

———. 1977j. "The Relations Between History and the History of Science." In his 1977b, pp. 127–61.

———. 1977k. "The Relations Between the History and the Philosophy of Science." In his 1977b, pp. 3–20.

———. 1977l. "Second Thoughts on Paradigms." In his 1977b, pp. 293–319. Repr. from Suppe, 1977b, pp. 459–82.

———. 1971. "Notes on Lakatos." In Buck and Cohen, 1971, pp. 137–46.

———. 1970a. "Logic of Discovery or Psychology of Research?" In Lakatos and Musgrave, 1970, pp. 1–23. Repr. in his 1977b, pp. 266–92.

———. 1970b. "Reflections on My Critics." In Lakatos and Musgrave, 1970, pp. 231–78.

———. 1970c. *The Structure of Scientific Revolutions*. 2d enl. ed. Vol. 2, no. 2, of International Encyclopedia of Unified Science. Chicago: University of Chicago Press.

———. 1963. "The Function of Dogma in Scientific Research." In *Scientific Change: Historical Studies in the Intellectual, Social and Technical Conditions for Scientific*

Discovery and Technical Invention, from Antiquity to Present, edited by A. C. Crombie, pp. 347–69. London: Heinemann, 1963.

Kuhn, Thomas, et al. 1977. "Discussion." In Suppe, 1977b, pp. 500–517.

Kunin, Leonard, and F. Stirton Weaver. 1971. "On the Structure of Scientific Revolutions in Economics." *History of Political Economy* 3, no. 2 (Fall): 391–97.

Kuttner, Robert. 1985. "The Poverty of Economics." *Atlantic Monthly* 255, no. 2 (February): 74–84.

Lachmann, Ludwig M. 1976a. "Austrian Economics in the Age of the Neo-Ricardian Counterrevolution." In Dolan, 1976b, pp. 215–23.

———. 1976b. "On the Austrian Capital Theory." In Dolan, 1976b, pp. 145–51.

———. 1976c. "On the Central Concept of Austrian Economics: Market Process." In Dolan, 1976c, pp. 126–32.

———. 1976d. "Toward a Critique of Macroeconomics." In Dolan, 1976b, pp. 152–59.

Lakatos, Imre. 1978a. *Mathematics, Science and Epistemology.* Vol. 2 of his *Philosophical Papers.* Edited by John Worrall and Gregory Currie. Cambridge: Cambridge University Press.

———. 1978b. *The Methodology of Scientific Research Programmes.* Vol. 1 of his *Philosophical Papers.* Edited by John Worrall and Gregory Currie. Cambridge: Cambridge University Press.

———. 1978c. "The Problem of Appraising Scientific Theories: Three Approaches." In his 1978b, pp. 107–20.

———. 1978d. "Understanding Toulmin." In his 1978b, pp. 224–43.

———. 1974a. "Popper on Demarcation and Induction." In his 1978b, pp. 139–67. Repr. from Schilpp, 1974, pp. 241–73.

———. 1974b. "The Role of Crucial Experiments in Science." *Studies in History and Philosophy of Science* 4:309–25.

———. 1974c. "Science and Pseudoscience." *Conceptus* 8:5–9.

———. 1971a. "History of Science and Its Rational Reconstructions." In Buck and Cohen, 1971, pp. 91–136. Repr. in Hacking, 1981b, pp. 107–25.

———. 1971b. "Replies to Critics." In Buck and Cohen, 1971, pp. 174–82.

———. 1970. "Falsification and the Methodology of Scientific Research Programmes." In Lakatos and Musgrave, 1970, pp. 91–196.

Lakatos, Imre, and Alan Musgrave, eds. 1970. *Criticism and the Growth of Knowledge.* London: Cambridge University Press.

———, eds. 1968. *Problems in the Philosophy of Science.* Vol. 3 of the *Proceedings of the International Colloquium in the Philosophy of Science, London, 1965.* Amsterdam: North-Holland.

Lancaster, Kelvin. 1962. "The Scope of Qualitative Economics." *Review of Economic Studies* 29:99–123.

Landreth, Harry. 1976. *History of Economic Theory.* Boston: Houghton Mifflin.

Lange, Oskar. 1945. "The Scope and Method of Economics." *Review of Economic Studies* 13, no. 1: 19–32.

Latsis, Spiro J. 1983. "The Role and Status of the Rationality Principle in the Social Sciences." In Cohen and Wartofsky, 1983, pp. 123–51.

———, ed. 1976a. *Method and Appraisal in Economics.* Cambridge: Cambridge University Press.

———. 1976b. "A Research Programme in Economics." In Latsis, 1976a, pp. 1–42.

———. 1972. "Situational Determinism in Economics." *British Journal for the Philosophy of Science* 23:207–45.

Laudan, Laurens. 1981. "A Problem-Solving Approach to Scientific Progress." In Hacking, 1981b, pp. 144–55.

———. 1979. "Historical Methodologies: An Overview and Manifesto." In Asquith and Kyburg, 1979, pp. 40–54.

———. 1977. "The Sources of Modern Methodology." In Butts and Hintikka, 1977, pp. 3–19.

———. 1976. "Two Dogmas of Methodology." *Philosophy of Science* 43, no. 4 (December): 585–97.

———. 1968. "Theories of Scientific Method from Plato to Mach: A Bibliographical Review." *History of Science* 7:1–63.

Lazarfeld, Paul F., and Wagner Thielens, Jr. 1958. *The Academic Mind: Social Scientists in a Time of Crisis.* Glencoe, Ill.: The Free Press.

Leamer, Edward E. 1985. "Self-Interpretation." *Economics and Philosophy* 1:295–302.

———. 1983. "Let's Take the Con Out of Econometrics." *American Economic Review* 73, no. 1 (March): 31–43.

Leary, David. 1982. "The Fate and Influence of John Stuart Mill's Proposed Science of Ethology." *Journal of the History of Ideas* 43, no. 1 (January–March): 153–62.

Lee, Dwight E., and Robert N. Beck. 1953–1954. "The Meaning of 'Historicism.' " *American Historical Review* 59:568–77.

Lee, Frederick. 1984. "Whatever Happened to the Full-Cost Principle (USA)?" In Wiles and Routh, 1984, pp. 233–39.

Leibenstein, Harvey. 1976. *Beyond Economic Man: A New Foundation for Microeconomics.* Cambridge: Harvard University Press.

Leijonhufvud, Axel. 1983a. "What Was the Matter with IS–LM?" In *Modern Macroeconomic Theory,* edited by Jean-Paul Fitoussi, pp. 64–90. Oxford: Basil Blackwell.

———. 1983b. "What Would Keynes Have Thought of Rational Expectations?" In *Keynes and the Modern World: Proceedings of the Keynes Centenary Conference, King's College, Cambridge,* edited by David Worswick and James Trevitheck, pp. 179–205. Cambridge: Cambridge University Press, 1983.

———. 1976. "Schools, 'Revolutions', and Research Programmes in Economic Theory." In Latsis, 1976a, pp. 65–108.

———. 1973. "Life Among the Econ." *Western Economic Journal* 11 (September): 327–37.

———. 1966, 1968. *On Keynesian Economics and the Economics of Keynes.* New York: Oxford University Press.

Leinfellner, Werner, and Eckehart Köhler, eds. 1974. *Developments in the Methodology of Social Science.* Dordrecht, The Netherlands: D. Reidel.

Lekachmann, Robert. 1976. *Economists at Bay: Why the Experts Will Never Solve Your Problems.* New York: McGraw-Hill.

Lenk, Hans. 1986. "Wiederkäuer des Weltgeistes? Aphorismen über Philosophen und Philosophie eine fast nächtliche, doch nicht ganz unernste Einführungsvorlesung." *Conceptus* 20, no. 51:3–25.

———. 1982. "Die 'Feyerabendglocke' des Szientismus." *Conceptus* 16:3–11. Repr. in Duerr, 1981, 2:26–41.

Leontief, Wassily. 1982. "Letters: Academic Economics." *Science* 217 (9 July): 104–5. Repr. in Marr and Raj, 1983, pp. 331–35.

———. 1971. "Theoretical Assumptions and Nonobserved Facts." *American Economic Review* 61, no. 1 (March): 1–7.

———. 1961. "The Problem of Quality and Quantity in Economics." In Lerner, 1961, pp. 117–28.

Lerner, Daniel, ed. 1961. *Quantity and Quality.* New York: The Free Press of Glencoe.

Lessa, Carlos. 1979. "Economic Policy: Science or Ideology?" *CEPAL Review* 8 (August): 119–44.

Letwin, William. 1963. *The Origins of Scientific Economics: English Economic Thought, 1660–1776.* London: Methuen.

Lewin, Arie Y., and Linda Duchan. 1971. "Women in Academia." *Science* 173, no. 4000 (3 September): 892–95.

Lindgren, J. Ralph, ed. 1967. *The Early Writings of Adam Smith.* New York: Augustus M. Kelley.

Linstromberg, R. C. 1969. "The Philosophy of Science and Alternative Approaches to Economic Thought." *Journal of Economic Issues* 3, no. 2 (June): 177–91.

Lipsey, Richard G. 1963 (1st ed.; 1979, 5th ed.). *An Introduction to Positive Economics.* London: Wiedenfeld and Nicolson.

Littlechild, Stephen. 1983. "Subjectivism and Method in Economics." In Wiseman, 1983a, pp. 38–49.

Loasby, Brian. 1983. "Knowledge, Learning and Enterprise." In Wiseman, 1983a, pp. 104–21.

———. 1984. "On Scientific Method." *Journal of Post Keynesian Economics* 6, no. 3 (Spring): 394–410.

Lombardini, Siro. 1985. "At the Roots of the Crisis in Economic Theory." *Economia Internazionale* 38, nos. 3/4 (August/November): 323–50.

Lorenz, Kunz. 1984. "Empirismus, logischer." *Enzyklopädie Philosophie und Wissenschaftstheorie.* 1984 ed.

Losee, John. 1983. "Whewell and Mill on the Relation Between Philosophy of Science and History of Science." *Studies in History of Science* 14, no. 2 (June): 113–26.

———. 1977. "Limitations of an Evolutionist Philosophy of Science." *Studies in History and Philosophy of Science* 8:349–52.

Lugg, Andrew. 1978. "Disagreement in Science." *Zeitschrift für allgemeine Wissenschaftstheorie* 9, no. 2: 276–92.

MacDougall, Donald. 1974. "In Praise of Economics." *Economic Journal* 84, no. 336 (December): 773–86.

Macfie, A. L. 1963. "Economics—Science, Ideology, Philosophy?" *Scottish Journal of Political Economy* 10:212–25.

Machan, Tibor R. 1982a. "Anarchosurrealism Revisited: Reply to Feyerabend's Comments." *Philosophy of the Social Sciences* 12, no. 2 (June): 197–99.

———. 1982b. "The Politics of Medicinal Anarchism." *Philosophy of the Social Sciences* 12, no. 2 (June): 183–89.

———. 1981. "The Non-Rational Domain and the Limits of Economic Analysis: Comment on McKenzie." *Southern Economic Journal* 47 (April): 1123–27.

Machlup, Fritz. 1978a. "Adolf Lowe's Instrumental Analysis." In his 1978m, pp. 505–12.

———. 1978b. "Are the Social Sciences Really Inferior?" In his 1978m, pp. 345–67.

———. 1978c. "Fact and Theory in Economics." In his 1978m, pp. 101–30.

———. 1978d. "Friedrich Hayek on Scientific and Scientistic Attitudes." In his 1978m, pp. 513–34.

———. 1978e. "Gunnar Myrdal on Concealed Value Judgements." In his 1978m, pp. 475–80.

———. 1978f. "The Ideal Type: A Bad Name for a Good Construct." In his 1978m, pp. 211–22.

———. 1978g. "Ideal Types, Reality, and Construction." In his 1978m, pp. 223–66.

———. 1978h. "If Matter Could Talk." In his 1978m, pp. 309–32.

———. 1978i. "The Inferiority Complex of the Social Sciences." In his 1978m, pp. 333–44.

———. 1978j. "John Neville Keynes' Scope and Method." In his 1978m, pp. 489–92.

———. 1978k. "Joseph Schumpeter's Economic Methodology." In his 1978m, pp. 461–74.

———. 1978l. "Homo Oeconomicus and His Classmates." In his 1978m, pp. 267–82.

———. 1978m. *Methodology of Economics and Other Social Sciences.* New York: Academic Press.

———. 1978n. "A Note on Models in Microeconomics." In his 1978m, pp. 75–100.

———. 1978o. "Operationalism and Pure Theory in Economics." In his 1978m, pp. 189–203.

———. 1978p. "Operational Concepts and Mental Constructs in Model and Theory Formation." In his 1978m, pp. 159–88.

———. 1978q. "Paul Samuelson on Theory and Realism." In his 1978m, pp. 481–84.

———. 1978r. "Positive and Normative Economics." In his 1978m, pp. 425–50.

———. 1978s. "The Problem of Verification in Economics." In his 1978m, pp. 137–58.

———. 1978t. "Spiro Latsis on Situational Determinism." In his 1978m, pp. 521–34. Repr. from *British Journal for the Philosophy of Science* 23 (1972):270–84.

———. 1978u. "Terence Hutchison's Reluctant Ultra-Empiricism." In his 1978m, pp. 493–504.

———. 1978v. "Theories of the Firm: Marginalist, Behavioral, Managerial." In his 1978m, pp. 391–424.

———. 1978w. "Three Writers on Social Theory: Madge, Rose, and Zetterberg." In his 1978m, pp. 485–88.

———. 1978x. "The Universal Bogey: Economic Man." In his 1978m, pp. 283–301.

———. 1978y. "What Is Meant by Methodology: A Selective Survey of the Literature." In his 1978m, pp. 5–62.

———. 1978z. "Why Bother with Methodology?" In his 1978m, pp. 63–70.

———. 1978aa. "Why Economists Disagree." In his 1978m, pp. 375–90.

———. 1971. "Academic Freedom." *Encyclopedia of Education.* 1971 ed.

———. 1967. *Essays in Economic Semantics.* New York: Norton.

———. 1953. "Do Economists Know Anything?" *American Scholar* 22, no. 2 (Spring): 167–82.

MacIver, Robert M. 1967. *Academic Freedom in Our Time.* New York: Gordion Press.

MacLennon, Barbara. 1972. "Jevons' Philosophy of Science." *Manchester School of Economic and Social Sciences* 60, no. 1: 53–71.

Magee, Bryan. 1985. *Philosophy and the Real World: An Introduction to Karl Popper.* La Salle, Ill.: Open Court.

Malkiel, Burton G. 1978. "Fritz Machlup as a Teacher." In Dreyer, 1978, pp. 13–45.

Mandelbaum, Maurice. 1967. "Historicism." *Encyclopedia of Philosophy.* 1967 ed.

———. 1966. "Societal Laws." In Dray, 1966, pp. 330–46.

Manno, Bruno. 1977. "Michael Polanyi, 1891–1976: A Remembrance." *Journal of Humanistic Psychology* 17, no. 3 (Summer): 65–70.

Marget, A. W. 1929. "Morgenstern on the Methodology of Economic Forecasting." *Journal of Political Economy* 37, no. 3 (June): 312–39.

Marr, William, and Baldev Raj, eds. 1983. *How Economists Explain: A Reader in Methodology.* Lanham, Md.: University Press of America.

Marschak, Jacob. 1943. "Money Illusion and Demand Analysis." *Review of Economic Statistics* 25:40–48.

Bibliography

———. 1941. "A Discussion of Methods in Economics." *Journal of Political Economy* 49 (June): 441–48.

Marsden, David. 1984. "Homo Economicus and the Labour Market." In Wiles and Routh, 1984, pp. 121–53; with comment by Lévy-Garboua, 154–58.

Marshall, Alfred. 1897. "The Old Generation of Economists and the New." *Quarterly Journal of Economists* (January): 113–35.

Mason, Edward. 1982. "The Harvard Department of Economics from the Beginning to World War II." *Quarterly Journal of Economics* 97, no. 3 (August): 383–434.

Mason, Will E. 1980–1981. "Some Negative Thoughts on Friedman's Positive Economics." *Journal of Post Keynesian Economics* 3:235–55.

Masterman, Margaret, 1970. "The Nature of a Paradigm." In Lakatos and Musgrave, 1970, pp. 59–89.

Matuszewski, T. 1980. "Misère de l'économique." *Canadian Journal of Economics* 13, no. 4 (November): 539–47; with English translation, "The Poverty of Economics," pp. 547–55.

Mayer, Thomas. 1980. "Economics as a Hard Science: Realistic Goal or Wishful Thinking?" *Economic Inquiry* 18 (April): 165–78.

———. 1975. "Selecting Economic Hypotheses by Goodness of Fit." *Economic Journal* 85 (December): 877–883.

Mayo, Deborah. 1981. "Testing Statistical Testing." In Pitt, 1981, pp. 175–203.

McClennan, E. F. 1981. "Constitutional Choice: Rawls vs. Harsanyi." In Pitt, 1981, pp. 93–109.

McCloskey, Donald N. 1985a. "A Conversation with Donald N. McCloskey About Rhetoric." *Eastern Economic Journal* 11, no. 4 (October–December): 293–96.

———. 1985b. *The Rhetoric of Economics.* Madison: University of Wisconsin Press.

———. 1984a. "The Literary Character of Economics." *Daedalus* 113, no. 3 (Summer): 97–119.

———. 1984b. "Reply to Caldwell and Coats." *Journal Economic Literature* 22, no. 2 (June): 579–80.

———. 1983. "The Rhetoric of Economics." *Journal of Economic Literature* 21 (June): 481–517.

———. 1976. "Does the Past Have Useful Economics?" *Journal of Economic Literature* 14, no. 2 (June): 434–61.

McEvoy, John G. 1975. "A 'Revolutionary' Philosophy of Science: Feyerabend and the Degeneration of Critical Rationalism into Skeptical Fallibilism." *Philosophy of Science* 42, no. 1 (March): 49–66.

McKenzie, Richard B. 1983. *The Limits of Economic Science.* Boston: Kluwer–Nijhoff.

———. 1981. "The Necessary Normative Context of Positive Economics." *Journal of Economic Issues* 15 (September): 703–19.

———. 1979. "The Non-Rational Domain and the Limits of Economics." *Southern Economic Journal* 46, no. 1 (July): 145–57.

———. 1978. "On the Methodological Boundaries of Economic Analysis." *Journal of Economic Issues* 12, no. 3 (September): 627–45.

———. 1977. "Where Is the Economics in Economic Education?" *Journal of Economic Education* 9, no. 1 (Fall): 5–13.

McMullin, Ernan. 1979. "The Ambiguity of 'Historicism.' " In Asquith and Kyburg, 1979, pp. 55–83.

———. 1976a. "The Fertility of Theory and the Unit of Appraisal in Science." In Cohen, Feyerabend, and Wartofsky, 1976, pp. 395–432.

———. 1976b. "History and Philosophy of Science: A Marriage of Convenience?" In

PSA 1974. Boston Studies in the Philosophy of Science, vol. 32, edited by R. S. Cohen, C. A. Hooker, et al., pp. 585–601. Dordrecht, The Netherlands: D. Reidel, 1976.

———. 1970. "The History and Philosophy of Science: A Taxonomy." In Stuewer, 1970, pp. 12–67.

———. 1969. "Philosophies of Nature." *New Scholasticism* 43, no. 1 (Winter): 29–74.

Medvedev, Zhores A. 1969. *The Rise and Fall of T. D. Lysenko.* Translated by I. Michael Lerner. New York: Columbia University Press.

Meek, Ronald L. 1967. *Economics and Ideology and Other Essays.* London: Chapman and Hall.

———. 1964. "Value-Judgements in Economics." *British Journal for the Philosophy of Science* 15, no. 58 (August): 89–96.

Meier, Alfred, and Daniel Mettler. 1985. "Auf der Suche nach einem neuen Paradigma der Wirtschaftspolitik." *Kyklos* 38, no. 2: 171–99.

Meissner, Werner, and Herman Wold. 1974. "The Foundation of Science on Cognitive Mini-Models, with Applications to the German *Methodenstreit* and the Advent of Econometrics." In Leinfellner and Köhler, 1974, pp. 111–46.

Melitz, Jack. 1965. "Friedman and Machlup on the Significance of Testing Economic Assumptions." *Journal of Political Economy* 73, no. 1 (February): 37–60.

Meltzer, Allan H. 1980. "Monetarism and the Crisis in Economics." *The Public Interest* (Special Issue: The Crisis in Economic Theory), pp. 35–45.

Menger, Carl. 1968–1970. *Gesammelte Werke.* Edited and with an introduction by F. A. Hayek. 2d ed. 4 vols. Tübingen: J.C.B. Mohr. Vol. 1, *Grundsätze der Volkswirtschaftslehre* (1871), 1968; vol. 2, *Untersuchungen über die Methode der Sozialwissenschaften und der politischen Ökonomie insbesondere* (1883), 1969; vol. 3, *Kleinere Schriften zur Methode und Geschichte der Volkswirtschaftslehre,* 1970; vol. 4, *Schriften über Geld und Währungspolitik,* 1970.

Menger, Karl. 1979. "A Counterpart of Occam's Razor." In his *Selected Papers in Logic and Foundations, Didactics, Economics,* pp. 105–34. Dordrecht, The Netherlands: D. Reidel, 1979.

Merrill, G. H. 1980. "Moderate Historicism and the Empirical Sense of 'Good Science.' " In Asquith and Giere, 1980, pp. 223–35.

Merton, Robert K. 1976. "The Ambivalence of Scientists." In Cohen, Feyerabend, and Wartofsky, 1976, pp. 433–55.

Methodology in Economics Symposium Issue: Part II. 1980. *Journal of Economic Issues* 14, no. 1 (March).

Methodology in Eoconomics Symposium: Part I. 1979. *Journal of Economic Issues* 14, no. 4 (December).

Meyer, Willi. 1975. "Values, Facts, and Science: On the Problem of Objectivity in Economics." *Zeitschrift für die gesamte Staatswissenschaft* 131 (July): 514–39.

———. 1972. "Wissenschaftstheorie und Erfahrung: Zur Überwindung des methodologischen Dogmatismus." *Zeitschrift für allgemeine Wissenschaftstheorie* 3, no. 2: 267–84.

Mill, James. 1966. "Whether Political Economy Is Useful." In *James Mill: Selected Economic Writings,* edited and with an introduction by Donald Winch, pp. 371–82. Chicago: University of Chicago Press.

Mill, John Stuart. 1984 (1867). "Inaugural Address Delivered to the University of St. Andrews" (on 1 February 1867). In his *Essays on Equality, Law, and Education,* vol. 21 of *Collected Works of John Stuart Mill,* edited by John M. Robson, pp. 213–57. Toronto: University of Toronto Press.

———. 1974 (1874). *Essays on Some Unsettled Questions of Political Economy.* 2d ed. Clifton, N.Y.: Kelley.

———. 1972. *Utilitarianism, On Liberty, and Considerations on Representative Government.* Edited by H. B. Acton. London: J. M. Dent.

———. 1965 (1843). *A System of Logic.* London: Longman's, Green.

———. 1929 (1848). *Principles of Political Economy with Some of Their Applications to Social Philosophy.* Edited by W. J. Ashley. London: Longmans.

Mini, Piero. 1974. *Philosophy and Economics.* Gainesville: The University Presses of Florida.

Mink, Louis O. 1967. "Comment on Stephen Toulmin's 'Conceptual Revolutions in Science.' " In Cohen and Wartofsky, 1967, pp. 348–55.

———. 1966. "The Autonomy of Historical Understanding." In Dray, 1966, pp. 160–92.

Mirowski, Philip. 1987a. "Book Review: *The Logic of Discovery.*" *Journal of Economic History* 42, no. 1 (March): 295–96.

———. 1987b. "Shall I Compare Thee to a Minkowski-Ricardo-Leontief-Metzler Matrix of the Masak-Hicks Type? Or, Rhetoric, Mathematics, and the Nature of Neoclassical Economic Theory." *Economics and Philosophy* 3, no. 1 (April): 67–95.

———. 1986a. "Institutions as a Solution Concept in a Game Theory Context." In his 1986d, pp. 241–63.

———. 1986b. "Introduction: Paradigms, Hard Cores, and Fugleman in Modern Economic Theory." In his 1986d, pp. 1–11.

———. 1986c. "Mathematical Formalism and Economic Explanation." In his 1986d, pp. 179–240.

———, ed. 1986d. *The Reconstruction of Economic Theory.* Boston: Kluwer-Nijhoff.

———. 1984. "Physics and the 'Marginalist Revolution.' " *Cambridge Journal of Economics* 8:361–79.

———. 1974. "The Role of Conversion Principles in Twentieth-Century Economic Theory." *Philosophy of the Social Sciences* 14:461–73.

Mises, Ludwig von, see von Mises.

Mishan, E. J. 1969. *Twenty-One Popular Economic Fallacies.* London: Allen Lane.

Mongin, Philippe. 1984. "Modèle rationnel ou modèle économique de la rationalité?" *Revue économique* 35, no. 1 (January): 9–63.

Moody, Ernest A. 1967. "William of Ockham." *Encyclopedia of Philosophy.* 1967 ed.

Moore, William J. 1973. "The Relative Quality of Graduate Programs in Economics, 1958–1972: Who Published and Who Perished." *Western Economic Journal* 11 (1973): 1–23.

———. 1972. "The Relative Quality of Economics Journals: A Suggested Rating System." *Western Economic Journal* 10, no. 2 (June): 156–69.

Morgan, Brian. 1981. "Sir John Hicks's Contribution to Economic Theory." In Shackleton and Locksley, 1981, pp. 108–40.

Morgenstern, Oskar. 1976a. "Experiment and Large Scale Computation in Economics." In his 1976d, pp. 405–39.

———. 1976b. "Limits to the Uses of Mathematics in Economics." In his 1976d, pp. 441–58.

———. 1976c. "Logistics and the Social Sciences." In his 1976d, pp. 389–404.

———. 1976d. *Selected Economic Writings of Oskar Morgenstern.* Edited by Andrew Schotter. New York: New York University Press.

———. 1972a. "Descriptive, Predictive and Normative Theory." *Kyklos* 225, no. 4: 669–714.

————. 1972b. "Thirteen Critical Points in Contemporary Economic Theory: An Interpretation." *Journal of Economic Literature* 10, no. 4 (December): 1163–89.

————. 1963. *On the Accuracy of Economic Observations.* 2d completely rev. ed. Princeton: Princeton University Press.

————. 1935. "Vollkommene Voraussicht und wirtschaftliches Gleichgewicht." *Zeitschrift für Nationalökonomie* 6, no. 3 (August): 337–57; translated by Frank H. Knight as "Perfect Foresight and Economic Equilibrium" in Morgenstern, 1976d, pp. 169–83.

————. 1934. *Die Grenzen der Wirtschaftspolitik.* Vienna: Julius Springer Verlag. Translated by Vera Smith as *The Limits of Economics.* London: W. Hodge, 1937.

————. 1928. *Wirtschaftsprognose, eine Untersuchung ihrer Voraussetzungen und Möglichkeiten.* Vienna: Julius Springer Verlag.

Morgenstern, Oskar, and Gerhard Schwödiauer. 1976. "Competition and Collusion in Bilateral Markets." *Zeitschrift für Nationalökonomie* 36:217–45.

Morishima, Michio. 1984a. "The Good and Bad Uses of Mathematics." In Wiles and Routh, 1984, pp. 51–73.

————. 1984b. "A Reply to Gorman." In Wiles and Routh, 1984, pp. 74–77.

Mortimore, G. W., and J. B. Maund. 1976. "Rationality in Belief." In Benn and Mortimore, 1976, pp. 11–33.

Moulines, C. Ulises. 1983. "On How the Distinction Between History and Philosophy of Science Should Not Be Drawn." *Erkenntnis* 19 (May): 285–96.

Mueller, Dennis. 1984. "Further Reflections on the Invisible-Hand Theorem." In Wiles and Routh, 1984, pp. 159–83.

Mulkay, Michael, and G. Nigel Gilbert. 1981. "Putting Philosophy to Work: Karl Popper's Influence on Scientific Practice." *Philosophy of the Social Sciences* 11:389–407.

Munitz, Milton K. 1981. *Contemporary Analytic Philosophy.* New York: Macmillan.

Musgrave, Alan. 1981. " 'Unreal Assumptions' in Economic Theory: The F-Twist Untwisted." *Kyklos* 34, no. 3: 377–87.

————. 1979. "Theorie, Erfahrung und wissenschaftlicher Fortschritt." In *Theorie und Erfahrung,* edited by Hans Albert and Kurt Stapf, pp. 21–53. Stuttgart: Klett.

————. 1976. "Method or Madness?" In Cohen, Feyerabend, and Wartofsky, 1976, pp. 457–91.

————. 1974. "Logical Versus Historical Theories of Confirmation." *British Journal for the Philosophy of Science* 25, no. 1 (March): 1–23.

————. 1973. "Falsification and Its Critics." In Suppes et al., 1973, pp. 393–406.

Myrdal, Gunnar. 1978. "Institutional Economics." *Journal of Economic Issues* 12, no. 4 (December): 771–83.

————. 1973 (1972). *Against the Stream: Critical Essays on Economics.* New York: Pantheon.

————. 1969. *Objectivity in Social Research.* New York: Pantheon.

————. 1958. *Value in Social Theory.* Edited by Paul Streeten. London: Routledge and Kegan Paul.

————. 1953. *The Political Element in the Development of Economic Theory.* Translated by Paul Streeten. London: Routledge and Kegan Paul.

————. 1933. "Das Zweck-Mittel-Denken in der Nationalökonomie." *Zeitschrift für Nationalökonomie* 4:305–29.

Nagel, Ernest. 1966. "Determinism in History." In Dray, 1966, pp. 347–82.

————. 1963. "Assumptions in Economic Theory." *American Economic Review: Supplement* 53, no. 2 (May): 211–19.

Nagel, Ernest, and James R. Newman. 1956. "Gödel's Proof." *Scientific American* 196, no. 6 (June): 71–86.

Naughton, J. 1978. "The Logic of Scientific Economics: A Response to Bray." *Journal of Economic Studies* 5, no. 2: 152–65.

Neild, Robert. 1984. "The Wider World and Economic Methodology." In Wiles and Routh, 1984, pp. 37–46; with comments by Hart, pp. 47–50.

Nelson, Richard R., and Sidney G. Winter. 1982. *An Evolutionary Theory of Economic Change.* Cambridge: Harvard University Press, Belknap Press.

Neumann, John von, see von Neumann.

Neurath, Otto. 1970 (1944). *Foundations of the Social Sciences.* Vol. 2, no. 1, of the International Encyclopedia of Unified Science. Chicago: University of Chicago Press.

Newton-Smith, W. H. 1981. *The Rationality of Science.* London: Routledge and Kegan Paul.

Ng, Yew-Kwang. 1972. "Value Judgements and Economists' Role in Policy Recommendation." *Economic Journal* 82, no. 327 (September): 1014–18.

Niehans, Jürg. 1981. "Economics: History, Doctrine, Science, Art." *Kyklos* 34, no. 2: 165–77.

Nipperdey, Thomas. 1985. *Deutsche Geschichte 1800–1866.* 3d rev. ed. München: C. H. Beck.

North, Douglass C. 1978. "Structure and Performance: The Task of Economic History." *Journal of Economic Literature* 26, no. 3 (September): 963–78.

———. 1977a. "Markets and Other Allocation Systems in History: The Challenge of Karl Polanyi." *Journal of European Economic History* 6, no. 3 (Winter): 703–16.

———. 1977b. "The New Economic History After Twenty Years." *American Behavioral Scientist* 21, no. 2 (November–December): 187–200.

———. 1965. "The State of Economic History." *American Economic Association Papers and Proceedings* 60 (May): 86–91.

Northrop, Dexter B. 1980. "Pursuing Scientific Truth." *Chemical and Engineering News* 58, no. 20 (May 19): 4.

Oakeshott, Michael. 1966. "Historical Continuity and Causal Analysis." In Dray, 1966, pp. 193–212.

O'Brien, D. P. 1983. "Theories of the History of Science: A Test Case." In Coats, 1983c, pp. 89–124.

———. 1976. "The Longevity of Adam Smith's Vision: Paradigms, Research Programmes and Falsifiability in the History of Economic Thought." *Scottish Journal of Political Economy* 23, no. 2 (June): 133–51.

———. 1974. "Whither Economics? An Inaugural Lecture." Durham: University of Durham.

Odagiri, Hiroyuki. 1984. "The Firm as a Collection of Human Resources." In Wiles and Routh, 1984, pp. 190–206; with comments by Hart, pp. 207–10.

O'Driscoll, Gerald Patrick, and Mario J. Rizzo. 1984. *The Economics of Time and Ignorance.* London: Basil Blackwell.

O'Driscoll, Gerald P., and Sudha R. Shenoy. 1976. "Inflation, Recession, and Stagflation." In Dolan, 1976b, pp. 185–211.

O'Toole, Margot. 1989. "Scientists Must Be Able to Disclose Colleagues' Mistakes Without Risking Their Own Jobs or Financial Support." *Chronicle of Higher Education,* 25 January, p. A44.

Papandreou, Andreas George. 1958. *Economics as a Science.* Philadelphia: Lippincott.

Papineau, David. 1980. "Review of *Knowledge and Ignorance in Economics* by T. W. Hutchison." *British Journal for the Philosophy of Science* 31:98–103.

Parrish, John B. 1967. "Rise of Economics as an Academic Discipline: The Formative Years to 1900." *Southern Economic Journal* 34, no. 1 (July): 1–16.

Parsons, Talcott. 1934. "Some Reflections on 'The Nature and Significance of Economics.' " *Quarterly Journal of Economics* 48:511–45.

Pasinetti, Luigi L. 1986. "Theory of Value–A Source of Alternative Paradigms in Economic Analysis." In Baranzini and Scazzieri, 1986a, pp. 409–31.

Passmore, John. 1974. "The Poverty of Historicism Revisited." *History and Theory,* Beiheft 14: *Essays on Historicism* 14, no. 4: 30–47.

———. 1967a. "Logical Positivism." *Encyclopedia of Philosophy.* 1967 ed.

———. 1967b. "Philosophy." *Encyclopedia of Philosophy.* 1967 ed.

———. 1966. "The Objectivity of History." In Dray, 1966, pp. 75–94.

———. 1960. "Popper's Account of Scientific Method." *Philosophy* 35:326–31.

Paul, Ellen Frankel. 1979. *Moral Revolution and Economic Science: The Demise of Laissez-Faire in Nineteenth-Century British Economy.* Westport, Conn.: Greenwood Press.

Pawlowski, Tadeusz. 1975. *Methodologische Probleme in den Geistes- und Sozialwissenschaften.* Translated from Polish into German by Georg Grzyb. Warszawa: PWN-Polnischer Verlag der Wissenschaften.

Peabody, Gerald E. 1971. "Scientific Paradigms and Economics: An Introduction." *Review of Radical Political Economics* 3, no. 2 (July): 1–16.

Perlman, Mark. 1978a. "Reflections on Methodology and Persuasion." In Dreyer, 1978, pp. 37–45.

———. 1978b. "Review of Hutchison's *Knowledge and Ignorance in Economics.*" *Journal of Economic Literature* 16 (June): 582–85.

Peston, Maurice. 1981. "Lionel Robbins: Methodology, Policy and Modern Theory." In Shackleton and Locksley, 1981, pp. 183–98.

Pfouts, Ralph W. 1973. "Some Proposals for a New Methodology in Economics." *Atlantic Economic Journal* 1, no. 1 (November): 13–22.

Pheby, John. 1988. *Methodology and Economics: A Critical Introduction.* London: Macmillan.

Phelps Brown, Sir E. H. 1980. "The Radical Reflections of an Economist." *Bianca Nazionale del Lavoro Quarterly Review* 132, no. 33: 3–14.

———. 1972. "The Underdevelopment of Economics." *Economic Journal* 82 (March): 1–10.

"Philosophy Comes Down from the Clouds." 1986. *The Economist,* 26 April, pp. 101–5.

"The Philosophy of Science." 1987. *The Economist,* 25 April, pp. 80–81.

Piron, Robert. 1962. "On 'The Methodology of Positive Economics': Comment." *Quarterly Journal of Economics* 76:664–66.

Pitt, Joseph C., ed. 1981. *Philosophy in Economics: Papers Deriving from and Related to a Workshop on Testability and Explanation in Economics Held at Virginia Polytechnic Institute and State University, 1979.* University of Western Ontario Series in Philosophy of Science, vol. 16. Dordrecht, The Netherlands: D. Reidel.

Pitt, Joseph C., and Marcello Pera, eds. 1987. *Rational Changes in Science: Essays on Scientific Reasoning.* Boston Studies in the Philosophy of Science, vol. 98. Dordrecht, The Netherlands: D. Reidel.

Poirier, Dale J. 1988. "Frequentist and Subjectivist Perspectives on the Problems of Model Building in Economics." *Journal of Economic Perspectives* 2, no. 1 (Winter): 121–44; with discussion (Rust, Pagan, Geweke, Poirier), 145–70.

Polanyi, Karl. 1968. "Our Obsolete Market Mentality." In *Primitive, Archaic, and*

Modern Economics: Essays of Karl Polanyi, edited by George Dalton, pp. 59–77. New York: Anchor Books, 1968.

Polanyi, Michael. 1969. *Knowing and Being.* Chicago: University of Chicago Press.

———. 1967 (1966). *The Tacit Dimension.* New York: Anchor Books.

———. 1963. "The Potential Theory of Adsorption." *Science* 141, no. 3585 (13 September): 1010–13.

———. 1962. "The Unaccountable Element in Science." *Philosophy* 37, no. 139 (January): 1–14.

———. 1958. *Personal Knowledge: Towards a Post-Critical Philosophy.* Chicago: University of Chicago Press.

———. 1957. "Scientific Outlook: Its Sickness and Cure." *Science* 125, no. 3245 (8 March): 480–84.

———. 1955. "Words, Conceptions and Science." *The Twentieth Century* 158 (July–December): 256–67.

———. 1952. "The Stability of Beliefs." *British Journal for the Philosophy of Science* 3, no. 2 (November): 218–19.

Polanyi, Michael, and Harry Prosch. 1975. *Meaning.* Chicago: University of Chicago Press.

Pollard, Sidney. 1987. "The Concept of the Industrial Revolution." Bielefeld University Working Paper.

———. 1968. *The Idea of Progress.* London: C. A. Watts.

Pope, David, and Robin Pope. 1972a. "In Defense of Predictionism." *Australian Economic Papers* 11, no. 19 (December): 232–238.

———. 1972b. "Predictionists, Assumptionists, and the Relatives of the Assumptionists." *Australian Economic Papers* 11, no. 19 (December): 224–28.

Popper, Sir Karl. 1988. "The Open Society and Its Enemies Revisited." *The Economist,* 23 April, pp. 25–28.

———. 1987a. "The Myth of the Framework." In Pitt and Pera, 1987, pp. 35–62.

———. 1987b. "Natural Selection and the Emergence of the Mind." In Radnitzky and Bartley, 1987, pp. 139–53.

———. 1987c. "Zur Theorie der Demokratie." *Der Spiegel,* 3 August, pp. 54–55.

———. 1984 (1976). *Ausgangspunkte: meine intellektuelle Entwicklung.* 3d ed. Hamburg: Hoffman und Campe Verlag.

———. 1983 (1956). *Realism and the Aim of Science.* Vol. 1 of his *The Postscript to The Logic of Scientific Discovery.* Edited by W. W. Bartley III. London: Hutchinson.

———. 1982a (1956). *The Open Universe: An Argument for Indeterminism.* Vol. 2 of his *The Postscript to The Logic of Scientific Discovery.* Edited by W. W. Bartley. London: Hutchinson.

———. 1982b (1956). *Quantum Theory and the Schism in Physics.* Vol. 3 of *The Postscript to The Logic of Scientific Discovery.* Edited by W. W. Bartley. London: Hutchinson.

———. 1979 (1972). *Objective Knowledge: An Evolutionary Approach.* Rev. ed. Oxford: Oxford University Press.

———. 1976a. "The Logic of the Social Sciences." In Adorno et al., 1976, pp. 87–104.

———. 1976b. *Logik der Forschung.* 6th rev. ed. Wien: J. Springer.

———. 1976c. "Reason or Revolution?" In Adorno et al., 1976, pp. 288–300.

———. 1975. "The Rationality of Scientific Revolutions." In *Problems of Scientific Revolutions,* edited by Rom Harré, pp. 72–101. Oxford: Clarendon Press, 1975. Repr. in Hacking, 1976b, pp. 80–106.

———. 1974a. "Autobiography of Karl Popper." In Schilpp, 1974, pp. 3–181.

————. 1974b. "Karl Popper: Replies to My Critics." In Schilpp, 1974, pp. 963–1197.

————. 1972a (1963). *Conjectures and Refutations*. 4th ed. London: Routledge and Kegan Paul.

————. 1972b (1959). *The Logic of Scientific Discovery*. 6th ed. London: Hutchinson.

————. 1972c. "Prediction and Prophecy in the Social Sciences." In his 1972a, pp. 336–46. Repr. in *Theories of History,* edited by Patrick Gardiner. New York: The Free Press, pp. 276–85.

————. 1971 (1965). *Das Elend des Historizismus*. 5th, rev. ed. Translated from 2d ed. Tübingen: J.C.B. Mohr.

————. 1970. "Normal Science and Its Dangers." In Lakatos and Musgrave, 1970. pp. 51–58.

————. 1968. "Remarks on the Problems of Demarcation and of Rationality." In Lakatos and Musgrave, 1968, pp. 88–102.

————. 1967. "La rationalité et la statut du principe de rationalité." In *Les fondements philosophiques des systèmes économiques: textes de Jacques Rueff et essais rédigés en son honneur,* edited by Emil M. Claassen, pp. 142–50. Paris: Payot, 1967.

————. 1966a (1962). *The Open Society and Its Enemies*. 5th rev. ed. Vol. 1, *The Spell of Plato*. Princeton: Princeton University Press.

————. 1966b (1962). *The Open Society and Its Enemies*. 5th rev. ed. Vol. 2, *The High Tide of Prophecy: Hegel, Marx, and the Aftermath*. Princeton: Princeton University Press.

————. 1964. "Über die Unwiderlegbarkeit philosophischer Theorien einschließlich jener, welche falsch sind." In *Club Voltaire Jahrbuch für kritische Aufklärung I,* pp. 271–79. München: Szczesny Verlag, 1964.

————. 1963. "The Demarcation Between Science and Metaphysics." In Schilpp, 1963, pp. 183–226.

————. 1960 (1957). *The Poverty of Historicism*. 2d ed. London: Routledge and Kegan Paul.

Postan, Michael M. 1968. "A Plague of Economists? On Some Current Myths, Errors, and Fallacies." *Encounter* 30 (January): 42–47.

The Public Interest. 1980. Special Issue: The Crisis in Economic Theory.

Putnam, Hilary. 1981. "The 'Corroboration' of Theories." In Hacking, 1981b, pp. 60–80.

Puu, Tonu. 1967. "Some Reflections on the Relation Between Economic Theory and Empirical Reality." *Swedish Journal of Economics* 69, no. 2 (June): 86–114.

Quine, Willard van Orman. 1976. "Two Dogmas of Empiricism." In Harding, 1976a, pp. 41–64.

————. 1962. "Paradox." *Scientific American* 206, no. 4 (April): 84–96.

Quinn, Philip. 1972. "Methodological Appraisal and Heuristic Advice." *Studies in the History and Philosophy of Science* 3:135–49.

"Radical Paradigms in Economics." 1971. *Review of Radical Political Economy* (Special Issue on Radical Paradigms in Economics) 3, no. 2 (July).

Radner, Michael, and Stephen Winokur, eds. 1970. *Analyses of Theories and Methods of Physics and Psychology*. Minnesota Studies in the Philosophy of Science, vol. 4. Minneapolis: University of Minnesota Press.

Radnitzky, Gerard. 1976. "Popperian Philosophy of Science as an Antidote Against Relativism." In Cohen, Feyerabend, and Wartofsky, 1976, pp. 505–46.

Radnitzky, Gerard, and Gunnar Andersson. 1978. *Progress and Rationality in Science,* Boston Studies in the Philosophy of Science, vol. 58. Dordrecht, The Netherlands. D. Reidel.

Radnitzky, Gerard, and William Bartley, eds. 1987. *Evolutionary Epistemology, Rationality, and the Sociology of Knowledge.* La Salle, Ill.: Open Court.

Ranson, Baldwin. 1980. "Rival Economic Epistemologies: The Logic of Marx, Marshall, and Keynes." *Journal of Economic Issues* 14 (March): 77–98.

Rappaport, Steven. 1986. "The Modal View and Defending Microeconomics." In *PSA 1986*, vol. 1, edited by Arthur Fire and Peter Machamer, pp. 289–97. East Lansing, Mich.: Philosophy of Science Association.

Ravetz, Jerome. 1985. "The History of Science." *Encyclopedia Britannica.* 1985 ed.

Reder, Melvin W. 1982. "Chicago Economics: Permanence and Change." *Journal of Economic Literature* 20, no. 1 (March): 1–38.

Redman, Barbara. 1976. "On Economic Theory and Explanation." *Journal of Behavioral Economics* 5 (Summer): 161–76.

Redman, Deborah A. Forthcoming. *Thirty Years of Rational Expectations: A Concise Introduction with a Comprehensive, Annotated Bibliography.*

———. 1989. *Economic Methodology: A Bibliography with References to Works in the Philosophy of Science (1860–1988).* Bibliographies and Indexes in Economics and Economic History, no. 9. Westport, Conn.: Greenwood Press.

———. 1981. "Adam Smith and the Scientific Nature of Economics." Master's thesis, Pennsylvania State University.

Reed, Mike. 1979. "Rationality and Neoclassical Economics." *Economic Forum* 10, no. 1 (Summer): 91–93.

Reese, William L. 1980. "Vienna Circle of Logical Positivists." *Dictionary of Philosophy and Religion.* 1980 ed.

Reichenbach, Hans. 1938. *Experience and Prediction: An Analysis of the Foundations and the Structure.* Chicago: University of Chicago Press.

Reynolds, L. 1976. "The Nature of Revolutions in Economics." *Intermountain Economic Review* 7, no. 1 (Spring): 25–33.

Rima, Ingrid. 1978. *Development of Economic Analysis.* 3d ed. Homewood, Ill.: Richard D. Irwin.

Ritter, Gerhard. 1946. "The German Professor in the Third Reich." *Review of Politics* 8:242–54.

Rivett, Kenneth. 1972. "Comment on Pope and Pope." *Australian Economic Papers* 11, no. 19 (December): 228–32.

———. 1970. " 'Suggest' Or 'Entail'?: The Derivation and Confirmation of Economic Hypotheses." *Australian Economic Papers* 9, no. 15 (December): 127–48.

Robbins, Lionel. 1983. "On Latsis's 'Method and Appraisal in Economics': A Review Essay." In Coats, 1983c, pp. 43–54.

———. 1984 (1932). *An Essay on the Nature and Significance of Economic Science.* 3d ed. Foreword by William Baumol. London: Macmillan.

———. 1981. "Economics and Political Economy." *American Economic Association Papers and Proceedings* 71, no. 2 (May): 1–10.

———. 1938. "Live and Dead Issues in the Methodology of Economics." *Economica,* n.s. 5 (August): 342–52.

Roberts, Marc J. 1974. "On the Nature and Condition of Social Science." *Daedalus* 103, no. 2 (Summer): 47–64.

Robinson, Joan. 1981a. "The Disintegration of Economics." In her 1981f, pp. 96–104.

———. 1981b. "Marxism: Religion and Science." In her 1981f, pp. 155–64.

———. 1981c. "Survey: 1950's." In her 1981f, pp. 105–111.

———. 1981d. "Thinking About Thinking." In her 1981f, pp. 54–63.

———. 1981e. "Time in Economic Theory." *Kyklos* 33, no. 2: 219–29.

———. 1981f. *What Are the Questions? and Other Essays.* Armonk, N.Y.: M. E. Sharpe.

———. 1977. "What Are the Questions?" *Journal of Economic Literature* 15, no. 4 (December): 1318–39.

———. 1973. "What Has Become of the Keynesian Revolution?" In *After Keynes,* edited by Joan Robinson, pp. 1–11. Oxford: Basil Blackwell, 1973.

———. 1972. "The Second Crisis of Economic Theory." *American Economic Review* 62, no. 2 (May): 1–10.

———. 1970. *Freedom and Necessity.* New York: Pantheon Books.

———. 1962. *Economic Philosophy.* London: C. A. Watts.

Rohatyn, D. A. 1975. *Naturalism and Deontology: An Essay on the Problem of Ethics.* The Hague: Mouton.

Rosefielde, Steven. 1976. "Economic Theory in the Excluded Middle Between Positivism and Rationalism." *Atlantic Economic Journal* 4, no. 2 (Spring): 1–9.

Rosenberg, Alexander. 1986a. "The Explanatory Role of Existence Proofs." *Ethics* 97 (October): 177–86.

———. 1986b. "Lakatosian Consolations for Economics." *Economics and Philosophy* 2, no. 1 (April): 127–39.

———. 1983. "If Economics Isn't Science, What Is It?" *Philosophical Forum* 14, nos. 3–4 (Spring–Summer): 296–314.

———. 1980. "A Skeptical History of Microeconomic Theory." *Theory and Decision* 12, no. 1 (March): 79–93. Slightly rev. version in Pitt, 1981, pp. 47–61.

———. 1979. "Can Economic Theory Explain Everything?" *Philosophy of the Social Sciences* 9 (December): 509–29.

———. 1978a. "The Puzzle of Economic Modelling." *Journal of Philosophy* 75, no. 11 (November): 679–83.

———. 1978b. "Weintraub's Aims: A Brief Rejoinder." *Economics and Philosophy* 3, no. 1 (April): 143–44.

———. 1976a. *Microeconomic Laws: A Philosophical Analysis.* Pittsburgh: University of Pittsburgh Press.

———. 1976b. "On the Interanimation of Micro- and Macroeconomics." *Philosophy of the Social Sciences* 6:35–53.

———. 1974. "Partial Interpretation and Microeconomics." In Leinfellner and Köhler, 1974, pp. 93–109.

———. 1972. "Friedman's 'Methodology' for Economics: A Critical Examination." *Philosophy of the Social Sciences* 2:15–29.

Rothacker, Erich. 1960. "Das Wort 'Historismus.' " *Zeitschrift für deutsche Wortforschung* 16:3–6.

Rothbard, Murray N. 1976a. "The Austrian Theory of Money." In Dolan, 1976b, pp. 160–84.

———. 1976b. "New Light on the Prehistory of the Austrian School." In Dolan, 1976b, pp. 52–74.

———. 1976c. "Praxeology: The Methodology of Austrian Economics." In Dolan, 1976b, pp. 19–39.

———. 1976d. "Praxeology, Value Judgements, and Public Policy." In Dolan, 1976b, pp. 89–111.

Rotwein, Eugene. 1980. "Friedman's Critics: A Critic's Reply to Boland." *Journal of Economic Literature* 18 (December): 1553–55.

———. 1973. "Empiricism and Economic Method: Several Views Considered." *Journal of Economic Issues* 7, no. 3 (September): 361–82.

———. 1962. "On 'The Methodology of Positive Economics': Reply." *Quarterly Journal of Economics* 76:666–68.

———. 1959. "On 'The Methodology of Positive Economics.' " *Quarterly Journal of Economics* 73, no. 4 (November): 554–75.

Rousseas, Stephen. 1981. "Wiles' Wily Weltanschauung." *Journal of Post Keynesian Economics* 3, no. 3 (Spring): 340–51.

———. 1973. "Paradigm Polishing Versus Critical Thought in Economics." *American Economist* 17 (Fall): 72–78.

Routh, Guy. 1984. "What to Teach Undergraduates." In Wiles and Routh, 1984, pp. 240–48; with comment by Morris Perlman, 249–59.

———. 1975. *The Origin of Economic Ideas*. London: Macmillan.

Rowe, Leo S. 1890. "Instruction in Public Law and Political Economy in German Universities." *Annals of the American Academy of Political and Social Science*, July, pp. 78–102.

Rowley, Robin, and Omar Hamouda. 1987. "Troublesome Probability and Economics." *Journal of Post Keynesian Economics* 10, no. 1 (Fall): 44–64.

Rudner, Richard. 1953. "The Scientist *Qua* Scientist Makes Value Judgements." *Philosophy of Science* 20, no. 1 (January): 1–6.

Sagal, Paul T. 1972. "Incommensurability Then and Now." *Zeitschrift für allgemeine Wissenschaftstheorie* 3, no. 2: 298–301.

Sakar, Husain. 1980. "Imre Lakatos's 'Meta-Methodology': An Appraisal." *Philosophy of the Social Sciences* 10:397–416.

Salant, Walter S. 1969. "Writing and Reading in Economics." *Journal of Political Economy* 77, no. 4 (July/August): 545–58.

Salanti, Andrea. 1982. "Neoclassical Tautologies and the Cambridge Controversies: Comment on Dow." *Journal of Post Keynesian Economics* 5, no. 1 (Fall): 128–31.

Salmon, Wesley C. 1973. "Confirmation." *Scientific American* 228, no. 5 (May): 75–83.

———. 1966. *The Foundations of Scientific Inference*. Pittsburgh: University of Pittsburgh Press.

Samuels, Warren, ed. 1980. *The Methodology of Economic Thought: Critical Papers from the Journal of Economic Issues*. New Brunswick, N.J.: Transaction Books.

———. 1974. "The History of Economic Thought as Intellectual History." *History of Political Economy* 6, no. 3 (Fall): 305–23.

Samuelson, Paul. 1983 (1947). *Foundations of Economic Analysis*. Enl. ed. Cambridge: Harvard University Press.

———. 1972. "Economics in a Golden Age: A Personal Memoir." In *The Twenthieth-Century Sciences: Studies in the Biography of Ideas*, pp. 155–70. New York: Norton, 1972. Repr. in E. C. Brown and Solow, 1983, pp. 1–14.

———. 1965a. "Economic Forecasting and Science." *Michigan Quarterly Review* 4 (October): 274–80.

———. 1965b. "Professor Samuelson on Theory and Realism: Reply." *American Economic Review* 55 (December): 1164–72.

———. 1964. "Theory and Realism: A Reply." *American Economic Review* 54 (September): 736–39.

———. 1963. "Problems of Methodology—Discussion." *American Economic Association Papers and Proceedings* 53 (May): 231–36.

Sargeant, J. R. 1963. "Are American Economists Better?" *Oxford Economic Papers*, n.s. 15, no. 1 (March): 1–7.

Sargent, Thomas. 1986. *Rational Expectations and Inflation*. New York: Harper and Row.

Schabas, Margaret. 1986. "An Assessment of the Scientific Standing of Economics." In Fire and Machamer, 1986, pp. 289–97.

Schagrin, Morton L. 1973. "On Being Unreasonable." *Philosophy of Science* 40, no. 1 (March): 1–9.

Scharnberg, Max. 1984. *The Myth of Paradigm-shift, Or How to Lie with Methodology.* Uppsala Studies in Education, no. 20. Stockholm: Almquist and Wiksell International.

Scheffler, Israel. 1972. "Vision and Revolution: A Postscript on Kuhn." *Philosophy of Science* 39, no. 3: 366–74.

Schilpp, Paul Arthur, ed. 1974. *The Philosophy of Karl Popper.* 2 vols. La Salle, Ill.: Open Court.

———, ed. 1963. *The Philosophy of Rudolf Carnap.* La Salle, Ill.: Open Court.

Schleichert, Hubert, ed. 1975. *Logischer Empirismus—der Wiener Kreis. Ausgewählte Texte mit einer Einleitung.* München: Wilhelm Fink Verlag.

Schmitt, Bernard. 1986. "The Process of Formation of Economics in Relation to the Other Sciences." In Baranzini and Scazzieri, 1986a, pp. 103–32.

Schneider, Beth E. 1987. "Graduate Women, Sexual Harassment, and University Policy." *Journal of Higher Education* 58, no. 1 (January/February): 46–65.

Schoeffler, Sidney. 1955. *The Failures of Economics: A Diagnostic Study.* Cambridge: Harvard University Press.

Schotter, Andrew. 1973. "Core Allocations and Competitive Equilibrium—A Survey." *Zeitschrift für Nationalökonomie* 33:281–313.

Schotter, Andrew, and Gerhard Schwödiauer. 1980. "Economics and the Theory of Games: A Survey." *Journal of Economic Literature,* June, pp. 479–527.

Schramm, Alfred. 1974. "Demarkation und rationale Rekonstruktion bei Imre Lakatos." *Conceptus* 8:10–16.

Schumacher, E. F. 1973. "Does Economics Help?" In *After Keynes,* edited by Joan Robinson, pp. 26–36. Oxford: Basil Blackwell, 1973.

Schumpeter, Joseph A. 1982 (posthumously). "The 'Crisis' in Economics—Fifty Years Ago." *Journal of Economic Literature* 10 (September): 1049–59.

———. 1970 (1908). *Das Wesen und der Hauptinhalt der theoretischen Nationalökonomie.* Berlin: Duncker und Humblot.

———. 1954. *History of Economic Analysis.* New York: Oxford University Press.

———. 1952 (1906). "Über die mathematische Methode der theoretischen Ökonomie." *Zeitschrift für Volkswirtschaft, Sozialpolitik und Verwaltung* 15:30–49. Repr. in his *Aufsätze zur Ökonomischen Theorie,* pp. 529–48. Tübingen: J.C.B. Mohr, 1952.

———. 1933. "The Common Sense of Econometrics." *Econometrica* 1 (1933): 5–12.

Scoon, Robert. 1943. "Professor Robbins' Definition of Economics." *Journal of Political Economy* 51:310–21.

Scriven, Michael. 1966. "Causes, Connections and Conditions in History." In Dray, 1966, pp. 238–64.

Scully, Malcolm G. 1986. "Women Account for Half of College Enrollment in U.S., 3 Other Nations." *Chronicle of Higher Education* 3, no. 3 (September): A1, A42.

Seager, Henry M. 1893. "Economics at Berlin and Vienna." *Journal of Political Economy* 1 (March): 236–62.

Sebba, Gregor. 1953. "The Development of the Concepts of Mechanism and Model in Physical Science and Economic Thought." *American Economic Association Papers and Proceedings* 43, no. 2 (May): 259–68.

Seligman, Ben B. 1969. "The Impact of Positivism on Economic Thought." *History of Political Economy* 1, no. 2 (Fall): 256–78.

Sen, Amartya. 1987. *On Ethics and Economics.* Oxford: Basil Blackwell.

————. 1977. "Rational Fools: A Critique of the Behavioral Foundations of Economic Theory." *Philosophy and Public Affaris* 6, no. 4 (Summer): 317–44. Repr. in his *Choice, Welfare and Measurement,* pp. 84–106. Oxford: Basil Blackwell, 1982.

Shackle, G.L.S. 1983. "The Bounds of Unknowledge." In Wiseman, 1983a, pp. 28–37.

————. 1982. "Means and Meaning in Economic Theory." *Scottish Journal of Political Economy* 29, no. 3 (November): 223–34.

————. 1972. *Epistemics and Economics: A Critique of Economic Doctrines.* Cambridge: Cambridge University Press.

————. 1967. *The Years of High Theory: Invention and Tradition in Economic Thought, 1926–1939.* Cambridge: Cambridge University Press.

Shackleton, J. R. and Gareth Locksley, eds. 1981. *Twelve Contemporary Economists.* London: Macmillan.

Shapere, Dudley. 1981. "Meaning and Scientific Change." In Hacking, 1981b, pp. 28–59.

————. 1971. "The Paradigm Concept." *Science* 172, no. 3984 (14 May): 706–9.

————. 1967. "Isaac Newton." *Encyclopedia of Philosophy.* 1967 ed.

————. 1964. "The Structure of Scientific Revolutions." *Philosophical Review* 73 (July): 383–94.

Shapin, Steven. 1982. "History of Science and Its Sociological Reconstructions." *History of Science* 20, no. 49 (September): 157–211.

Shearer, J. O. 1939. "[Review of Hutchison's] *The Significance and Basic Postulates of Economic Theory.*" *Economic Record* 15 (June): 136–38.

Shearmur, Jeremy. 1983. "Subjectivism, Falsification, and Positive Economics." In Wiseman, 1983a, pp. 65–86.

Shubik, Martin. 1982, 1984. *Game Theory in the Social Sciences.* 2 vols. Cambridge: The MIT Press. Vol. 1, *Concepts and Solutions,* 1982; vol. 2, *A Game-Theoretic Approach to Political Economy,* 1984.

————. 1970. "A Curmudgeon's Guide to Microeconomics." *Journal of Economic Literature* 8, no. 1 (March): 405–34.

Shupak, Mark B. 1962. "The Predictive Accuracy of Empirical Demand Analysis." *Economic Journal* 72 (September): 550–75.

Silberberg, Eugene. 1978. *The Structure of Economics: A Mathematical Analysis.* New York: McGraw-Hill.

Simon, Herbert A. 1983. "Rational Decision Making in Business Organizations." Nobel lecture, 1978. In Marr and Raj, 1983, pp. 281–315.

————. 1976. "From Substantive to Procedural Rationality." In Laksis, 1976a, pp. 129–48.

————. 1960. *The New Science of Management Decision.* Rev. ed. Englewood Cliffs, N.J.: Prentice-Hall.

Skinner, Andrew S. 1983. "Adam Smith: Rhetoric and the Communication of Ideas." In Coats, 1983c, pp. 71–88.

————. 1965. "Economics and History—The Scottish Enlightenment." *Scottish Journal of Political Economy* 12 (February): 1–22.

Skouras, Thanos. 1981. "The Economics of Joan Robinson." In Shackleton and Locksley, 1981, pp. 199–218.

Smeaton, W. A. and Stafford Finlay. 1962. "Book Review: Towards an Historiography of Science." *Annals of Science* 18:125–29.

Smith, Adam. 1983 (1748–1749). *Lectures on Rhetoric and Belles Lettres.* Edited by J. C. Bryce. Oxford: Clarendon Press.

————. 1967 (ca. 1750). "The History of Astronomy." In Lindgren, 1967, pp. 53–109.

Smith, Adam [pseud.]. 1972. "The Last Days of Cowboy Capitalism." *Atlantic Monthly* 230, no. 3 (September): 43–55.

Smyth, R. L., ed. 1962. *Essays in Economic Method: Selected Papers Read to Section F of the British Association for the Advancement of Science, 1860–1913.* Introduction by T. W. Hutchison. London: Duckworth.

Sneed, Joseph D. 1981. *The Logical Structure of Mathematical Physics.* Dordrecht, The Netherlands: D. Reidel.

Snow, C. P. 1960, 1961. *Science and Government.* Cambridge: Harvard University Press.

Solow, R. M. 1985. "Economic History and Economics." *American Economic Association Papers and Proceedings* 75 (May): 328–31.

————. 1971. "Science and Ideology in Economics." *The Public Interest* 23 (Spring), pp. 94–107.

Somers, Amy. 1982. "Sexual Harassment in Academe: Legal Issues and Definitions." *Journal of Social Issues* 38, no. 4: 23–32.

Sowell, Thomas. 1974. *Classical Economics Reconsidered.* Princeton: Princeton University Press.

Spellman, William, and Bruce Gabriel. 1978. "Graduate Students in Economics 1940–74." *American Economic Review* 68 (March): 182–87.

Spengler, Joseph J. 1968a. "Economics: Its History, Themes, Approaches." *Journal of Economic Issues* 2, no. 1 (March): 42–49.

————. 1968b. "Exogenous and Endogenous Influences in the Formation of Post-1870 Economic Thought: A Sociology of the Knowledge Approach." In Eagley, 1968, pp. 159–87; with discussion, pp. 188–90.

————. 1961. "Quantification in Economics: Its History." In Lerner, 1961, pp. 129–211.

Spinner, Helmut F. 1982. *Ist der Kritische Rationalismus am Ende?* Weinheim: Beltz.

————. 1974. *Pluralismus als Erkenntnismodell.* Frankfurt am Main: Suhrkamp.

Stanfield, J. Ron. 1979. *Economic Thought and Social Change.* Carbondale and Edwardsville: Southern Illinois University Press.

————. 1974. "Kuhnian Scientific Revolutions and the Keynesian Revolution." *Journal of Economic Issues* 8 (March): 97–109.

Stark, Werner. 1967. "The Sociology of Knowledge." *Encyclopedia of Philosophy.* 1967 ed.

Steedman, P. H. 1982. "Should Debbie Do Shale? A Playful Polemic in Honor of Paul Feyerabend." *Educational Studies* 13, no. 2 (Summer): 240–51.

Stegmüller, Wolfgang. 1978, 1987. *Hauptströmungen der Gegenwartsphilosophie. Eine kritische Einführung.* 3 vols. Vol. 1 (6th ed.), 1978; vol. 2 (8th ed.), 1987; vol. 3 (8th ed.), 1987. Stuttgart: Kröner.

————. 1986. *Theorie und Erfahrung.* 3. Teilband. *Die Entwicklung des neuen Strukturalismus seit 1973.* Berlin: Springer-Verlag.

Stegmüller, Wolfgang, W. Balzer, and W. Spohn, eds. 1982. *Philosophy of Economics.* Berlin: Springer-Verlag.

Stewart, I.M.T. 1979. *Reasoning and Method in Economics.* Toronto: McGraw-Hill.

Stigler, George J. 1983. "The Process and Progress of Economists." *Journal of Political Economy* 91 (August): 529–45.

————. 1969. "Does Economics Have a Useful Past?" *History of Political Economy* 1, no. 2 (Fall): 217–30.

Stigler, George J., and Claire Friedland. 1975. "The Citation Practices of Doctorates in Economics." *Journal of Political Economy* 83, no. 3 (June): 477–507.

Stone, Sir Richard. 1980. "Political Economy, Economics and Beyond." *Economic Journal* 90 (December): 719–36.

———. 1978. *Keynes, Political Arithmetic and Econometrics.* Keynes Lecture 1978. Proceedings of the British Academy, vol. 64. London: Oxford University Press.

Stonier, Alfred W. 1939. "Review of Hutchison's *The Significance and Basic Postulates of Economic Theory.*" *Economic Journal* 49 (March): 114–15.

Stove, David C. 1982. *Popper and After: Four Modern Irrationalists.* Oxford: Pergamon Press.

Strasnick, Steven. 1981. "Neo-Utilitarian Ethics and the Ordinal Representation Assumption." In Pit, 1981, pp. 63–92.

Streissler, Erich W. 1972. "To What Extent Was the Austrian School Marginalist?" *History of Political Economy* 4 (Fall): 426–41. Repr. in Black, Coats, and Goodwin, 1973, pp. 160–75.

———. 1970. *Pitfalls in Econometric Forecasting.* London: Institute of Economic Affairs.

———, ed. 1970 [1969]. *Roads to Freedom.* London: Routledge and Kegan Paul.

Stuewer, Roger H., ed. 1970. *Historical and Philosophical Perspectives of Science.* Minnesota Studies in the Philosophy of Science, vol. 5. Minneapolis: University of Minnesota Press.

Stuhlhofer, Franz. 1986. "August Weismann—Ein 'Vorläufer' Poppers." *Conceptus* 20, no. 50: 99–100.

Suchting, W. A. 1972. "Marx, Popper, and 'Historicism.' " *Inquiry* 15:235–66.

Suppe, Frederick. 1979. "Theory Structure." In Asquith and Kyburg, 1979, pp. 317–38.

———. 1977a. "Exemplars, Theories and Disciplinary Matrixes." In his 1977b, pp. 483–99.

———. ed. 1977b. *The Structure of Scientific Theories.* 2d ed. Urbana: University of Illinois Press.

———. 1976. "Theoretical Laws." In *Formal Methods in the Methodology of Empirical Science: Proceedings of the Conference for Formal Methods in the Methodology of Empirical Sciences, Warshaw, June 17–21, 1974,* edited by Marian Prezelecki, Klemens Szaniawski, and Ryszard Wojcicki, pp. 247–67. Dordrecht, The Netherlands: D. Reidel, 1976.

———. 1974. "Theories and Phenomena." In Leinfellner and Köhler, 1974, pp. 45–91.

Suppes, Patrick. 1979. "The Role of Formal Methods in the Philosophy of Science." In Asquith and Kyburg, 1979, pp. 16–27.

Suppes, Patrick, et al., eds. 1973. *Logic, Methodology and Philosophy of Science.* Amsterdam: North-Holland.

Swamy, P.A.V.D., R. K. Conway, and P. von zur Muehlen. 1985. "The Foundations of Econometrics—Are There Any?" *Econometric Reviews* 4, no. 1: 1–61; with discussion, pp. 62–119.

Sweezy, Paul M. 1971. "Toward a Critique of Economics." *Review of Radical Political Economics* 3, no. 2 (July): 59–66.

Swinburne, R. G. 1971. "The Paradoxes of Confirmation—A Survey." *American Philosophical Quarterly* 8, no. 4 (October): 318–29.

Tarascio, Vincent J. 1975. "Intellectual History and the Social Sciences: The Problem of Methodological Pluralism." *Social Science Quarterly* 56, no. 1 (June): 37–54.

———. 1971. "Some Recent Developments in the History of Economic Thought in the United States." *History of Political Economy* 3, no. 2 (Fall): 419–97.

Tarascio, Vincent J., and Bruce Caldwell. 1979. "Theory Choice in Economics: Philosophy and Practice." *Journal of Economic Issues* 13 (December): 983–1006.

Tarski, Alfred. 1969. "Truth and Proof." *Scientific American* 220, no. 6 (June): 63–77.

Thackray, Arnold. 1985. "The Historican and the Progress of Science." *Science, Technology, and Human Values* 10, no. 1 (Winter): pp. 17–27.

———. 1970. "Science: Has Its Present Past a Future?" In Stuewer, 1970, pp. 112–27.

Thoben, H. 1982. "Mechanistic and Organistic Analogies in Economics Reconsidered." *Kyklos* 35, no. 2: 292–306.

Thompson, Herbert. 1965. "Adam Smith's Philosophy of Science." *Quarterly Journal of Economics* 79, no. 2 (May): 212–33.

Thornton, Robert J., and Jon T. Innes. 1988. "The Status of Master's Programs in Economics." *Journal of Economic Perspectives* 2, no. 1 (Winter): 171–78.

Thurow, Lester C. 1983. *Dangerous Currents: The State of Economics.* Oxford: Oxford University Press.

Tinter, Gerhard. 1953. "The Definition of Econometrics." *Econometrica* 21, no. 1 (January): 31–40.

———. 1968. *Methodology of Mathematical Economics and Econometrics.* Chicago: University of Chicago Press.

Tisdell, Clem. 1975. "Concepts of Rationality in Economics" *Philosophy of the Social Sciences* 5:259–72.

Tobin, James. 1984. *Asset Accumulation and Economic Activity: Reflections on Contemporary Macroeconomic Theory.* Oxford: Basil Blackwell.

Topitsch, Ernst. 1972. *Logik der Sozialwissenschaften.* Köln: Kiepenheuer und Witsch.

Toulmin, Stephen E. 1985. "Pluralism and Responsibility in Post-Modern Science." *Science, Technology, and Human Values* 10, no. 1 (Winter): 28–37.

———. 1982. "The Construal of Reality: Criticism in Modern and Postmodern Science." *Critical Inquiry* 9, no. 1 (Autumn): 93–110.

———. 1977a. "From Form to Function: Philosophy and History of Science in the 1950's and Now." *Daedalus* 106, no. 3 (Summer): 143–62.

———. 1977b. "The Structure of Scientific Theories." In Suppe, 1977b, pp. 600–614.

———. 1976. "History, Praxis and the 'Third World.' " In Cohen, Feyerabend, and Wartofsky, 1976, pp. 655–75.

———. 1972a. "The Historical Background to the Anti-Science Movement." In *Civilization and Science in Conflict or Collaboration?*, pp. 23–32. Amsterdam: Ciba Foundation, 1972.

———. 1972b. *Human Understanding,* vol. 1. Princeton: Princeton University Press.

———. 1971. "Rediscovering History." *Encounter* 36, no. 1 (January): 53–64.

———. 1969. "From Logical Analysis to Conceptual History." In Achinstein and Barker, 1969, pp. 25–53.

———. 1967. "Conceptual Revolutions in Science." In Cohen and Wartofsky, 1967, pp. 331–47.

———. 1961. *Foresight and Understanding.* New York: Harper and Row.

———. 1960. *The Philosophy of Science.* London: Hutchinson.

———. 1955. "Is There a Fundamental Problem in Ethics?" *Australian Journal of Philosophy* 33, no. 1 (May): 1–19.

Urbach, Peter. 1978. "Is Any of Popper's Arguments Against Historicism Valid?" *British Journal for the Philosophy of Science* 29:117–30.

Veblen, Thorstein. 1948a. "The Higher Learning." In *The Portable Veblen,* edited by Max Lerner, pp. 507–38. New York: Viking Press, 1948.

———. 1948b. "Why Is Economics Not an Evolutionary Science?" In *The Portable Veblen,* edited by Max Lerner, pp. 215–40. New York: Viking Press, 1948.

Veit-Brause, Irmline. 1985. "Review of Thomas Nipperdey's *Deutsche Geschichte 1800–1866.*" *History and Theory* 24:209–21.

Vickers, Douglas. 1983. "Formalism, Finance and Decisions in Real Economic Time." In Coats, 1983c, pp. 247–64.

Viner, Jacob. 1963. "The Economist in History." *American Economic Review Papers and Proceedings* 53, no. 2 (May): 1–22; with discussion (G. J. Stigler), 23–25.

———. 1958. "A Modest Proposal for Some Stress on Scholarship in Graduate Training." In his *The Long View and the Short,* pp. 369–81. Glencoe, Ill.: The Free Press, 1958.

von Mises, Ludwig. 1978 (posthumously). *The Ultimate Foundation of Economic Science.* 2d ed. Kansas City, Kans.: Sheed Andrews and McMeel.

von Neumann, John, and Oskar Morgenstern. 1944. *Theory of Games and Economic Behavior.* 3d ed. New York: Wiley.

Vroey, Michel de. 1975. "The Transition from Classical to Neoclassical Economics: A Scientific Revolution." *Journal of Economic Issues* 9:415ff. Repr. in Samuels, 1980, pp. 297–321.

Wade, Thomas. 1977. "Thomas S. Kuhn: Revolutionary Theorist of Science." *Science* 197, no. 4299 (8 July): 143–45.

Walsh, W. H. 1966. "The Limits of Scientific History." In Dray, 1966, pp. 54–74.

Ward, Benjamin. 1972. *What's Wrong with Economics?* London: Macmillan.

Wartofsky, Marx W. 1976. "The Relation Between Philosophy of Science and History of Science." In Cohen, Feyerabend, and Wartofsky, 1976, pp. 717–37.

Watts, M. 1981. "The Non-Rational Domain and the Limits of Economic Analysis: Comment on McKenzie." *Southern Economic Journal* 47 (April): 1120–22.

Weaver, James H., and Jon D. Wisman. 1978. "Smith, Marx, and Malthus—Ghosts Who Haunt Our Future." *The Futurist* 12. no. 2 (April): 93–104.

Weber, Max. 1973. *Soziologie, universalgeschichtliche Analysen, Politik.* Edited by Johannes Winckelmann. 5th rev. ed. Stuttgart: Alfred Kröner Verlag.

Wedeking, Gary. 1976. "Duhem, Quine and Grünbaum on Falsification." In Harding, 1976a, pp. 176–83.

Weinheimer, Heinz. 1986. *Rationalität und Begründung: Das Grundlagenproblem in der Philosophie Karl Poppers.* Mainzer philosophische Forschungen, vol. 30. Bonn: Bouvier Verlag.

Weintraub, E. Roy. 1987. "Rosenberg's 'Lakatosian Consolations for Economists.' " *Economics and Philosophy* 3, no. 1 (April): 139–42.

———. 1985. *General Equilibrium Analysis.* Cambridge: Cambridge University Press.

———. 1982–1983. "Substantive Mountains and Methodological Molehills." *Journal of Post Keynesian Economics* 5, no. 2 (Winter): 295–303.

Weisskopf, Walter. 1979. "The Methodology Is the Ideology: From a Newtonian to a Heisenbergian Paradigm in Economics." *Journal of Economic Issues* 13 (December): 869–84.

———. 1977. "Normative and Ideological Elements in Social and Economic Thought." *Journal of Economic Issues* 11 (March): 103–17.

———. 1973. "The Image of Man in Economics." *Social Research* (Autumn): 547–63.

———. 1955. *The Psychology of Economics.* London: Routledge and Kegan Paul.

Whitaker, J. K. 1975. "John Stuart Mill's Methodology." *Journal of Political Economy* 83, no. 5 (October): 1033–50.

White, John. 1982. *Rejection.* Reading, Mass.: Addison-Wesley.

Whitrow, G. J. 1967. "Albert Einstein." *Encyclopedia of Philosophy.* 1967 ed.

Whittaker, Edmund. 1940. "Review of Hutchison's *The Significance and Basic Postulates of Economic Theory.*" *American Economic Review* 30 (March): 128.

"Why Scientific Fact Is Sometimes Fiction." 1987. *The Economist,* 28 February, pp. 103–4.

Wiener, Norbert. 1964. *God and Golem, Inc.: A Comment on Certain Points Where Cybernetics Impinges on Religion.* Cambridge: MIT Press.

Wilber, Charles K. 1979. "Empirical Verification and Theory Selection: The Keynesian–Monetarist Debate." *Journal of Economic Issues* 13 (December): 973–81.

Wilber, Charles, and Jon Wisman. 1975. "The Chicago School: Positivism or Ideal Type." *Journal of Economic Issues* 9 (December): 665–79.

Wilber, Charles, and Robert S. Harrison. 1978. "The Methodological Basis of Institutional Economics: Pattern Model, Storytelling, and Holism." *Journal of Economic Issues* 12, no. 1 (March): 61–89.

Wilde, Louis L. 1981. "On the Use of Laboratory Experiments in Economics." In Pitt, 1981, pp. 137–48.

Wiles, Peter. 1984a. "Epilogue: The Role of Theory." In Wiles and Routh, 1984, pp. 293–325.

———. 1984b. "Whatever Happened to the Full-Cost Principle (UK)?" In Wiles and Routh, 1984, pp. 211–21; with comment by Austin Robinson, 222–32.

———. 1981. "Methodology and Ideology: Reply." *Journal of Post Keynesian Economics* 3, no. 3 (Spring): 352–58.

———. 1979–1980. "Ideology, Methodology, and Neoclassical Economics." *Journal of Post Keynesian Economics* 2, no. 2 (Winter): 155–80.

Wiles, Peter, and Guy Routh, eds. 1984. *Economics in Disarray.* Oxford: Basil Blackwell.

Wilkins, Burleigh Taylor. 1978. *Has History Any Meaning? A Critique of Popper's Philosophy of History.* Hassocks, Eng.: Harvester Press.

Willes, Mark H. 1980. " 'Rational Expectations' as a Counterrevolution." *The Public Interest* (Special Issue: The Crisis in Economic Theory), pp. 81–96.

Winch, D. N. 1962. "What Price the History of Economic Thought?" *Scottish Journal of Political Economy* 9, no. 3 (November): 193–204.

Wiseman, Jack, ed. 1983a. *Beyond Positive Economics?* London: The British Association for the Advancement of Science.

———. 1983(b). "Beyond Positive Economics—Dream and Reality." In his 1983a, pp. 13–27.

Wisman, Jon D. 1980. "The Sociology of Knowledge as a Tool for Research into the History of Economic Thought." *American Journal of Economics and Sociology* 39, no. 1 (January): 83–94.

———. 1979. "Legitimation, Ideology-Critique, and Economics." *Social Research* 46 (Summer): 291–320.

———. 1978. "The Naturalist Turn of Orthodox Economics: A Study of Methodological Misunderstanding." *Review of Social Economy* 36 (December): 263–84.

Wold, Herman O. 1969. "Mergers of Economics and Philosophy of Science." *Synthese* 20, no. 4 (December): 427–82.

Wolfson, R. J. 1981. "New Consumer Theory and the Relations Between Goods." In Pitt, 1981, pp. 33–46.

Wolozin, Harold. 1974. "Lying and Economic Dogma." *Review of Existential Psychology and Psychiatry* 13, no. 2: 196–203.

"Women Pay for Success." 1978. *The Times Higher Education Supplement* 744 (6 February): 8.

Wong. Stanley. 1978. *The Foundations of Paul Samuelson's Revealed Preference Theory.* London: Routledge and Kegan Paul.

———. 1973. "The F-Twist and the Methodology of Paul Samuelson." *American Economic Review* 63 (June): 312–25.

Woodbury, Stephan A. 1979. "Methodological Controversy in Labor Economics." *Journal of Economic Issues* 13 (December): 933–72.

Woodman, Harold D. 1972. "Economic History and Economic Theory: The New Economic History in America." *Journal of Interdisciplinary History* 3:323–50.

Worland, Stephen T. 1972. "Radical Political Economy as a Scientific Revolution." *Southern Economic Journal* 39, no. 2 (October): 274–84.

Worrall, John. 1976. "Imre Lakatos (1922–1974): Philosopher of Mathematics and Philosopher of Science." In Cohen, Feyerabend, and Wartofsky, 1976, pp. 1–8.

Worswick, G.D.N. 1972. "Is Progress in Economic Science Possible?" *Economic Journal* 82 (March): 73–100.

Worswick, David, and James Trevitheck. 1983. *Keynes and the Modern World: Proceedings of the Keynes Centenary Conference, King's College, Cambridge.* Cambridge: Cambridge University Press.

Zinam, Oleg. 1978. "Search for a Logic of Change Economic Theories: Evolution, Revolutions, Paradigmatic Shifts and Dialectic." *Revista Internazionale di Scienze Economiche e Commerciali* 25 (February): 156–88.

Zweig, Michael. 1971. "Bourgeois and Radical Paradigms in Economics." *Review of Radical Political Economics* 3, no. 2 (July): 43–58.

Author Index

Subject Index

Ad hoc theories/hypotheses, 33, 39, 49
  defense of, 46
Against Method, 44, 45, 51, 52
  dedication to Lakatos, 49, 50, 71–72 *n.* 40
Aggregation problem, 110
America. *See* U.S.A.
American Economic Association (AEA), 154,
  155, 162, 165, 172
  founding of, 176 *n.* 11
Anti-intellectualism, 159, 167
Anti-Semitism, 73 *n.* 42, 175 *n.* 8
Anything goes, 45, 46, 73 *n.* 43
A priorist tradition, 138 *n.* 16
Astrology, pseudoscience of, 62 *n.* 11
Astronomy, 44, 114, 127, 150
  and Adam Smith, 91, 98 *n.* 8, 98 *n.* 9
Atomism, 61 *n.* 11
Auschwitz, 15
Austria, 44, 114, 148. *See also,* Vienna; Vi-
  enna Circle; Vienna, University of
Austrian optimism, 114
Austrians (school of economics), 114, 138 *n.*
  16, 148, 166
Authority, of science, 12, 14, 144
Axioms, axiomatization, 9, 10, 48
  of economics, 94
  of mathematics, 118
  of paradigms, 145

Balance of payments, 106
Basic statements, 75 *n.* 48, 107, 124
  fallibility of, 34
Beliefs, 116
  closed system of, 15
  collective, 22
  exposure of, 53, 54
  prejudiced, 79

rationality of, 53
scientific, 14
Berkeley, University of, 11 *n.* 3, 44
Berlin, 154, 159, 160
Biology, 3, 13, 20, 21, 22 *n.* 2, 80, 104, 106,
  150, 170
Bold conjectures. *See* Conjectures
Boston University, 11 *n.* 3
Bristol, University of, 44
Britain, 35, 176 *n.* 11
Brookings Institution, 155
Brown University, 155
Buchenwald, 15
Budapest, University of, 12

Cambridge, University of, 11 *n.* 3, 35, 152,
  178 *n.* 15
Canterbury University College, 28
Ceteris paribus, 169, 175
Chemistry, 3, 12, 13, 68 *n.* 28, 150, 175
Chicago, University of, 81, 133
Clergy. *See* Theology
Cliometrics, 127, 128
Cold War, 78
Columbia University, 81, 164
Confirmation, confirmability, 8, 53, 58 *n.* 4
  in economics, 122
Conjectures, 33, 36, 41. *See also* Hypotheses
  bold, 31, 53
Consensus, 86, 133, 146–50, 166, 167, 172
Context of discovery, 82. *See also* Discovery,
  scientific
Convention, 14, 34, 59 *n.* 7, 124
Cornell University, 81
Correspondence rules (operational rules, dic-
  tionary), 9, 10, 18
Corroboration, 32, 40, 67 *n.* 26, 124

Social sciences (*continued*)
  and critical tradition, 131–32
  and development, like the natural sciences,
    107
  difference between natural sciences and,
    94, 95
  economics' isolation from other, 126
  economics more scientific than other, 128
  foundation of, 109, 128
  grounded in methodological individualism,
    109
  and Kuhn, 145, 146, 150, 151, 180 *n.* 23
  less complicated than natural sciences, 114
  less successful than natural sciences, 104, 108
  objectivity in, 130, 131
  physics-type, 110
  and Popper, 142
  queen of, 104, 110
  and research support in, 159
  and theory absorption in, 134 *n.* 3
  unclear writing in, 129, 130
  wertfrei, 125
  youthful state of, 108
Social Security Administration, 171
Sociology, 13, 21, 22 *n.* 2, 85, 136 *n.* 8, 149, 151
  and economics, 152, 155, 166
  and Lakatos, 87
  and physics, 153
  and Popper, 104, 109, 135 *n.* 6
  of science/knowledge, 6, 12, 16, 18, 19, 82,
    86, 132
Statements. *See also* Basic statements
  analytic, 94
  synthetic, 94
Structuralism. *See* German structuralism
The Structure of Scientific Revolutions
    (Kuhn), 16, 17, 19–21, 23–24 *n.* 9, 26 *n.*
    9, 26 *n.* 15, 84 *n.* 8, 145
Subjective knowledge, 9, 13
Subjectivity, 115
Supply-side economics, 148
Switzerland, 148

Tacit knowledge, 12–14. *See also* Personal
    knowledge
Testing in economics, 120–24
  by Darwinian struggle for survival, 127
Theology, 176 *n.* 11
Theoretical terms, 10
Theory absorption, 134 *n.* 3
Theory appraisal (comparison, evaluation), 3,
    12, 35
  in economics, 129, 146
  Keynes on, 129
  Kuhn's five criteria of, 20
  Lakatos and, 38, 41

Theory choice, 14, 22
Theory-ladenness (theory-dependence) of ob-
    servation, 47, 79, 81, 82, 122, 140 *n.* 20.
    See also Observation–theory distinction
The Theory of Moral Sentiments (Smith), 98
    n. 7, 98 *n.* 9
Tolerance, 51
  spirit of, 168
Trends, economic, 105–7, 122, 124
Trial by jury, 34
Tu quoque argument, 53
Twardowski school (= Lwów-Warszawa-
    Schule) of philosophy. *See* philosophy

Uncertainty, 149
Unemployment. *See* Employment
Union for Radical Political Economy
    (URPE), 145
Unit of appraisal, 32, 67 *n.* 26
U.S.A., 58 *n.* 5, 110, 145
  Congress of, 170
  economics of, 148, 152–65, 176 *n.* 11, 177
    n. 14
  and emigrants from Europe, 56 *n.* 3, 57 *n.*
    3, 139 *n.* 17, 175 *n.* 8
Unity of science thesis, 104
Universal method, 4
UCLA, 11 *n.* 3

Value(s), 64 *n.* 13, 171, 172, 181
Verifiability, verification, 8, 9, 14, 30, 53, 58
    n. 4, 62 *n.* 11, 64 *n.* 13, 75 *n.* 48
Verisimilitude, 32, 33, 40, 67 *n.* 26, 120
Victoria University, 13
Vienna, 27, 35, 44, 123
Vienna Circle (*Wiener Kreis*), 7–10, 15, 22–23
    n. 3, 48, 53, 87, 168
  and Carnap, 58 *n.* 5
  and Machlup, 102 *n.* 20
  and Popper school, 27, 28, 56 *n.* 1, 56 *n.* 2,
    57–58 *n.* 4, 59–60 *n.* 7, 61–62 *n.* 11, 70 *n.*
    37, 72 *n.* 40, 93
  and Wittgenstein, 59 *n.* 7
Vienna, University of, 27, 44, 154
Vietnam, 145
Visual doubles (reversible figures), 18, 24 *n.*
    11, 82. *See also* Gestalt switch

The Wealth of Nations (Smith), 98 *n.* 7, 98 *n.* 9
Women, 160–62, 166
  admission to degrees at Cambridge, 161
  number of Ph.D.s in economics, 162

Yale University, 11 *n.* 3, 44, 133

Zero method, 112, 124. *See also* Method
  and rationality, 112